3
Microbiology Monographs

Series Editor: Alexander Steinbüchel

Magnetoreception and Magnetosomes in Bacteria

Volume Editor: Dirk Schüler

With 90 Figures, 8 in Color

Springer

Volume Editor:

Professor Dr. Dirk Schüler
Max-Planck-Institut für Marine Mikrobiologie
Celsiusstraße 1
28359 Bremen
Gremany

Present address:
Department Biologie I
Maria-Ward-Str. 1a
80638 München
Germany
e-mail: dirk.schueler@erz.uni-muenchen.de

Series Editor:

Professor Dr. Alexander Steinbüchel
Institut für Molekulare Mikrobiologie und Biotechnologie
Westfälische Wilhelms-Universität
Corrensstraße 3
48149 Münster
Germany
e-mail: steinbu@uni-muenster.de

ISSN 1862-5576

DOI 10.1007/11741862

ISBN 978-3-642-07224-6

e-ISBN 978-3-540-37468-8

Springer is a part of Springer Science+Business Media

springer.com

Editor: Dr. Christina Eckey, Heidelberg
Desk Editor: Dr. Jutta Lindenborn, Heidelberg
Cover design: WMXDesign GmbH, Heidelberg

Preface

Magnetoreception and Magnetosomes in Bacteria is published as Vol. 3 of the series *Microbiology Monographs*, which is devoted to hot topics and fields in which important, recent progress has been made. This volume focus on magnetosomes, prokaryotic organelles which are responsible for magnetotaxis, the orientation and migration of cells along the geomagnetic field lines.

The magnetotactic bacteria, a diverse group of aquatic prokaryotes that orient and swim along geomagnetic field lines, were discovered over thirty years ago by Richard P. Blakemore (1975). This completely unexpected discovery inspired a number of fundamental and pioneering studies by a rather small community of researchers including Blakemore, Richard Frankel, Dennis Bazylinski, Marcos Farina, Joseph Kirschvink, Stephen Mann, Tadashi Matsunaga, Nikolai Petersen, Ralph Wolfe, and others. During this early research period, it was established that magnetotactic bacteria occur world-wide, from high mountain lakes to the deep ocean, display a remarkable diversity, and are metabolically versatile. A characteristic of all magnetotactic bacteria is the presence of magnetosomes. In most strains, the magnetosome mineral is magnetite (Fe_3O_4), although some species from marine or brackish water habitats make magnetosomes with the mineral greigite (Fe_3S_4). The crystal sizes, compositions, and shapes are remarkably consistent within each bacterial species or strain. In most species, magnetosomes are arranged in one or multiple chains, and comprise a permanent magnetic dipole in the cell. Because the chain is fixed in position within the cell, the cell passively orients in the magnetic field and migrates along geomagnetic field lines as it swims. Magnetotaxis is thought to work in conjunction with aerotaxis to increase the efficiency of the bacteria in finding and maintaining position at a preferred concentration of oxygen in vertical oxygen concentration gradients in aquatic environments. Moreover, magnetosomes and magnetotactic bacteria have gained interest in fields such as in palaeomagnetism research, geobiology, or as potential biomarkers in astrobiology.

Many of these early milestone discoveries have been already extensively covered in a number of excellent monographs, which were published mostly in the 1980s and early 1990s. However, since then a number of important novel developments in the research on magnetotactic bacteria have emerged, which are reviewed here.

For example, a key question regarding magnetotactic bacteria since their discovery has been how they form and organize their magnetosomes. It was predicted early on that magnetosome synthesis is under genetic control but research progress was hampered by the fastidiousness of most magnetotactic bacteria that makes them difficult to isolate and grow in pure culture. Now, however, tractable genetic systems have been devised for several magnetotactic bacteria species. Secondly, the genomes of several magnetotactic bacteria have been sequenced, either partially or completely. These developments have led to the identification of a large genomic island containing many of the genes involved in magnetosome formation and positioning in the cell. Many of these genes encode for proteins associated with the magnetosome membrane and are organized in clusters or operons within the "magnetosome island". Finally, new cryogenic techniques have been developed for use with electron tomography that enable 3-dimensional visualization of cytoskeletal structures of magnetic bacteria as never before, and which have provided exciting insights into the cell biology of magnetosome formation and assembly. Also, potential nanobiotechnological applications of the magnetosome crystals, which have magnetic and crystalline characteristics unmatched by their inorganic counterparts, are reviewed in this book, as well as the current knowledge on their magnetic microstructure and mineralogy. Related topics, such as the impact of biogenic magnetic crystals in geobiology, palaeomagnetism, and magnetoreception, are covered in several chapters by experts from various interdisciplinary fields of research.

Magnetotactic bacteria and magnetosomes have certainly developed from a somewhat exotic subject into a bona fide field of microbiological research. The recent developments described here have opened a host of new questions that beg to be answered. The aim of this book is to provide a broad survey of this highly multidisciplinary field, and identify new directions and inspire future research on these fascinating organisms.

Finally, the editor wants to thank all authors of the chapters, the series editor, Alexander Steinbüchel, for help and encouragement, Jutta Lindenborn of Springer for her suggestions and support, Richard Frankel for his advice on this preface and many invaluable discussions, and the many people who worked behind the scenes in making this book a reality.

Bremen, August 2006 Dirk Schüler

References

Blakemore, R (1975) Magnetotactic bacteria. Science 190:377–379

Contents

Microbiol Monogr (3)
D. Schüler: Magnetoreception and Magnetosomes in Bacteria
DOI 10.1007/7171_2006_036/Published online: 8 September 2006

Magneto-Aerotaxis

Richard B. Frankel[1] (✉) · Timothy J. Williams[2] · Dennis A. Bazylinski[3]

[1]Department of Physics, California Polytechnic State University, 1 Grand Avenue, San Luis Obispo, CA 93407, USA
rfrankel@calpoly.edu

[2]Department of Biochemistry, Biophysics and Molecular Biology, Iowa State University, Ames, IA 50011, USA

[3]School of Life Sciences, University of Nevada, Las Vegas, 4505 Maryland Parkway, Las Vegas, NV 89154-4004, USA

Abstract Magnetotactic bacteria orient and migrate along geomagnetic field lines. Magneto-aerotaxis increases the efficiency of respiring cells to efficiently find and maintain position at a preferred microaerobic oxygen concentration. Magneto-aerotaxis could also facilitate access to regions of higher nutrient and electron acceptor concentration via periodic excursions above and below the preferred oxygen concentration level.

1 Introduction

1.1 History

The terms "magnetotaxis" and "magnetotactic bacteria" were first used by Richard P. Blakemore in his landmark 1975 paper (Blakemore 1975) announcing the discovery of aquatic bacteria from Woods Hole, Massachusetts that migrated northward in water drops along magnetic field lines. Using transmission electron microscopy he found that the cells were roughly coccoid with two bundles of seven flagella each on one side of the cell. He also found the cells contained chains of elongated, electron-dense, iron-rich crystals, later shown to consist of magnetite (Fe_3O_4) (Frankel et al. 1979). The crystals were contained in intracytoplasmic vesicles arranged adjacent to the cytoplasmic membrane in the cell. He noted the probable relationship of the chains of crystals and magnetotaxis in the cocci and other magnetotactic bacteria recovered from aquatic sediments in Woods Hole. Blakemore (1975) postulated "Perhaps the iron-rich cell inclusions serve as magnetic dipoles which convey a magnetic moment on the cells, thus orienting the cells in magnetic fields. Magnetotaxis would result if, within each cell, a fixed spatial relationship existed between the orienting mechanism and cell propulsion". In a sense, all subsequent research on magnetotactic bacteria follows from these and other original observations in that paper. In this review, we describe and discuss recent work on magnetotaxis.

1.2 General Features of Magnetotactic Bacteria

Magnetotactic bacteria inhabit water columns or sediments with vertical chemical concentration stratification, where they occur predominantly at the oxic–anoxic interface (OAI) and the anoxic regions of the habitat or both (Bazylinski et al. 1995; Bazylinski and Moskowitz 1997; Simmons et al. 2004). All known magnetotactic bacteria phylogenetically belong to the domain Bacteria and are associated with different subgroups of the Proteobacteria and the Nitrospira phylum (Spring and Bazylinski 2000; Simmons et al. 2004). They represent a diverse group of microorganisms with respect to morphology and physiology (Bazylinski and Frankel 2004).

The magnetotactic bacteria are difficult to isolate and cultivate (Bazylinski and Frankel 2004) and thus there are relatively few axenic cultures of these organisms. Most cultured strains belong to the genus *Magnetospirillum*. Currently recognized species include *M. magnetotacticum* strain MS-1 (Blakemore et al. 1979; Maratea and Blakemore 1981; Schleifer et al. 1991), *M. gryphiswaldense* (Schleifer et al. 1991) and *M. magneticum* strain AMB-1

(Matsunaga et al. 1991). Several other freshwater magnetotactic spirilla in pure culture have not yet been completely described (Schüler et al. 1999). Other species of cultured magnetotactic bacteria include a variety of as yet incompletely characterized organisms: the marine vibrios, strains MV-1 (Bazylinski et al. 1988) and MV-2; a marine coccus, strain MC-1 (DeLong et al. 1993; Meldrum et al. 1993a); and a marine spirillum, strain MMS-1 (formerly MV-4) (Bazylinski and Frankel 2000; Meldrum et al. 1993b). There is also an anaerobic, sulfate-reducing, rod-shaped magnetotactic bacterium named *Desulfovibrio magneticus* strain RS-1 (Sakaguchi et al. 1993, 2002). These cultured organisms, except *D. magneticus*, are facultatively anaerobic or obligate microaerophiles. All are chemoorganoheterotrophic although the marine strains can also grow chemolithoautotrophically (Bazylinski et al. 2004; Williams et al. 2006). The genomes of several strains, including *M. magnetotacticum* strain MS-1 and strain MC-1, have been partially sequenced while that of *M. magneticum* strain AMB-1 (Matsunaga et al. 2005) has been recently completed.

Several uncultured, morphologically conspicuous, magnetotactic bacteria have also been examined in some detail. A very large, rod-shaped bacterium, *Candidatus* Magnetobacterium bavaricum, has been found to inhabit the OAI in the sediments of calcareous freshwater lakes in Bavaria (Spring et al. 1993; Spring and Bazylinski 2000). Cells of this organism biomineralize multiple chains of tooth-shaped crystals of magnetite. A multicellular bacterium, referred to as the many-celled magnetotactic prokaryote (MMP) (Rogers et al. 1990), biomineralizes crystals of iron sulfides (Mann et al. 1990; Farina et al. 1990; Pósfai et al. 1998) and is comprised of about 20–30 cells in a roughly spherical arrangement that moves as an entire unit. There is evidence that suggests that the MMP is a sulfate-reducing bacterium (DeLong et al. 1993) and organisms like it have been found in marine and brackish aquatic habitats around the world.

All studied magnetotactic bacteria are motile by means of flagella and have a cell wall structure characteristic of Gram-negative bacteria (Bazylinski and Frankel 2004). The arrangement of flagella varies between species/strains and can be either polarly monotrichous, bipolar, or in tufts (lophotrichous). The MMP is peritrichously flagellated as a unit but not as individual cells, which are multi-flagellated on only one side (Rogers et al. 1990). It is the only magnetotactic bacterium whose external surface is covered with flagella. Like other flagellated bacteria, magnetotactic bacteria propel themselves through the water by rotating their helical flagella. Because of their magnetosomes, magnetotactic bacteria passively orient and actively migrate along the local magnetic field **B**, which in natural environments is the geomagnetic field. Reported swimming speeds (Table 1) vary between species/strains, from ca. 40 to 1000 μm/s. In general, the magnetotactic spirilla are at the slower end (<100 μm/s) (Maratea and Blakemore 1981) and the magnetotactic cocci are at the faster end of the range at >100 μm/s (Blakemore 1975; Moench 1988; Cox et al. 2002).

Table 1 Lengths, swimming speeds, and magnetic dipole moments of selected motile microorganisms. Magnetotactic bacterial species are designated with an * in front of their name

Organism	Average cell length µm	Observed swimming speed µm/s	Magnetic moment Am^2	Refs.
* *Magnetospirillum magnetotacticum* strain MS-1	3	44	5.0×10^{-16}	Dunin-Borkowski et al. 1998
* *Candidatus* Magnetobacterium bavaricum	9	40	3.2×10^{-14}	Spring et al. 1993
* Many-celled Magnetotactic Prokaryote (MMP)	8	170/100		Greenberg et al. 2005
* *Candidatus* Bilophococcus magneticus	1	69		Moench 1988
* Unidentified Woods Hole Coccus	1	159	7.0×10^{-16}	Kalmijn 1980
* Coccus "ARB-1"	1	1000		Cox et al. 2002
* Strain MV-1	2		7.0×10^{-16}	Dunin-Borkowski et al. 1998
* Morro Bay greigite-containing Rod	3.6		9.0×10^{-16}	Kasama et al. 2006
Anisonema platysomum (protist)	20		7.0×10^{-13}	Torres de Araujo et al. 1986
Escherichia coli	2	20		Berg 1999
Psuedomonas aeruginosa	1.5	55		Garcia-Pichel 1989
Vibrio comma	4	200		Garcia-Pichel 1989
Thiovulum majus	15	600		Garcia-Pichel 1989

1.3 Detection of Magnetotactic Bacteria

Magnetotactic bacteria can be detected and roughly enumerated in environmental samples using phase contrast or differential interference contrast microscopy and a bar magnet or Helmholtz coil pair. In this method, a drop of water and sediment from an environmental sample is placed directly on a microscope slide or on a cover slip which is placed on a rubber o-ring with the drop on the underside (called a hanging drop). The bar magnet is placed on the microscope stage near the drop so the axis of the magnet is parallel to the plane of the slide or cover slip and oriented radial to the drop. The magnetic field (**B**) at the drop should be at least a few gauss. Magnetotactic bacteria in the drop will swim persistently toward or away from the bar magnet and accumulate along the edge of the drop, close to the near pole of the bar magnet or on the other side of the drop farthest away from the near pole. If the magnet is rotated 180°, the bacteria will rotate and swim away from their position toward the opposite side of the drop, i.e., they swim in the same direction relative to **B**. Another 180° rotation of the bar magnet will cause the bacteria to return to the original position at the edge of the drop. Bacteria that swim toward the "south" magnetic pole of the bar magnet, i.e., swim parallel to **B**, are said to have North-seeking (NS) polarity because they would swim northward in the geomagnetic field; bacteria that swim away from the "south" magnetic pole or toward the "north" magnetic pole, i.e., swim antiparallel to **B**, are said to have South-seeking (SS) polarity (Fig. 1). Using this assay, it has been found that magnetotactic bacteria from Northern hemisphere habitats are predominantly NS whereas those from Southern hemisphere habitats are predominantly SS (Blakemore 1975; Blakemore et al. 1980; Kirschvink 1980; Nogueira and Lins de Barros 1995). It should be noted that because of the rapid diffusion of oxygen from the air into the drop, this assay is carried out under oxic conditions. A device, known as a bacteriodrome, in which the magnetic field rotates in the horizontal plane at constant angular velocity, is also useful for detecting bacteria in environmental samples and for measuring some of their magnetic properties (Hanzlik et al. 2002).

Magnetotaxis involves passive orientation and active swimming along the field by bacteria. Cells are not appreciably pulled or pushed by the field which is demonstrated by the fact that killed cells in suspension also orient but do not move along the field. While many magnetotactic bacteria swim persistently in one direction relative to the field under oxic conditions, they are able to reverse direction without turning around under anoxic conditions (Frankel et al. 1997). Other bacteria, particularly magnetotactic spirilla, migrate in both directions along the field with occasional spontaneous reversals of the swimming direction without turning around under both oxic and anoxic conditions (Blakemore 1982; Spormann and Wolfe 1984; Frankel et al. 1997). It

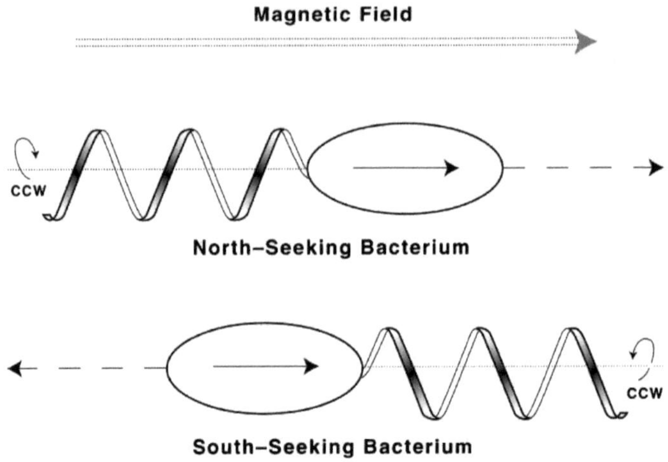

Fig. 1 Transmission electron micrograph of *Magnetospirillum magnetotacticum* showing the chain of magnetosomes. The magnetite crystals incorporated in the magnetosomes have a cuboctahedral morphology and are about 42 nm in diameter. The magnetosome chain is fixed in the cell and the interaction between the magnetic dipole moment associated with the chain and the local magnetic field causes the cell to be oriented along the magnetic field lines. Rotation of the cellular flagella (not shown) causes the cell to migrate along the field lines

should be noted that magnetotaxis is a misnomer, i.e., cells do not swim towards or away from the stimulus (the magnetic field) unlike in other forms of taxis known in bacteria (e.g. phototaxis).

2 Magnetosomes

All magnetotactic bacteria contain magnetosomes, which are intracellular structures comprising magnetic iron mineral crystals enveloped by a phospholipid bilayer membrane (Gorby et al. 1988). The magnetosome membrane is presumably a structural entity that is anchored to the mineral particles at particular locations in the cell, as well as the locus of biological control over the nucleation and growth of the mineral crystal (Scheffel et al. 2005; Komeili et al. 2004, 2005). The magnetosome magnetic mineral phase consists of magnetite, Fe_3O_4, or greigite, Fe_3S_4. The magnetosome crystals are typically of order 35 to 120 nm in length, which is within the permanent single-magnetic-domain (SD) size range for both minerals, although magnetite crystals with lengths up to 250 nm are known (Spring et al. 1998; McCartney et al. 2001; Lins et al. 2005). In the majority of magnetotactic bacteria, the magnetosomes are organized in one or more straight chains of various lengths parallel to the

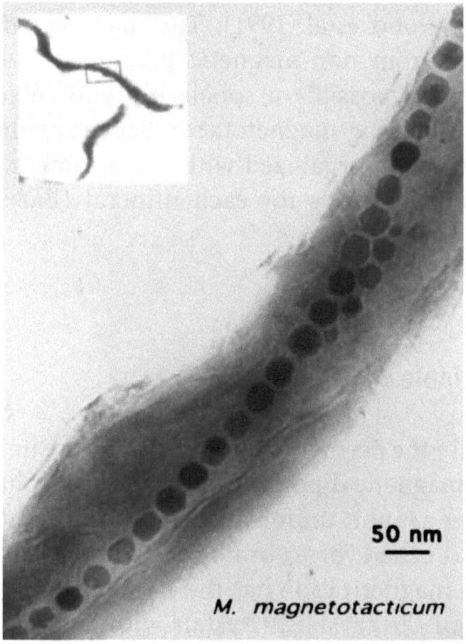

Fig. 2 Schematic representation of magnetotactic NS and SS polarity showing the two possible orientations of the cell's magnetic dipole with respect to the cellular poles (i.e., the flagellum). In both polarities, the magnetic dipole orients parallel to the magnetic field. If both NS and SS cells rotate their flagella ccw under oxic conditions, the cell with NS polarity will migrate parallel to the magnetic field, whereas the cell with SS polarity will migrate antiparallel to the magnetic field. Migration directions under oxic conditions are indicated by *dashed lines*. Under anoxic conditions, the cells switch their flagellar rotation to the opposite sense, and the cells migrate opposite to the direction shown without turning around. The flagellum is a left-handed helix. Just as a left-handed screw advances when turned ccw and retracts when turned cw, ccw and cw flagellar rotation pushes and pulls the cell, respectively

axis of motility of the cell (Fig. 2). Clusters of separate magnetosomes occur in some species, usually at the side of the cell where the flagella are inserted. The narrow size range and consistent morphologies of the magnetosome crystals in each species or strain are clear indications that the magnetotactic bacteria exert a high degree of control over the processes of magnetosome formation. Recent progress in elucidating the biomineralization process and the construction of the magnetosome chain in magnetotactic bacteria will be presented elsewhere in this volume.

All known freshwater magnetotactic bacteria and some marine, estuarine and salt marsh strains have magnetite magnetosomes. Other strains in the latter habitats have greigite magnetosomes. While none of the latter are available in pure culture, recognized greigite-bearing magnetotactic bacteria include the MMP (Mann et al. 1990) and a variety of relatively large, rod-shaped bacteria (Heywood et al. 1991). The magnetosome greigite crystals are thought to form from non-magnetic precursors including mackinawite (tetragonal FeS) and possibly a sphalerite-type cubic FeS (Pósfai et al. 1998). Some greigite-bearing magnetotactic bacteria contain magnetite and greigite magnetosomes, co-organized within the same magnetosome chains but with distinct morphologies for each mineral (Bazylinski et al. 1993b, 1995).

3 Cellular Magnetic Dipole

Magnetosomes within the permanent SD size range are uniformly magnetized with the maximum magnetic dipole moment per unit volume. Magnetic crystals larger than SD size are non-uniformly magnetized because of the formation of domain walls or so-called vortex or flower configurations (McCartney et al. 2001). Non-uniform magnetization has the effect of significantly reducing the magnetic moments of the crystals. Crystals with lengths below about 35 nm are superparamagnetic (SPM). Although SPM particles are SD, thermally induced reversals of their magnetic moments result in a time-averaged moment of zero. Thus, by controlling particle size, magnetotactic bacteria optimize the magnetic dipole moment per magnetosome. For magnetosomes arranged in a chain, as in *M. magnetotacticum*, magnetostatic interactions between the SD crystals cause the magnetic moments to spontaneously orient parallel to each other along the chain direction (Frankel 1984; Frankel and Blakemore 1980). This results in a permanent magnetic dipole for the entire chain with a magnetization approaching its saturation value (0.6 T). Since the chain of magnetosomes is fixed within the cell, the entire cell is oriented in the magnetic field by the torque exerted on the magnetic dipole, causing the cell to migrate along the magnetic field as it swims. The permanent magnetic structure of magnetosome chains has been demonstrated by electron holography (Dunin-Borkowski et al. 1998), and by pulsed magnetic field remanence measurements on individual cells (Penninga et al. 1995; Hanzlik et al. 2002).

Reported and estimated magnetic moments of several organisms are shown in Table 1. For the smaller organisms the moments are ca. 1.0×10^{-15} Am^2, and the corresponding magnetic energy in the geomagnetic field of 50 μ Tesla is 5.0×10^{-20} J. This value is greater than thermal energy at room temperature, 4.1×10^{-21} J. The average orientation of a cell along the magnetic field as it swims is determined by the ratio of magnetic to thermal energy (Frankel 1984). For a ratio of 10, the average projection of the magnetic dipole on the magnetic field, $< \cos\Theta > = 0.9$, which means the cell can migrate along the field at 90% of its forward speed. Thus, a magnetotactic bacterium is, in effect, a self-propelled magnetic compass needle.

4 Magnetotaxis

4.1 Adaptiveness of Magnetotaxis

The original model of magnetotaxis was based on the assumption that all magnetotactic bacteria have a polar preference to their swimming direction and are microaerophiles indigenous in sediments (Blakemore 1975; Blakemore and Frankel 1981). The geomagnetic field is inclined downward from horizontal in the Northern Hemisphere and upward in the Southern hemisphere, with the magnitude of inclination increasing from the equator to the poles. NS cells swimming northward in the Northern hemisphere and SS cells swimming southward in the Southern hemisphere would migrate downward towards the sediments along the inclined geomagnetic field lines. Thus, polar magnetotaxis appeared to guide cells in each hemisphere downward to less oxygenated regions of aquatic habitats. Once cells have reached their preferred microhabitat they would presumably stop swimming and adhere to sediment particles until conditions changed, as for example, when additional oxygen was introduced, or when disturbance of the sediments caused them to be displaced into the water column. This theory is supported by the predominant occurrence of NS polar magnetotactic bacteria in the Northern hemisphere and SS polar magnetotactic bacteria in the Southern hemisphere, as determined by the magnetotaxis assay under oxic conditions (Blakemore 1975; Blakemore et al. 1980; Nogueira and Lins de Barros 1995). Because of the negative and positive sign of the geomagnetic field inclination in the Northern and Southern hemispheres, respectively, polar magnetotactic bacteria in both hemispheres therefore swim downward toward the sediments under oxic conditions.

4.2 Magneto-Aerotaxis

The discovery of large populations of magnetotactic bacteria at the OAI in the water columns of certain chemically stratified aquatic habitats, and the isolation of obligately microaerophilic, coccoid, magnetotactic bacteria strains in pure culture, has led to a revised view of magnetotaxis (Frankel et al. 1997). The original model did not completely explain how bacteria in the anoxic zone of a water column benefit from magnetotaxis, nor did it explain how the polar magnetotactic cocci such as strain MC-1 form horizontal microaerophilic bands in semi-solid oxygen gradient media instead of accumulating and growing at the bottom of the tube. Bands of strain MC-1 and *M. magnetotacticum* were studied in oxygen concentration gradients in thin, flattened capillaries. When the head space gas was switched from air to pure

N_2, the bands moved up the capillary, eventually to the meniscus. When the N_2 was replaced with air, the bands moved back to their original position. Pure O_2 caused the bands to move further down the capillary. This shows that magnetotaxis and aerotaxis work together in these magnetotactic bacteria. The behavior observed in strain MC-1 and *M. magnetotacticum* has been denoted "magneto-aerotaxis" (Frankel et al. 1997).

Two different magneto-aerotactic mechanisms, polar and axial, have been proposed for strain MC-1 and *M. magnetotacticum*, respectively (Frankel et al. 1997). The magnetotactic bacteria, including magnetotactic cocci in addition to strain MC-1, which swim persistently in one direction along the magnetic field **B** in the hanging drop assay, are polar magneto-aerotactic (NS or SS). Those, including the magnetotactic spirilla in addition to *M. magnetotacticum*, which do not show a polar preference in their swimming direction and swim in either direction along **B** with frequent, spontaneous reversals of swimming direction, are axial magneto-aerotactic. The distinction between NS and SS does not apply to axial magneto-aerotactic bacteria.

4.2.1 Polar Magneto-Aerotaxis

The large majority of naturally occurring magnetotactic bacteria, including many magnetotactic marine and freshwater spirilla, display polar magnetotaxis. Although NS cells swim persistently parallel to **B** under oxic conditions it was demonstrated that under reducing conditions they swim antiparallel to **B** without turning around (Frankel et al. 1997). This suggests that the sense of flagellar rotation (presumably ccw) is unchanged as long as the cells remain under oxic conditions, and furthermore, that the opposite sense of flagellar rotation (cw) occurs under reducing conditions and likewise remains unchanged as long as the cells remain under reducing conditions. Thus, instead of a temporal sensory mechanism, polar magneto-aerotactic cells have a two-state sensory mechanism that determines the sense of flagellar rotation and consequently swimming direction relative to **B** (Fig. 3). Under higher than optimal oxygen tensions, the cell is presumably in an "oxidized state" and cw flagellar rotation causes the cell to migrate persistently parallel to **B**, i.e., downward in the Northern hemisphere. Under reducing conditions, or suboptimal oxygen concentrations, the cell switches to a "reduced state" in which cw flagellar rotation causes the cell to migrate antiparallel to **B** (upward in the Northern hemisphere). The two-state sensing mechanism results in an efficient aerotactic response, provided that the oxygen-gradient is oriented vertically so that it is more or less antiparallel to **B**, guiding the cell back toward the optimal oxygen concentration from either reducing or oxidizing conditions. This is especially important because adaptation, which would lead to spontaneous reversals of the swimming direction, is never observed in controlled experiments with the cocci. This model accounts for the fact that cells

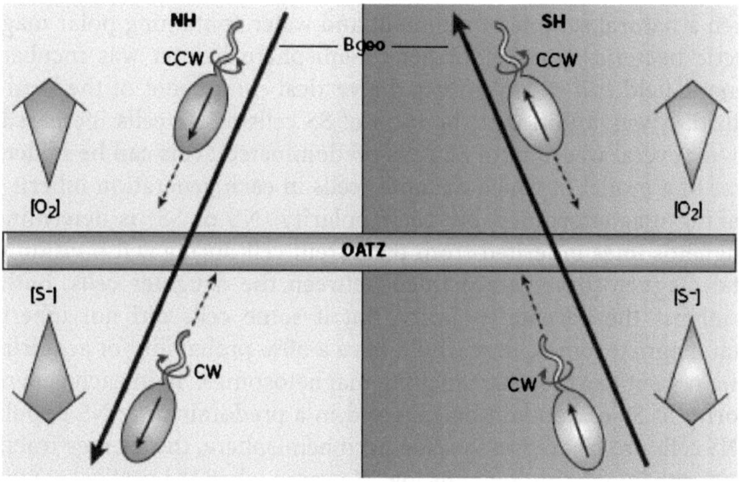

Fig. 3 Schematic showing how polar magneto-aerotaxis keeps cells at the preferred oxygen concentration at the oxic–anoxic interface (OATZ) in chemically stratified water columns and sediments (NH, Northern hemisphere; SH, Southern hemisphere; Bgeo, geomagnetic field). In both hemispheres, cells at higher than optimal oxygen concentration ('oxidized state') swim forward by rotating their flagella counter clockwise (ccw: see Fig. 1), until they reach a lower than optimal oxygen concentration ('reduced state') that switches the sense of flagellar rotation to clockwise (cw), causing the cell to back up without turning around. Note that the geomagnetic field selects for cells with polarity such that ccw flagellar rotation causes cells to swim downward along the magnetic field lines in both hemispheres

swim away from an aerotactic band when the magnetic field is reversed. In this situation, cells do not encounter the redox condition that switches them into the other state and hence do not reverse their swimming direction. It also accounts for the fact that in a capillary with both ends open NS polar bacteria only form a stable band at the end for which the oxygen gradient and B are antiparallel. Unlike the axial cells, polar cells have been observed to stop swimming and remain stationary by attachment to a solid surface or other cells at the optimum oxygen concentration, resuming swimming when the oxygen concentration changes. Finally, in some polar strains exposure to light of short wavelengths (< 500 nm) can switch the cell into the "oxidized state" even in reducing conditions for which the oxygen concentration is suboptimal (Frankel et al. 1997).

The polar magneto-aerotaxis model would also apply to SS polar magneto-aerotactic bacteria if it is assumed that their flagellar rotation is also ccw in the "oxidized" state, and cw in the "reduced" state. In flat capillaries with both ends open, SS bacteria would also form only a single band but at the end of the capillary for which the magnetic field is parallel to the oxygen concentration gradient, i.e., at the other end from that at which the NS band forms.

When a natural sample of sediment and water containing polar magneto-aerotactic bacteria from a Northern hemisphere habitat was incubated in a magnetic field coil which inverts the vertical component of the local magnetic field, it was found that the ratio of SS cells to NS cells increased with time over several weeks until SS cells predominated. This can be understood in terms of a model in which daughter cells in each generation inherit genes for making magnetosomes, but their polarity (NS or SS) is determined by the magnetosomes inherited from the parent cell during cell division. If the parental magnetosomes are divided between the daughter cells, both cells could inherit the parental polarity. But if some cells did not inherit any parental magnetosomes, they would have a 50% probability of acquiring the opposite polarity as they start making magnetosomes. So in each generation, a minority of SS cells might be expected in a predominantly NS population. Since NS cells are favored in the Northern hemisphere, the average fraction of SS cells in the population remains low. However, when the vertical component of the magnetic field is inverted, the SS cells are favored and they eventually become the majority polarity in the population. This process might also occur in a given location during reversals or excursions of the geomagnetic field. A further indication that cell polarity is not determined genetically comes from the fact that SS cells can result when NS cells are pulsed with magnetic fields greater than the coercive force of the magnetosome chain (ca. 300 gauss), with the magnetic pulse oriented opposite to the local background magnetic field.

4.2.2 Axial Magneto-Aerotaxis

The aerotactic, axial magnetotactic spirilla appear to locate and remain at a preferred or optimal oxygen concentration, at which the proton motive force generated by the cell is maximal (Zhulin et al. 1996; Taylor et al. 1999), by means of a temporal sensory mechanism that occurs in many non-magnetotactic, chemotactic bacteria (Berg 1983, 1999). Cells sample the oxygen concentration as they swim and compare the present concentration with that in the recent past. The change in oxygen concentration with time is connected to the probability of switching the sense of flagellar rotation (cw or ccw) and hence the direction of migration. Axial magneto-aerotactic cells moving away from the optimal oxygen concentration toward higher or lower oxygen concentration have an increased probability of reversing the sense of flagellar rotation and hence the direction of migration along **B** which causes them to return to the band. Cells moving toward the optimum oxygen concentration have a decreased probability of reversing the sense of flagellar rotation. At constant oxygen concentration band formation does not occur and the cells revert to an intermediate probability of reversal; this is known as adaptation. In the axial magneto-aerotactic model, the bacteria must be actively motile in order to quickly measure and respond to local concentration gradients. Since the cells use the magnetic field to provide an axis but not a direction of motility, the relative orientation of **B** and the concentration gradient is unimportant to aerotactic band formation. The combination of a passive alignment along inclined geomagnetic field lines with an active, temporal, aerotactic response provides axial magneto-aerotactic organisms with an efficient mechanism to find the OAI in habitats with vertical, chemical gradient stratification.

4.2.3 Redoxtaxis

It has been suggested that the polar magneto-aerotaxis model could be extended to a more complex redoxtaxis in habitats in which rapid chemical oxidation of reduced chemical species such as sulfur near the OAI results in separated pools of reductants and oxidants (Spring and Bazylinski 2000). For some magnetotactic bacteria, it might be necessary to perform excursions to anoxic zones of their habitat in order to accumulate reduced sulfur compounds. In this situation, polar magnetotaxis could efficiently guide bacteria, either downward to accumulate reduced sulfur species or upward to oxidize stored sulfur with oxygen. The "oxidized state" would result from the almost complete consumption of stored sulfur or another electron donor, and the cells would swim parallel to **B** toward deeper anoxic layers where they could replenish the depleted stock of electron donor using nitrate or other compounds as an alternative electron acceptor. Finally, they would reach a "reduced state" in which the electron acceptor is depleted. In this state the cells would swim antiparallel to **B** to return to the microoxic zone where oxygen is available to them as an electron acceptor. The advantage of polar magnetotaxis is that an oxygen concentration gradient is not necessary for efficient orientation in the anoxic zone, thereby enabling a rapid return of the cell along relatively large distances to the preferred microoxic conditions. A further benefit would be that cells avoid the waste of energy by constant movement along gradients, but instead can attach to particles in preferred microniches until they reach an unfavorable internal redox state that triggers a magnetotactic response either parallel or antiparallel to the geomagnetic field lines. In any case, greater than optimal concentrations of oxygen would switch cells immediately to the "oxidized state" provoking the typical down-seeking response of magnetotactic bacteria under oxic conditions.

Cells of MC-1, like other uncultivated magnetotactic cocci, are small (ca. 1 μm diameter) with twin, multiflagellar bundles on one side of the cell. Magnetotactic cocci have been reported to swim at speeds in excess of 100 μm/s (about 100 body lengths per second) (e.g., Moench 1988; Cox et al. 2002). In $[O_2]$ gradients in flat, thin capillaries, cells of MC-1 form microaerophilic bands of cells (Frankel et al. 1997). Some cells within the band make long, straight traverses through the band whereas others stop swimming and attach to the walls of the capillary or to each other at the OAI. Cells thus appear to alternate between active swimming and sessile behavior.

Cells of strain MC-1 grow chemolithoautotrophically with sulfide and other reduced sulfur sources as electron donors and molecular oxygen as the terminal electron acceptor (Williams et al. 2006). In addition, these cells also fix atmospheric dinitrogen (Bazylinski, unpublished data). This is presumably true for other magnetotactic cocci that inhabit the OAI in many marine and brackish habitats (Simmons et al. 2006). However, oxidation of S^{2-} by O_2 is autocatalytic, so an inverse $[O_2]/[S^{2-}]$ double gradient (from the downward diffusion of O_2 from air at the surface and the upward diffusion of S^{2-} from the anaerobic zone through the action of sulfate-reducing bacteria) will form even without the presence of bacteria. Consumption of S^{2-} and O_2 by bacteria at the OAI makes the gradients steeper. The coexistence or overlap region (both O_2 and S^{2-} present together) is only a few hundred μm deep (Schultz and Jorgensen 2001) and has very low (< 1 μM) concentrations of both O_2 and S^{2-}. Thus, cells have to contend with relatively low nutrient concentrations, as well as diffusion-limited flux of S^{2-} from below and O_2 from above into the overlap region.

Nutrient limitation is a fact of life in many marine habitats, and results in predominantly small, fast swimming cells (Mitchell 1991). Smaller cells require lower amounts of nutrients to grow and their higher surface to volume ratio ($S/V \sim 1/R$), increases their rate of nutrient uptake relative to their nutrient requirement. This is especially advantageous in low nutrient conditions. However, consumption of nutrients results in a greater local depletion because of diffusion limitation. Cells can solve this problem by swimming and relying on chemotaxis to find areas of locally higher nutrient concentration. At minimum, cells have to swim fast and straight enough to outrun nutrient diffusion (about 30 μm/s for 1 s) (Purcell 1977). However, small cells lose their heading in times of the order of milliseconds from buffeting by Brownian motion. One solution is swimming faster so as to get farther before going off course, which is presumably the reason why small cells that swim fast are the rule in marine environments (Mitchell 1991). However, faster swimming also burns more cellular energy because the viscous drag on cells depends on their velocity, so swimming must result in increased access to nutrients.

Cells of strain MC-1 and similar marine magnetotactic cocci with bilophotrichous flagellation are fast swimmers, yet have their magnetic dipole to keep their heading. As noted above, fast swimming perhaps allows them to make traverses from one side of the overlap region to the other to sequentially access higher concentrations of S^{2-} and O_2. However, small cells such as the cocci have low carrying capacity so they have to make shorter, more frequent, traversals than larger cells. In this case, the horizontal chemical stratification could guarantee a payoff that would cover the cost of fast swimming. Then why do cells of strain MC-1 sometimes stop swimming, as seen in the bands in the flat capillaries? The answer might involve the N_2-fixing enzyme nitrogenase. Nitrogen fixation is energy demanding and only occurs at O_2 concentrations less than about 5 μM (Zhulin et al. 1996). As noted above, the O_2 concentration in the overlap region is less than that so cells can fix N_2 there. If a cell is fixing N_2, its energy balance might improve if it stops swimming altogether.

Cells of *M. magnetotacticum*, like all other magnetospirilla, have a single flagellum at both poles of the cell and swim at about 40 μm/s, forwards and backwards with equal facility. Cultivated cells grow heterotrophically on certain organic acids (e.g., succinic acid) as an electron source with O_2 or nitrate as the terminal electron acceptor (Bazylinski and Blakemore 1983). When O_2 is the only electron acceptor available in $[O_2]$ gradients, cells form microaerophilic bands, seeking a preferred O_2 concentration that presumably maximizes the proton motive force generated by transfer of electrons (Zhulin et al. 1996; Taylor et al. 1999). Cells are in constant motion making straight-line excursions above and below the band. However, because there is no autocatalytic oxidation of electron donor by acceptor, access to nutrients is mostly limited by the diffusion of O_2 and electron source and consumption by the cells. In this situation, cells need only outrun diffusion in order to access increased concentrations of electron donor and acceptor below and above the preferred O_2 concentration, respectively. There is no need to incur the cost of faster swimming because the cellular magnetic dipole allows cells to maintain their heading, minimizing the straight run time for temporal chemotaxis (Berg 1983, 1999). Cells of the magnetospirilla, like those of strain MC-1, also fix N_2, but since they do not expend as much energy swimming as does MC-1, they likely do not need to stop swimming to conserve energy for N_2 fixation.

It should be noted that the situation for cells in situ in natural environments for the magnetospirilla might be more complex than that for the magnetotactic cocci. The fact that cells of magnetotactic spirilla collected from natural environments often display polar magnetotaxis in the hanging drop assay might indicate this. Many of the magnetotactic cocci collected from natural environments contain sulfur-rich globules suggesting they are actively oxidizing S^{2-} at the OAI. Many of the cultivated magnetospirilla possess genes encoding for cbbM, a type II ribulose-1,5-bisphosphate carboxylase/oxygenase, a key enzyme of the Calvin–Benson–Bassham cycle for autotrophy (Bazylinski and Williams, 2006, in this volume). Thus, the magnetospirilla might be able to grow chemolithoautotrophically like strain MC-1 and may also use inorganic electron donors as well as organic ones.

4.2.4
Deviations from the Magneto-Aerotaxis Models

Polar magneto-aerotaxis has been observed in some of the freshwater spirilla (D. Schüler, 2006, personal communication), bacteria that are nominally axial magneto-aerotactic. Magnetic polarity was most pronounced in strains that were freshly isolated but was gradually lost upon repeated subcultivation. Polar magnetotaxis has also been observed in cells of *M. gryphiswaldense* (D. Schüler, 2006, personal communication) and *M. magnetotacticum* (D.A. Bazylinski, unpublished data) grown in semi-solid [O_2]-gradient medium and in highly reduced medium under the microscope. However, these experiments were not entirely reproducible and thus the trigger that causes axially magneto-aerotactic cells to switch into polarly magneto-aerotactic cells is not known. Since the difference between axial and polar magneto-aerotaxis at the molecular level is not known, it is possible that the two models represent the endpoints of a continuum of responses.

The predominance of freshwater, south-seeking, magnetotactic cocci in a pond in the Northern hemisphere was reported by Cox et al. (2002) without discussion. Simmons et al. (2006) recently observed a population of uncultured, marine magnetotactic bacterium, collected from the anoxic zone of a coastal pond in the Northern hemisphere, that were primarily SS under oxic conditions in the hanging drop assay. Other, polar magnetotactic, bacteria in the sample were generally NS as expected although on occasion the ratio of SS to NS cells was greater than 0.1. Since the SS cells were not identified, it is not clear whether they are microaerophiles, leaving open the possibility that they use the magnetic field to find a preferred position in a vertical concentration gradient of a molecule or ion other than O_2 or at a specific oxidation–reduction potential. If the organism turns out to be microaerophilic, then the SS response is difficult to understand on the basis of the magneto-aerotaxis models. However, since the cells do not migrate up to the surface of the pond, something must cause them to reverse direction and swim downward in the water column. Alternatively, they may not be actively swimming and may be attached to particles. The solution to this intriguing mystery will probably require examination of the motility of the cells in an oxygen concentration gradient.

Finally, the magneto-aerotaxis model comprises passive magnetic orientation and active swimming due to flagellar rotation with the rotation sense determined by oxygen or redox sensing. On the basis of analysis of kinematics in magnetic fields, Greenberg et al. (2005) have proposed that the MMP may have magnetoreception, i.e., a magnetic field-sensing mechanism.

4.2.5 Bacterial Hemerythrins, [O_2]-Sensing, and Magneto-Aerotaxis

Hemerythrins are a group of O_2-handling proteins originally identified in certain marine invertebrates including sipunculids, priapulids, annelids, and brachiopods (Dunn et al. 1977; Vergote et al. 2004). Many prokaryotes are known to have open reading frames (ORFs) that encode for putative hemerythrins including proteins with hemerythrin-like domains. On the basis of the large number of these ORFs encoding for hemerythrin-like proteins identified in genomes, magnetotactic bacteria appear to contain the highest number of hemerythrin-like proteins among the prokaryotes. The genomes of *M. magnetotacticum*, *M. magneticum* and strain MC-1 each contain approximately 30 or more ORFs that encode for putative proteins with hemerythrin-like domains. None of these proteins have been characterized, however. Given that magnetotactic bacteria occur predominantly at the OAI and/or anoxic regions of the water column, O_2-binding proteins such as hemerythrins may serve as a sensory mechanism for O_2, and thus play a key role in magneto-aerotaxis.

Hemerythrin domains contain a sequence motif that includes five histidine residues and two carboxylate ligands that coordinate two iron atoms; reversible O_2-binding occurs at the diiron site located in a hydrophobic pocket of the protein (Stenkamp et al. 1985). Thus, these hemerythrins, both eukaryotic and prokaryotic, share certain conserved amino acid residues associated with the diiron site, in the form of the motifs H... HxxxE... HxxxH... HxxxxD (where H = histidine, E = glutamate, D = aspartate, and x_n = conserved spacer region) (Stenkamp et al. 1985; C.E. French, 2006, personal communication). These putative hemerythrins include short ($\leq$ 200 amino acid residues) single-domain proteins, such as the hemerythrin-like protein McHr (131 amino acid residues) of the methanotrophic bacterium *Methylococcus capsulatus* (Karlsen et al. 2005), as well as longer proteins in which the hemerythrin-like domain is associated with one or more other domains (especially those involved in signal transduction), such as the multidomain protein DcrH (959 residues) from the bacterial sulfate-reducing bacterium *Desulfovibrio vulgaris* (Xiong et al. 2000). For magnetotactic bacteria, some ORFs that encode for putative hemerythrin-like proteins are located within the magnetosome membrane protein gene islands in strain MC-1 (Mmc1DRAFT_1515 from draft genome) and *M. gryphiswaldense* (ORF12, ORF13; Schübbe et al. 2003; Ullrich et al. 2005). Two adjacent ORFs that encode putative proteins with hemerythrin-like domains have been identified in the genome of the magnetotactic vibrio strain MV-1, although it is not known if these ORFs are situated within the magnetosome island (D.A. Bazylinski, unpublished). One of these MV-1 hemerythrin-like proteins is of a single-domain kind (202 residues). The other (748 residues) contains multiple domains, including two histidine kinase-like domains (the second of these a putative histidine kinase-like ATPase) followed by a hemerythrin-like domain and a carboxy terminus signal receiver domain (Fig. 4). Other putative multi-domain proteins from other magnetotactic bacteria also include hemerythrin domains associated with signal transduction domains (e.g., histidine kinases, methyl-accepting chemotaxis proteins).

In marine invertebrates, hemerythrin is used for O_2 transport between tissues (Stenkamp et al. 1985). The function of hemerythrins in prokaryotes is unclear, and they may perform disparate functions in different or-

Fig. 4 Domain structure of a putative hemerythrin-like protein predicted from an ORF identified in the genome of the magnetotactic vibrio strain MV-1, based on translation of ORF, showing the following putative domains: HisKA (histidine kinase A), HATPase (histidine kinase-like ATPase), hemerythrin, and REC (signal receiver domain). Polypeptide is 748 amino acid residues long

ganisms. The putative chemotaxis protein DcrH from *D. vulgaris* contains a hemerythrin-like domain at the carboxy terminus, and has been suggested to have a role in O_2-sensing (Xiong et al. 2000). The hemerythrin-like McHr protein from *Meth. capsulatus*, may furnish oxygen-dependent enzymes with O_2 (Karlsen et al. 2005). It has also been suggested that hemerythrin is part of a detoxification mechanism for bacteria that have a low tolerance for O_2 (anaerobes, microaerophiles) (Xiong et al. 2000). Many motile bacteria are exposed to variable $[O_2]$, and, like magnetotactic bacteria, may selectively migrate to anoxic and oxic conditions (such as to obtain electron donors and acceptors, respectively), so hemerythrins may serve to differentially bind and release O_2 (C.E. French, 2006, personal communication). For magnetotactic bacteria migrating within and through the OAI, hemerythrins may serve to bind O_2 when the cell is exposed to elevated O_2 concentrations, and then release the O_2 when the cell descends into anoxic conditions (C.E. French, 2006, personal communication). Multi-domain proteins with both signal transduction and hemerythrin domains suggests a role in O_2-sensing, as proposed for DcrH in *D. vulgaris* (Xiong et al. 2000). Even single-domain hemerythrins may serve a sensory function, if they are co-transcribed and/or acting with signal transduction proteins. Given the prevalence of hemerythrin-like ORFs in the known genomes of magnetotactic bacteria, including those within the magnetosome protein gene island (Ullrich et al. 2005), hemerythrins may play a role in magneto-aerotaxis (including directing flagellar rotation). However, this has yet to be determined.

The genomes of *M. magnetotacticum*, *M. magneticum*, and strain MC-1 show numerous ORFs that encode for putative proteins with PAS domains, providing many potential candidate genes for the identification of aero-, redox-, and (perhaps) phototaxis in these bacteria. In bacteria, PAS domains are responsible for sensing stimuli such as $[O_2]$, redox potential, and light (Taylor and Zhulin 1999; Repik et al. 2000; Watts et al. 2006). For example, the aerotaxis receptor (Aer) responds to oxygen concentration in the environment, and is the first step in the intracellular pathway that governs the sense of flagellar rotation in *Escherichia coli* (Watts et al. 2006). As mentioned above, the polarly magneto-aerotactic coccus, strain MC-1, display a negative phototaxis in response to short-wavelength light, but the mechanism is unknown. It is difficult to infer the precise identity of the stimulus that the

PAS-containing protein is sensitive to based on amino acid sequence alone. This is also the case for the numerous ORFs that encode putative methyl-accepting chemotaxis proteins in *M. magnetotacticum*, *M. magneticum*, and MC-1, including putative hemerythrins.

In *Magnetospirillum* species and strain MC-1, the genes for the proteins implicated in magnetosome biosynthesis are located within a genomic island. In *M. gryphiswaldense* the magnetosome genes are located within a hypervariable 130-kb stretch of the genome within the magnetosome island (Schübbe et al. 2003; Ulrich et al. 2005). In *M. gryphiswaldense*, the genes for MamA, MamB, MamJ and MamK are located on the *mamAB* operon, and the genes for MamC and MamD are located on the *mamDC* operon (Schübbe et al. 2003; Ulrich et al. 2005). Functions for magnetosome-membrane associated proteins have been determined for MamJ and MamK. MamJ was demonstrated to be essential for the assembly of magnetosome chains in *M. gryphidwaldense*, probably through interaction with MamK (Scheffel et al. 2005); and MamK appears to be involved in the formation of a network of actin-like filaments that comprise the magnetosomal cytoskeleton and is responsible for the linear chain-like alignment of magnetosomes within the cell (Komeili et al. 2005). The presence of hemerythrin-like genes in the magnetosome islands may imply some interaction between magnetosome synthesis and O_2-handling mediated by hemerythrins, but this remains to be elucidated. In *M. gryphiswaldense*, two putative hemerythrin ORFs (ORF12, OR13) are located between the *mamAB* and *mamDC* operons (Schübbe et al. 2003).

5 Questions about Magnetotaxis

Magnetotactic bacteria have solved the problem of constructing an internal, permanent, magnetic dipole that is sufficiently robust so that a cell will be oriented along the geomagnetic field as it swims, yet be no longer than the length of the cell itself (ca. 1–2 μm). The solution, the magnetosome chain, is very elegant and efficient in that it makes maximum use of the minimum amount of magnetite, assuming that cells want to maximize the ratio of magnetic moment to volume of magnetite. Since magnetite is four times more magnetic than the same volume of greigite, why do some magnetotactic bacteria biomineralize greigite? This question is particularly acute for those cells that contain both magnetite and greigite magnetosomes co-organized in the same chains. Why not all magnetite? Also, SD magnetite is a clear winner over multidomain magnetite for making permanent magnets. So why are there magnetotactic bacteria that make magnetosomes up to 250 nm in length, larger than SD, hence with a lower magnetic moment per unit volume? Since arranging magnetosomes in chains is so efficient, why do some species have dispersed clusters of magnetosomes? From the magnetism point of view, this is not as efficient as alignment in chains, because it requires the cell to align the long axes of all magnetosomes parallel to each other. Might there be other, non-magnetic roles for magnetite in cells? Possibilities include the storage and sequestration of iron as an electron acceptor/donor reserve although the iron present in magnetosomes has never been shown to be utilized by cells. Moreover, there is evidence that some species cannot utilize the iron in magnetite-containing magnetosomes and will continue to synthesize magnetosomes and limit their growth under iron-limiting conditions (Dubbels et al. 2004). Also, magnetite crystals can disproportionate H_2O_2, and probably oxygen radicals produced during aerobic respiration, suggesting magnetite magnetosomes could be an elementary catalase, or have another catalytic role. Cu, a potentially toxic element, is sometimes found in greigite magnetosomes, which suggests a possible detoxification role (Bazylinski et al. 1993a). Free iron ions within the cell are also toxic through the generation of highly reactive and toxic oxygen species such as hydroxyl radicals (Halliwell and Gutteridge 1984). The toxic effect of these ions could be eliminated by concentrating the free iron ions in a relatively inert mineral like magnetite. But this doesn't explain why the cell takes up so much iron in the first place.

There are also questions about magnetotaxis itself (Frankel and Bazylinski 2004). There are many microaerophilic organisms, including non-magnetic mutants of magnetotactic bacteria, which form aerotactic bands without the aid of magnetism. Simulations of axial magnetotactic bacteria confirm the fact that magneto-aerotaxis is more efficient than aerotaxis alone for finding the optimal $[O_2]$, meaning magnetotactic bacteria would find the optimal concentration before non-magnetic aerotactic bacteria with the same swimming speed, but only at high inclinations of the geomagnetic field. Many polar magnetotactic bacteria are fast swimmers, ca. 100 body lengths per second or more, so the efficiency argument may hold over a greater range of geomagnetic inclination for these organisms. Nevertheless, the question of whether aerotactic efficiency alone is sufficient to account for the persistence of magnetotaxis in bacteria over geologic time scales is still open.

Acknowledgements We thank B. L. Cox, C. E. French, S. L. Simmons, and D. Schüler for discussions. DAB was supported by US National Science Foundation Grant EAR-0311950.

References

Bazylinski DA, Blakemore RP (1983) Denitrification and assimilatory nitrate reduction in *Aquaspirillum magnetotacticum*. Appl Environ Microbiol 46:1118–1124

Bazylinski DA, Frankel RB (2000) Biologically controlled mineralization of magnetic iron minerals by magnetotactic bacteria. In: Lovley DR (ed) Environmental microbe-metal interactions. ASM Press, Washington, DC, p 109–144

Bazylinski DA, Frankel RB (2004) Magnetosome formation in prokaryotes. Nature Rev Microbiol 2:217–230
Bazylinski DA, Moskowitz BM (1997) Microbial biomineralization of magnetic iron minerals: microbiology, magnetism, and environmental significance. In: Banfield JF, Nealson KH (eds) Geomicrobiology: interactions between microbes and minerals (Rev Mineral Vol 35). Mineralogical Society of America, Washington, DC, p 181–223
Bazylinski DA, Dean AJ, Williams TJ, Kimble-Long L, Middleton SL, Dubbels BL (2004) Chemolithoautotrophy in the marine, magnetotactic bacterial strains MV-1 and MV-2. Arch Microbiol 182:373–387
Bazylinski DA, Frankel RB, Heywood BR, Mann S, King JW, Donaghay PL, Hanson AK (1995) Controlled biomineralization of magnetite (Fe_3O_4) and greigite (Fe_3S_4) in a magnetotactic bacterium. Appl Environ Microbiol 61:3232–3239
Bazylinski DA, Frankel RB, Jannasch HW (1988) Anaerobic magnetite production by a marine, magnetotactic bacterium. Nature 334:518–519
Bazylinski DA, Garratt-Reed AJ, Abedi A, Frankel RB (1993a) Copper association with iron sulfide magnetosomes in a magnetotactic bacterium. Arch Microbiol 160:35–42
Bazylinski DA, Heywood BR, Mann S, Frankel RB (1993b) Fe_3O_4 and Fe_3S_4 in a bacterium. Nature 366:218
Berg H (1983) Random walks in biology. Princeton University Press, Princeton, NJ
Berg H (1999) Motile behavior of bacteria, Physics Today on the web (http://www.aip.org/pt/vol-53/iss-1/berg.htm)
Blakemore RP (1975) Magnetotactic bacteria. Science 190:377–379
Blakemore RP (1982) Magnetotactic bacteria. Annu Rev Microbiol 36:217–238
Blakemore RP, Frankel RB (1981) Magnetic navigation in bacteria. Sci Am 245(6):58–65
Blakemore RP, Frankel RB, Kalmijn AJ (1980) South-seeking magnetotactic bacteria in the southern hemisphere. Nature 236:384–386
Blakemore RP, Maratea D, Wolfe RS (1979) Isolation and pure culture of a freshwater magnetic spirillum in chemically defined medium. J Bacteriol 140:720–729
Cox BL, Popa R, Bazylinski DA, Lanoil B, Douglas S, Belz A, Engler DL, Nealson KH (2002) Organization and elemental analysis of P-, S-, and Fe-rich inclusions in a population of freshwater magnetococci. Geomicrobiol J 19:387–406
DeLong EF, Frankel RB, Bazylinski DA (1993) Multiple evolutionary origins of magnetotaxis in bacteria. Science 259:803–806
Dubbels BL, DiSpirito, Morton JD, Semrau JD, Neto JNE, Bazylinski DA (2004) Evidence for a copper-dependent iron transport system in the marine, magnetotactic bacterium strain MV-1. Microbiology 150:2931–2945
Dunin-Borkowski RE, McCartney MR, Frankel RB, Bazylinski DA, Posfai M, Buseck PR (1998) Magnetic microstructure of magnetotactic bacteria by electron holography. Science 282:1868–1870
Dunn JB, Addison AW, Bruce RE, Loehr JS, Loehr TM (1977) Comparison of hemerythrins from four species of sipunculids by optical absorption, circular dichroism, fluorescence emission, and resonance Raman spectroscopy. Biochem 16:1743–1749
Farina M, Esquivel DMS, Lins de Barros HGP (1990) Magnetic iron-sulphur crystals from a magnetotactic microorganism. Nature 343:256–258
Frankel RB (1984) Magnetic guidance of organisms. Annu Rev Biophys Bioeng 13:85–103
Frankel RB, Bazylinski DA (2004) Magnetosome mysteries. ASM News 70:176–183
Frankel RB, Blakemore RP (1980) Navigational compass in magnetic bacteria. J Magn Magn Mater 15–18:1562–1565
Frankel RB, Bazylinski DA, Johnson M, Taylor BL (1997) Magneto-aerotaxis in marine, coccoid bacteria. Biophys J 73:994–1000

Frankel RB, Blakemore RP, Wolfe RS (1979) Magnetite in freshwater magnetic bacteria. Science 203:1355–1357

Garcia-Pichel F (1989) rapid bacteria swimming measured in swarming cells of Thiovulum majus. J Bacteriol 171:3560–3563

Gorby YA, Beveridge TJ, Blakemore RP (1988) Characterization of the bacterial magnetosome membrane. J Bacteriol 170:834–841

Greenberg M, Canter K, Mahler I, Thornheim A (2005) Observation of magnetoreceptive behavior in a multicellular, magnetotactic, prokaryote in higher than geomagnetic fields. Biophys J 88:1496–1499

Halliwell B, Gutteridge JM (1984) Oxygen toxicity, oxygen radicals, transition metals and disease. Biochem J 219:1–14

Hanzlik M, Winklhofer M, Petersen N (2002) Pulsed-field-remanence measurements on individual magnetotactic bacteria. J Magn Magn Mater 248:258–267

Heywood BR, Bazylinski DA, Garratt-Reed AJ, Mann S, Frankel RB (1990) Controlled biosynthesis of greigite in magnetotactic bacteria. Naturwiss 77:536–538

Kalmijn AJ (1981) Biophysics of geomagnetic field detection. IEEE Trans Magn Mag 17:1113–1124

Karlsen OA, Ramsevik L, Bruseth LJ, Larsen Ø, Brenner A, Berven FS, Jensen HB, Lillehaug JR (2005) Characterization of a prokaryotic haemerythrin from the methanotrophic bacterium *Methylococcus capsulatus* (Bath). FEBS J 272:2428–2440

Kasama T, Posfai M, Chong RKK, Finlayson AP, Dunin-Borkowski RE, Frankel RB (2006) Magnetic microstructure of iron sulfide crystals in magnetotactic bacteria from off-axis electron holography. Physica B (in press)

Kirschvink JL (1980) South-seeking magnetic bacteria. J Exp Biol 86:345–347

Komeili A, Li Z, Newman DK, Jensen GJ (2005) Magnetosomes are invaginations organized by the actin-like protein MamK. Science 311:242–245

Komeili A, Vali H, Beveridge TJ, Newman DK (2004) Magnetosome vesicles are present prior to magnetite formation and MamA is required for their activation. Proc Natl Acad Sci USA 101:3839–3844

Lins U, McCartney MR, Farina M, Buseck PR, Frankel RB (2005) Crystal habits and magnetic microstructures of magnetosomes in coccoid magnetotactic bacteria. Appl Environ Microbiol 71:4902–4905

Mann S, Sparks NHC, Frankel RB, Bazylinski DA, Jannasch HW (1990) Biomineralization of ferrimagnetic greigite (Fe_3S_4) and iron pyrite (FeS_2) in a magnetotactic bacterium. Nature 343:258–261

Maratea D, Blakemore RP (1981) *Aquaspirillum magnetotacticum* sp. nov., a magnetic spirillum. Int J Syst Bacteriol 31:452–455

Matsunaga T, Sakaguchi T, Tadokoro F (1991) Magnetite formation by a magnetic bacterium capable of growing aerobically. Appl Microbiol Biotechnol 35:651–655

Matsunaga T, Okamura Y, Fukuda Y, Wahyudi AT, Murase Y, Takeyama H (2005) Complete genome sequence of the facultative anaerobic magnetotactic bacterium *Magnetospirillum* sp. strain AMB-1. DNA Res 12:157–166

McCartney MR, Lins U, Farina M, Buseck PR, Frankel RB (2001) Magnetic mirostructure of bacterial magnetite by electron holography. Eur J Mineral 13:685–689

Meldrum FC, Mann S, Heywood BR, Frankel RB, Bazylinski DA (1993a) Electron microscope study of magnetosomes in a cultured coccoid magnetotactic bacterium. Proc Roy Soc Lond B 251:231–236

Meldrum FC, Mann S, Heywood BR, Frankel RB, Bazylinski DA (1993b) Electron microscope study of magnetosomes in two cultured vibrioid magnetotactic bacteria. Proc Roy Soc Lond B 251:237–242

Mitchell JG (1991) The influence of cell size on marine bacterial motility and energetics. Microb Ecol 22:227–238

Moench TT (1988) *Bilophococus magnetotacticus*, gen. nov. sp. nov., a motile, magnetic coccus. Antonie van Leeuwenhoek 54:483–496

Nogueira FS, Lins de Barros HGP (1995) Study of the motion of magnetotactic bacteria. Eur Biophys J 24:13–21

Penninga I, de Waard H, Moskowitz BM, Bazylinski DA, Frankel RB (1995) Remanence measurements on individual magnetotactic bacteria using pulsed magnetic fields. J Magn Magn Mater 149:279–286

Pósfai M, Buseck PR, Bazylinski DA, Frankel RB (1998) Reaction sequence of iron sulfides in bacteria and their use as biomarkers. Science 280:880–883

Purcell EM (1977) Life at low Reynolds number. Am J Phys 45:3–11

Repik A, Rebbapragada A, Johnson MS, Haznedar JO, Zhulin IB, Taylor BL (2000) PAS domain residues involved in signal transduction by the Aer redox sensor of *Escherichia coli*. Mol Microbiol 36:806–816

Rogers FG, Blakemore RP, Blakemore NA, Frankel RB, Bazylinski DA, Maratea D, Rogers C (1990) Intercellular structure in a many-celled magnetotactic procaryote. Arch Microbiol 154:18–22

Sakaguchi T, Arakaki A, Matsunaga T (2002) *Desulfovibrio magneticus* sp. nov., a novel sulfate-reducing bacterium that produces intracellular single-domain-sized magnetite particles. Int J Syst Evol Microbiol 52:215–221

Sakaguchi T, Burgess JG, Matunaga T (1993) Magnetite formation by a sulphate reducing bacterium. Nature 365:47–49

Scheffel A, Gruska M, Faive D, Linaroudisn A, Plitzko JM, Schüler D (2006) An acidic protein aligns magnetosomes along a filamentous structure in magnetotactic bacteria. Nature 440:110–114

Schleifer K-H, Schüler D, Spring S, Weizenegger M, Amann R, Ludwig W, Kohler M (1991) The genus *Magnetospirillum* gen. nov., description of *Magnetospirillum gryphiswaldense* sp. nov. and transfer of *Aquaspirillum magnetotacticum* to *Magnetospirillum magnetotacticum* comb. nov. Syst Appl Microbiol 14:379–385

Schübbe S, Kube M, Scheffel A, Wawer C, Heyen U, Meyerdierks A, Madkour MH, Mayer F, Reinhardt R, Schüler D (2003) Characterization of a spontaneous nonmagnetic mutant of *Magnetospirillum gryphiswaldense* reveals a large deletion comprising a putative magnetosome island. J Bacteriol 185:5779–5790

Schüler D, Spring S, Bazylinski DA (1999) Improved technique for the isolation of magnetotactic spirilla from freshwater sediment and their phylogenetic characterization. Syst Appl Microbiol 22:466–471

Schultz HN, Jorgensen BB (2001) Big bacteria. Annu Rev Microbiol 55:105–137

Simmons SL, Bazylinski DA, Edwards KJ (2006) South-seeking magnetotactic bacteria in the northern hemisphere. Science 311:371–374

Simmons SL, Sievert SM, Frankel RB, Bazylinski DA, Edwards KJ (2004) Spatiotemporal distribution of marine magnetotactic bacteria in a seasonally stratified coastal pond. Appl Environ Microbiol 70:6230–6239

Spormann AM, Wolfe RS (1984) Chemotactic, magnetotactic, and tactile behavior in a magnetic spirillum. FEMS Microbiol Lett 22:171–177

Spring S, Bazylinski DA (2000) Magnetotactic bacteria. In: The prokaryotes. Published on the web at http://www.springer-ny.com/, Springer, New York

Spring S, Amann R, Ludwig W, Schleifer K-H, van Gemerden H, Petersen N (1993) Dominating role of an unusual magnetotactic bacterium in the microaerobic zone of a freshwater sediment. Appl Environ Microbiol 59:2397–2403

Spring S, Lins U, Amann R, Schleifer K-H, Ferreira LCS, Esquivel DMS, Farina M (1998) Phylogenetic affiliation and ultrastructure of uncultured magnetic bacteria with unusually large magnetosomes. Arch Microbiol 169:136–147

Stenkamp RE, Sieker LC, Jensen LH, McCallum JD, Sanders-Loehr J (1985) Active site structures of deoxyhemerythrin and oxyhemerythrin. Proc Natl Acad Sci USA 82:713–716

Taylor BL, Zhulin IB, Johnson MS (1999) Aerotaxis and other energy-sensing behavior in bacteria. Annu Rev Microbiol 53:103–128

Torres de Araujo FF, Pires MA, Frankel RB, Bicudo CE (1986) Magnetite and magnetotaxis in magnetotactic bacteria. Biophys J 50:375–378

Ullrich S, Kube M, Schübbe S, Reinhardt R, Schüler D (2005) A hypervariable 130-kilobase genomic region of *Magnetospirillum gryphiswaldense* comprises a magnetosome island which undergoes frequent rearrangements during stationary growth. J Bacteriol 187:7176–7184

Vergote D, Sautière P-E, Vandenbulcke F, Vieau D, Mitta G, Macagno ER, Salzet M (2004) Up-regulation of neurohemerythrin expression in the central nervous system of the medicinal leech, *Hirudo medicinalis*, following septic injury. J Biol Chem 279:43828–43837

Watts KJ, Johnson MS, Taylor BL (2006) Minimal requirements for oxygen sensing by the aerotaxis receptor Aer. Mol Microbiol 59:1317–1326

Williams TJ, Zhang CL, Scott JH, Bazylinski DA (2006) Evidence for autotrophy via the reverse tricarboxylic acid cycle in the marine magnetotactic coccus strain MC-1. Appl Environ Microbiol 72:1322–1329

Xiong J, Kurtz DM Jr, Ai J, Sanders-Loehr J (2000) A hemerythrin-like domain in a bacterial chemotaxis protein. Biochem 39:5117–5125

Zhulin IB, Bespalov VA, Johnson MS, Taylor BL (1996) Oxygen taxis and proton motive force in *Azospirillum brasilense*. J Bacteriol 178:5199–5204

Microbiol Monogr (3)
D. Schüler: Magnetoreception and Magnetosomes in Bacteria
DOI 10.1007/7171_037/Published online: 30 August 2006

Diversity and Taxonomy of Magnetotactic Bacteria

Rudolf Amann (✉) · Jörg Peplies · Dirk Schüler

Max-Planck-Institut für Marine Mikrobiologie, Celsiusstr. 1, 28359 Bremen, Germany
ramann@mpi-bremen.de

Abstract Studies of the diversity of magnetotactic bacteria (MTB) benefit from the unique advantage that MTB can be readily separated from sediment particles and other bacteria based on their magnetotaxis. This is the reason why current knowledge on MTB diversity relies to a lesser extent on the isolation and characterization of pure cultures, the classical tool of microbiology, than in other groups of microorganisms. Microscopy of magnetotactic enrichments retrieved from various environmental samples has consistently revealed significant morphological and ultrastructural diversity of MTB. However, of the many morphotypes detected, including spirilla, cocci, vibrios, ovoid, rod-shaped and even multicellular bacteria, only few bacteria could so far be brought into pure culture. The taxonomy of MTB is therefore heavily based on comparative sequence analysis of their 16S rRNA genes which can be investigated without prior cultivation. Based on 16S rRNA sequence similarity MTB are polyphyletic. Most of the MTB pure cultures and many of the so far uncultured phylotypes cluster within the *Alphaproteobacteria*, but MTB have also been affiliated to *Deltaproteobacteria*, to the phylum *Nitrospira*, and, tentatively, also to *Gammaproteobacteria*.

1 Introduction

The studies on the diversity of magnetotactic bacteria (MTB) benefit from the unique advantage that MTB can be readily separated from sediment particles and other bacteria based on their magnetotaxis. Because of this, current knowledge on MTB diversity relies to a lesser extent on the isolation and characterization of pure cultures, the classical tool of microbiology, than in other groups of microorganisms. Also, the microscopy of magnetotactic enrichments retrieved from various environmental samples has consistently revealed significant morphological and ultrastructural diversity of MTB. However, of the many morphotypes detected, including spirilla, cocci, vibrios, ovoid, rod-shaped and even multicellular bacteria (Fig. 1), only few bacteria could so far be brought into pure culture. The taxonomy of MTB is therefore heavily based on comparative sequence analysis of their 16S rRNA genes which can be investigated without prior cultivation (Amann et al. 1995). Based on 16S rRNA sequence similarity MTB are polyphyletic. Most of the MTB pure cultures and many of the so far uncultured phylotypes cluster within the *Alphaproteobacteria*, but MTB have also been affiliated to *Deltaproteobacteria*, to the phylum *Nitrospira*, and, tentatively, also to *Gammaproteobacteria* (Fig. 2). In the following we summarize the current knowledge on MTB diversity along the groupings suggested by 16S rRNA.

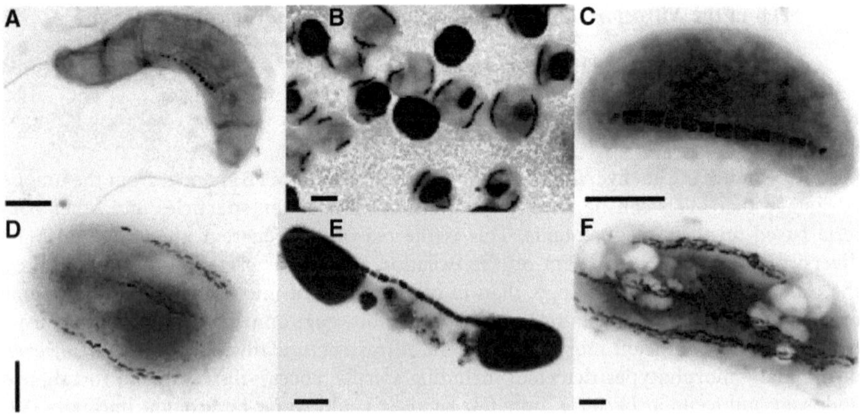

Fig. 1 Transmission electron micrographs showing characteristic cell morphologies of various magnetotactic bacteria. The diversity of abundant morphotypes includes spirilla (**A**), cocci (**B**), vibrios (**C**), ovoid forms (**D**), and large rods (**E**,**F**). *Scale bar* represents 0.5 μm (part (**D**) courtesy of Christine Flies)

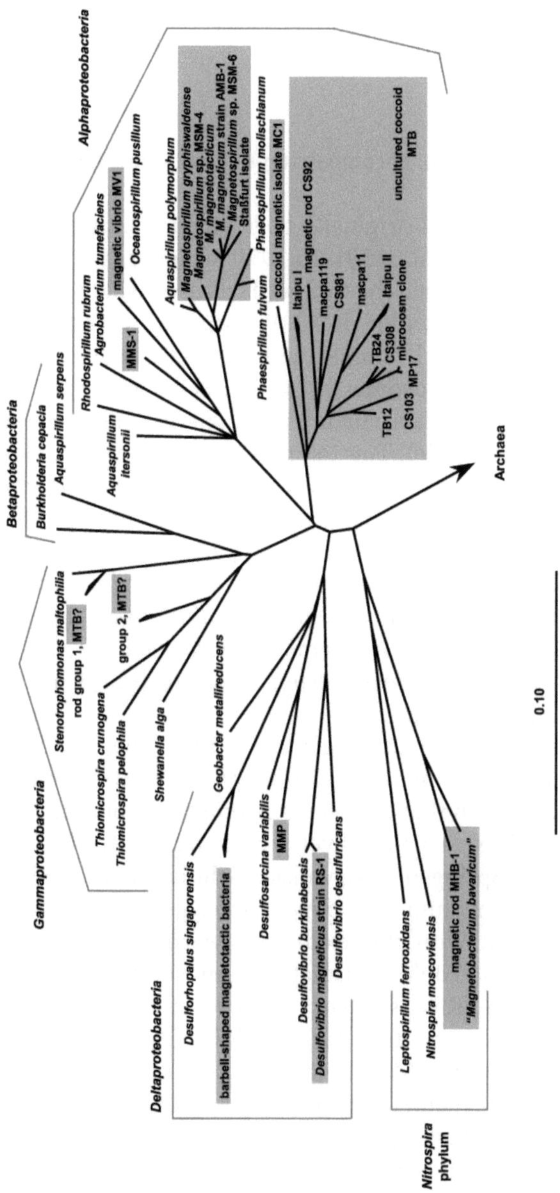

Fig. 2 16S rRNA-based tree reconstruction showing the phylogeny of magnetotactic bacteria (*grey boxes*). The tree is based on neighbor joining analyses and was corrected according to the results of maximum likelihood and maximum parsimony methods using the ARB program package (Ludwig et al. 2004). Multifurcations were drawn whenever branching orders were not stable. The tree topology is based on almost full-length 16S rRNA sequences. The partial sequences of strains CS103, MC-1, MV-1, and MMP have been added with the parsimony tool without allowing changes of the overall tree topology (Ludwig et al. 2004). *Bar* = 10% estimated sequence divergence

2 Alphaproteobacterial MTB

2.1 The Genus *Magnetospirillum*, Encompassing Heterotrophic Culturable MTB

The first MTB isolate had originally been named "*Aquaspirillum magnetotacticum*" (Blakemore et al. 1979). In 1991 Schleifer, Schüler and coworkers (Schleifer et al. 1991) studied the two pure cultures available at that time. Both are magnetospirilla of a width of 0.2–0.7 μm and a length of 1–3 μm. The ultrastructure is also highly similar with respect to the arrangement (single chain of up to 60 magnetosomes), size (diameter approximately 40–45 nm) and cubo-octahedral crystal structure of magnetosomes as well as flagellation (single flagella at each pole). The 16S rRNA sequences were closely related to each other (94%) and clustered with that of *Alphaproteobacteria* whereas the 16S rRNA sequence of the type species of the genus *Aquaspirillum*, *A. serpens*, clusters with that of *Betaproteobacteria*. In parallel, the 16S rRNA sequence of *A. magnetotacticum* was also determined by Eden and coworkers (Eden et al. 1991). Consequently, the genus *Magnetospirillum* and the two species *Magnetospirillum magnetotacticum* (formerly "*Aquaspirillum magnetotacticum*") and *Magnetospirillum gryphiswaldense* were created (Schleifer et al. 1991; Schüler and Schleifer 2005).

In 1993 it was shown (Burgess et al. 1993) that the 16S rRNAs of two facultatively anaerobic strains of magnetic spirilla (AMB-1 and MGT-1) share 98–99% similarity with that of *M. magnetotacticum* but only 95–96% to that of *M. gryphiswaldense*. Further diversity of magnetospirilla was revealed by a study of Schüler, Spring and Bazylinski (Schüler et al. 1999) in which a new two-layer isolation medium with opposing oxygen and sulfide gradients was used for cultivation. With this technique seven strains of microaerophilic magnetotactic spirilla could be isolated from one freshwater pond in Iowa, USA. While the 16S rRNA sequences of five of the isolates (MSM-1,-6,-7,-8,-9) were very similar to either *M. gryphiswaldense* or *M. magnetotacticum* (> 99.7%), two (MSM-3, MSM-4) may represent an additional group. In a recent study, novel magnetotactic spirilla strains were isolated from various freshwater habitats including a ditch and several ponds in Northern Germany (Flies et al. 2005). Again, 16S rRNA analysis affiliated them all with the genus *Magnetospirillum* with highest similarity to strain MSM-6. Although by now only two species have been validly named, the diversity within this genus of culturable MTB is significant. It is likely that an in-depth taxonomic study would result in the valid description of additional species.

The 16S rRNA sequence similarity between *Magnetospirillum gryphiswaldense* and *M. magnetotacticum* of 96% is similar to that shared by *Magnetospirillum* spp. and members of the photoorganoheterotrophic, nonmagnetotactic genus *Phaeospirillum* (formerly *Rhodospirillum*) (Fig. 2). Furthermore, with 98% 16S rRNA similarity the non-magnetotactic *Aquaspirillum polymorphum* is significantly closer to *M. gryphiswaldense* than *M. magnetotacticum*. Recently, more bacteria have been isolated which are based on a 16S rRNA sequence similarity of about 95%, morphology, and physiology similar to *Magnetospirillum* spp., but lack the capability to form magnetosomes (Coates et al. 1999; Shinoda et al. 2000). Based on the different phenotypes in otherwise closely related and physiologically similar strains of *Magnetospirillum* it is tempting to speculate whether either *A. polymorphum* represents a magnetospirillum strain that recently lost genes essential for magnetotaxis, or *M. gryphiswaldense* has recently acquired the set of genes responsible for magnetosome formation. If the former scenario applies, *A. polymorphum* should be renamed *Magnetospirillum polymorphum*. Further taxonomic investigations are required to clarify the evolutionary relationships of these magnetic and non-magnetic spirilli.

Interestingly, spirillum-shaped morphotypes represent only a minority of the mostly coccoid to rod-shaped MTB in primary enrichments obtained from aquatic environments based on magnetotaxis. However, during subsequent attempts of pure culture retrieval the magnetospirilla outcompete the other morphotypes (Spring and Schleifer 1995). Many of the MTB have therefore only been phylogenetically identified based on cultivation-independent 16S rRNA-based comparative sequence analysis and fluorescence in situ hybridization (FISH) of single bacterial cells with 16S rRNA-targeted oligonucleotide probes (Amann et al. 1995).

2.2
Other Cultivated Alphaproteobacterial MTB

Dennis Bazylinski was the first to isolate MTB other than *Magnetospirillum* strains in pure culture, which included a magnetic vibrio (strain MV-1) (Bazylinski et al. 1988), and a magnetic coccus (strain MC-1). Strain MC-1 represents the only so far cultured magnetotactic coccus and was isolated from brackish water collected from the Pettaquamscutt Estuary (Rhode Island, USA). The name *Magnetococcus marinus* is being proposed for this strain (D.A. Bazylinski, personal communication 2006). Comparative 16S rRNA sequence analysis affiliated the two isolates, which both contain iron oxide magnetosomes and can grow either chemoheterotrophically or chemolithoautotrophically, with the *Alphaproteobacteria* (DeLong et al. 1993) (Fig. 2). Isolate MC-1 is grouping with other, yet uncultured magnetococci (discussed in detail below), whereas MV-1 is closer to *Magnetospirillum*.

Two novel magnetotactic marine spirillum strains designated MMS-1 and MMS-2 were isolated (see chapter by Bazylinski and Williams, this volume). Although not fully characterized, preliminary analysis suggest that MMS-1 and MMS-2 represent a new genus within the *Alphaproteobacteria* and have no close phylogenetic relatives (T.J. Williams and D.A. Bazylinski, personal communication 2006). The new isolate MMS1 warrants further taxonomic studies since it is a new phylotype that based on a 16S rRNA sequence similarity of only 90% species with MV-1 could represent a new genus of culturable MTB.

2.3 Uncultured Alphaproteobacterial MTB

When upper sediment layers of Lake Chiemsee—a large, mesotrophic freshwater lake in Upper Bavaria, Germany—was stored on a laboratory shelf protected from direct light for several weeks, high numbers of magnetotactic bacteria enriched right beneath the water-sediment interface (Spring et al. 1992). Magnetotactic enrichments encompassed four distinct morphotypes: abundant cocci, two big rods of distinct morphology (one slightly bent and therefore originally referred to as "big vibrio" (Spring et al. 1992)), and small vibrios. Partial 16S rRNA genes were PCR-amplified directly from the enriched cells, singularized by cloning in *E. coli*, and sequenced. Most of the retrieved sequences grouped with 16S rRNA sequences of *Alphaproteobacteria*. FISH with three 16S rRNA-targeted oligonucleotide probes constructed complementary to signature regions of the most frequent alphaproteobacterial sequences (CS, for Chiemsee, 103, 308, and 310) identified three discrete subpopulations of the cocci. Simultaneous applications of two differentially labeled probes showed differences in abundance and taxis: magnetococci of genotype CS308 dominated over coccal genotypes CS103 and CS310, and, under the influence of a magnetic field, cells of genotype CS103 were predominantly entrapped nearest to the agarose solution-air interface.

In 1994, Spring and coworkers used the cultivation-independent approach to retrieve another three partial and seven almost full length 16S rRNA gene sequences from freshwater sediments of various sites in Germany (Spring et al. 1994). By FISH all sequences were assigned to magnetotactic bacteria, nine to magnetotactic cocci and one to the rod-shaped magnetotactic phylotype CS92. The magnetotactic rod shared a 16S rRNA similarity of 90–92% with the magnetotactic cocci, indicating affiliation in a new genus. Most cocci shared 16S rRNA similarities of less than 97%, suggesting that they represent different species. All sequences grouped with those earlier retrieved from the uncultured Chiemsee magnetotactic cocci (Spring et al. 1992).

Another rRNA-based study on the phylogeny and in situ identification of MTB addressed enrichments from the Itaipu lagoon near Rio de Janeiro (Spring et al. 1998). These were dominated by coccoid-to-ovoid morphotypes. Some of the cells produced unusually large magnetosomes that with a length of 200 nm and a width of 160 nm are almost twice as big as those found in other magnetotactic bacteria (Farina et al. 1994). Sequencing of 16S rRNA genes revealed two clusters (Itaipu I and II) of closely related sequences within the lineage of magnetotactic cocci (Spring et al. 1992, 1994; Thornhill et al. 1995). In order to link at high resolution the ultrastructure of the enriched cells with their 16S rRNA sequence a new methodology was applied. In situ hybridizations were performed with digoxigenin- and fluorescein-labeled polynucleotide probes on ultrathin sections of embedded magnetotactic bacteria. For probe synthesis one representative clone of each of the two closely related 16S rRNA clusters was used as a template for in vitro transcription of a 230 nucleotide long variable region at the 5′ end of the 16S rRNA. A bound polynucleotide probe was detected by incubation of the sections with gold-labeled antibodies specific for fluorescein or digoxigenin. The gold labels could then be detected in the electron microscope. This enabled for the first time a detailed description of the ultrastructure of in situ identified single MTB: the unusually large magnetosomes were only found in ovoid magnetotactic bacteria of the Itaipu I 16S RNA genotype.

Cox and coworkers investigated the diversity of magnetotactic cocci in Baldwin Lake (Los Angeles) by restriction fragment length patterns (RFLP) analysis (Cox et al. 2002). They found several 16S rRNA sequences, which reportedly had high similarities to known magnetotactic cocci from the database. In addition, they identified six sequences, which formed a monophyletic cluster (ARB-1 cluster) related to, but distinct from other magnetotactic bacteria (89% similarity to the magnetotactic rod CS92). Unfortunately, these sequences are currently not accessible in public databases.

Recently, Flies and coworkers (Flies et al. 2005) have investigated the diversity of magnetotactic bacteria in various microcosms with freshwater and marine sediments from Germany and Sweden by DGGE and amplified ribosomal DNA restriction analysis (ARDRA) of the 16S rRNA genes. Initially, the sediments contained a highly diverse population of magnetotactic bacteria displaying a variety of different morphotypes. However, the magnetotactic population in the microcosms underwent a rapid succession, which usually resulted after several weeks of incubation in the dominance of a magnetotactic coccus affiliated with the *Alphaproteobacteria.*

While most MTB 16S rRNA sequences were identified after magnetic enrichment, sequences putatively originating from marine MTB have been repeatedly found without magnetic manipulation. For instance, Riemann and coworkers retrieved two sequences when investigating the bacterial community composition in the Arabian Sea by DGGE analysis (Riemann et al. 1999). Both sequences were nearly identical to each other and closely related (95% similarity) to that of an uncultivated magnetotactic coccus from a freshwater habitat. This indicates that MTB may also occur in the water column in significant numbers. Alternatively, these sequences may represent closely related non-magnetotactic species.

3
A Gammaproteobacterial Greigite-Producing Rod?

In 2004, Edwards and coworkers reported preliminary evidence for the existence of magnetotactic gammaproteobacteria in a seasonally stratified coastal salt pond (Simmons et al. 2004). At the sulfide-rich base of the chemocline in the meromictic Salt Pond, Falmouth, Massachusetts, the authors have found morphologically conspicuous, large (approximately 5 μm long, 3 μm wide) slow-moving rods. These were shown by transmission electron microscopy to contain irregular shaped electron-dense inclusions resembling magnetosomes. A single crystal electron diffraction pattern corresponded to that of greigite. A 16S rDNA library obtained from a sample highly enriched in the slow-moving rod contained 42% sequences with affiliation to *Gammaproteobacteria*. Two clone groups dominated, one with 99% similarity to *Stenotrophomonas maltophila*, and a second with about 90% similarity to *Thiomicrospira* spp. To further confirm that the newly identified large rod-shaped greigite-producing bacterium is a gammaproteobacterium the authors performed FISH with the group-specific 23S RNA-targeted probe GAM42a (Manz et al. 1992). Binding of this probe does, however, not allow to unambiguously link either one of the two dominant or one of the other diverse gammaproteobacterial sequences to the large magnetotactic rod. Furthermore, the group-specific probe GAM42a is outdated and should only be used as part of probe sets, but not as a stand-alone tool for phylogenetic assignments (Musat et al. 2006). Although further experiments are clearly required to unambiguously prove the existence of gammaproteobacterial MTB, these findings might suggest that the diversity of uncultivated MTB goes beyond previously identified groups.

4
Deltaproteobacterial MTB: The Isolate *Desulfovibrio magneticus* RS-1, and Two Yet-Uncultured Unusual MTB

The magnetosomes of most MTB contain magnetite (Fe_3O_4), but some MTB collected from sulfidic, brackish-to-marine aquatic habitats are made of greigite (Fe_3S_4). The 16S rRNA sequence retrieved from an uncultured many-celled, magnetotactic prokaryote (MMP) with iron sulfide magnetosomes collected at various coastal sites in New England was specifically related to the dissimilatory sulfate-reducing bacteria within the *Deltaproteobacteria* (DeLong et al. 1993) (see also chapter by Farina et al., this volume). The closest relative at that time was *Desulfosarcina variabilis* with a 16S rRNA similarity of 91% (DeLong et al. 1993). The assignment of MMP to *Deltaproteobacteria* together with the earlier assignments of *Magnetospirillum* spp., isolates MC-1 & MV-1, and the uncultured magnetococci to multiple groups within the *Alphaproteobacteria* was the first clear evidence for a polyphyletic origin of MTB. The authors also argue that their findings suggest that magnetotaxis based on iron oxide and iron sulfide magnetosomes evolved independently. They state that the biochemical basis for biomineralization and magnetosome formation for iron oxide-type and iron sulfide type bacteria are likely fundamentally different and speculate that in two independent phylogenetic groups of bacteria analogous solutions for the problem of effective cell positioning along physicochemical gradients were found based on intracellular particles with permanent magnetic dipole moments (DeLong et al. 1993).

In 1995 the group of Matsunaga (Kawaguchi et al. 1995) reported the comparative 16S rRNA sequence analysis of a sulfate-reducing MTB pure culture, RS-1, originally described in 1993 (Sakaguchi et al. 1993). This isolate affiliates with the genus *Desulfovibrio* of the *Deltaproteobacteria*. It was accordingly named as *Desulfovibrio magneticus* strain RS-1 and represented the first bacterium outside the *Alphaproteobacteria* that contains magnetite inclusions (Sakaguchi et al. 1993, 2002). It therefore disrupts the correlation between the alpha and deltaproteobacterial magnetotactic bacteria and iron oxide (magnetite) and iron sulfide (greigite) magnetosomes, respectively, suggested by DeLong and coworkers (DeLong et al. 1993).

A barbell-shaped population was recently reported from a narrow layer immediately below the oxygen-sulfide interface in Salt Pond (Falmouth, MA) when enriching for South-seeking magnetotactic bacterium in the Northern hemisphere (Simmons et al. 2006). This morphotype consists of short chains of 2 to 5 cocci. It was shown by 16S rRNA-based sequencing and FISH with a specific oligonucleotide probe to be another deltaproteobacterium affiliated with the genus *Desulforhopalus*.

5
MTB in the *Nitrospira* Phylum: "*Magnetobacterium bavaricum*" and More

MTB are not restricted to *Proteobacteria*. A large (approximately 5–10 μm long, and 1.5 μm wide) rod-shaped magnetic bacterium has been enriched from the calcareous sediments of a few freshwater lakes in Upper Bavaria, including Lake Chiemsee (Vali et al. 1987). Based on its 16S rRNA this conspicuous morphotype, tentatively named "*Magnetobacterium bavaricum*", was affiliated with the *Nitrospira* phylum (Spring et al. 1993). This independent bacterial phylum encompasses few other isolates including the iron-oxidizer *Leptospirillum ferrooxidans* and the chemolithoautotrophic nitrite oxidizer *Nitrospira moscoviensis*. The magnetosomes of "*M. bavaricum*" were shown to consist of the iron oxide magnetite (N. Petersen, personal communication 2006). It contains up to 1000 hook-shaped magnetosomes with a length of 110–150 nm, often arranged in several chains. The large cells are gram-negative, contain sulfur globules, and are mobile by one polar tuft of flagella.

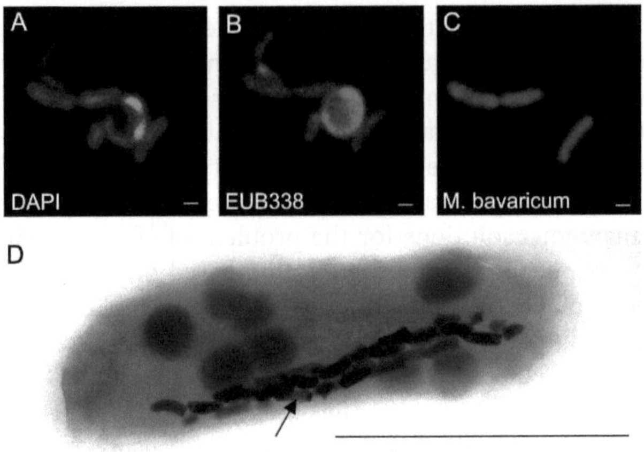

Fig. 3 Micrograph images of MHB-1, a novel magnetotactic rod of the *Nitrospira* phylum, which is closely related to "*Magnetobacterium bavaricum*": DAPI stain (**A**), cells hybridized with a bacterial probe EUB338 (**B**) and a probe for "*M. bavaricum*" (**C**), electron micrograph of MHB-1 (**D**). *Scale bar* represents 1 μm; *arrow* indicates magnetosomes (redrawn after Flies et al. 2005)

As is the case for many other MTB, microbiologists were hitherto unable to grow "*M. bavaricum*" in pure culture. Recently, it was, however, shown that the occurrence of MTB from the *Nitrospira* phylum is apparently not restricted to Bavaria. A conspicuous rod (MHB-1) was magnetically collected from sediment of Waller See, a lake nearby Bremen (Flies et al. 2005). The magnetosomes from MHB-1 display the same bullet-shaped crystal morphology like those from "*M. bavaricum*" (Fig. 3) and are aligned in multiple chains. The cells hybridized with the FISH probe originally used for the identification of "*M. bavaricum*" (Spring et al. 1993). However, unlike the latter organism, MHB-1 has fewer magnetosomes, which form a single bundle. Based on a 16S rRNA sequence similarity of 91% to "*M. bavaricum*", the genotype MTB-1 is a candidate for a new species, "*Magnetobacterium bremense*", if not a new genus, indicating that there is also significant diversity within the MTB of the phylum *Nitrospira*.

6 Final Remarks on MTB Phylogeny

Based on 16S rRNA sequences MTB are phylogenetically diverse, with representatives in *Alphaproteobacteria*, *Deltaproteobacteria* and in the *Nitrospira* phylum. In addition several recent studies suggested that MTB diversity is far from being fully explored. Considering the recent discovery of a potentially mobile magnetosome island (Schübbe et al. 2003; Ullrich et al. 2005) this is, however, not ruling out that biomineralization of magnetite magnetosomes is still monophyletic, and spread by lateral gene transfer. This hypothesis could be investigated by comparative sequence analysis of the genes involved in the magnetosome formation. Did lateral gene transfer, e.g., from the alphaproteobacterial magnetotactic bacteria to "*M. bavaricum*" contribute to the spreading of magnetite-based magnetotaxis? Or, have mechanisms of magnetosome formation independently developed in the different phylogenetic groups? Studies of this type will not necessarily rely on cultured strains since large DNA fragments are now routinely retrieved from the environment (Stein et al. 1996). The analysis of large contiguous sequences harboring such islands might be an extremely powerful approach to gain further insights in the genetic diversity of biomineralization mechanism of MTB.

Acknowledgements The authors were supported by the Max Planck Society, the Fonds der chemischen Industrie, the Deutsche Forschungsgemeinschaft, and the German BMBF.

References

Amann R, Ludwig W, Schleifer KH (1995) Phylogenetic identification and in situ detection of individual microbial cells without cultivation. Microbiol Rev 59:143–169

Bazylinski DA, Frankel RB, Jannasch HW (1988) Anaerobic magnetite production by a marine magnetotactic bacterium. Nature 334(6182):518–519

Blakemore RP, Maratea D, Wolfe RS (1979) Isolation and pure culture of a freshwater magnetic spirillum in chemically defined medium. J Bacteriol 140:720–729

Burgess J, Kawaguchi R, Sakaguchi T, Thornhill RH, Matsunaga T (1993) Evolutionary relationships among *Magnetospirillum* strains inferred from phylogenetic analysis of 16S rRNA sequences. J Bacteriol 175(20):6689–6694

Coates JD et al. (1999) Ubiquity and diversity of dissimilatory (per)chlorate-reducing bacteria. Appl Environ Microbiol 65(12):5234–5241

Cox L et al. (2002) Organization and elemental aalysis of P-, S-, and Fe-rich inclusions in a population of freshwater magnetococci. Geomicrobiol J 19:387–406

DeLong EF, Frankel RB, Bazylinski DA (1993) Multiple evolutionary origins of magnetotaxis in bacteria. Science 259(5096):803–806

Eden PA, Schmidt TM, Blakemore RP, Pace NR (1991) Phylogenetic analysis of *Aquaspirillum magnetotacticum* using polymerase chain reaction-amplified 16S rRNA-specific DNA. Int J Syst Bact 41(2):324–325

Farina M, Kachar B, Lins U, Broderick R, De Barros Henrique L (1994) The observation of large magnetite (Fe-3O-4) crystals from magnetotactic bacteria by electron and atomic force microscopy. J Microscopy 173(1):1–8

Flies C, Peplies J, Schüler D (2005) A combined approach for the characterization of uncultivated magnetotactic bacteria from various aquatic environments. Appl Environ Microbiol 71:2723–2731

Kawaguchi R et al. (1995) Phylogenetic analysis of a novel sulfate-reducing magnetic bacterium, Rs-1, demonstrates its membership of the deltaproteobacteria. FEMS Microbiol Lett 126(3):277–282

Ludwig W et al. (2004) ARB: a software environment for sequence data. Nucleic Acids Res 32(4):1363–1371

Musat N et al. (2006) Microbial community structure of sandy intertidal sediments in the North Sea, Sylt-Romo Basin, Wadden Sea. Syst Appl Microbiol 29(4):333–348

Riemann L et al. (1999) Bacterial community composition during two consecutive NE Monsoon periods in the Arabian Sea studied by denaturing gradient gel electrophoresis (DGGE) of rRNA genes. Deep-Sea Research Part II-Topical Studies in Oceanography 46(8–9):1791–1811

Sakaguchi T, Burgess JG, Matsunaga T (1993) Magnetite formation by a sulphate-reducing bacterium. Nature 365:47–49

Sakaguchi T, Arakaki A, Matsunaga T (2002) *Desulfovibrio magneticus* sp. nov., a novel sulfate-reducing bacterium that produces intracellular single-domain-sized magnetite particles. Int J Syst Evol Microbiol 52(1):215–221

Schleifer KH et al. (1991) The genus *Magnetospirillum* gen. nov., description of *Magnetospirillum gryphiswaldense* sp. nov. and transfer of *Aquaspirillum magnetotacticum* to *Magnetospirillum magnetotacticum* comb. nov. Syst Appl Microbiol 14:379–385

Schübbe S et al. (2003) Characterization of a spontaneous non-magnetic mutant of *Magnetospirillum gryphiswaldense* reveals a large deletion comprising a putative magnetosome island. J Bacteriol 185(19):5779–5790

Schüler D, Spring S, Bazylinski DA (1999) Improved technique for the isolation of magnetotactic spirilla from a freshwater sediment and their phylogenetic characterization. Syst Appl Microbiol 22(3):466–471

Schüler D, Schleifer KH (2005) In: Brenner DJ, Krieg NR, Staley JT (eds) The Genus *Magnetospirillum*. Bergey's Manual of Determinative Bacteriology. Springer, Berlin Heidelberg New York, p 28–31

Shinoda Y, Sakai Y, Ue M, Hiraishi A, Kato N (2000) Isolation and characterization of a new denitrifying spirillum capable of anaerobic degradation of phenol. Appl Environ Microbiol 66(4):1286–1291

Simmons SL, Bazylinski DA, Edwards KJ (2006) South-seeking magnetotactic bacteria in the Northern Hemisphere. Science 311(5759):371–374

Spring S, Amann R, Ludwig W, Schleifer KH, Petersen N (1992) Phylogenetic diversity and identification of non-culturable magnetotactic bacteria. Syst Appl Microbiol 15(1):116–122

Spring S et al. (1993) Dominating role of an unusual magnetotactic bacterium in the microaerobic zone of a freshwater sediment. Appl Environ Microbiol 59(8):2397–2403

Spring S et al. (1994) Phylogenetic analysis of uncultured magnetotactic bacteria from the alpha-subclass of proteobacteria. Syst Appl Microbiol 17(4):501–508

Spring S, Schleifer KH (1995) Diversity of magnetotactic bacteria. Syst Appl Microbiol 18(2):147–153

Spring S et al. (1998) Phylogenetic affiliation and ultrastructure of uncultured magnetic bacteria with unusually large magnetosomes. Arch Microbiol 169(2):136–147

Stein JL, Marsh TL, Wu KY, Shizuya H, DeLong EF (1996) Characterization of uncultivated prokaryotes: isolation and analysis of a 40-kilobase-pair genome fragment from a planktonic marine archaeon. J Bacteriol 178:591–599

Thornhill RH, Burgess JG, Matsunaga T (1995) PCR For direct detection of indigenous uncultured magnetic cocci in sediment and phylogenetic analysis of amplified 16s ribosomal DNA. Appl Environ Microbiol 61(2):495–500

Ullrich S, Kube M, Schübbe S, Reinhardt R, Schüler D (2005) A hypervariable 130 kb genomic region of *Magnetospirillum gryphiswaldense* comprises a magnetosome island, which undergoes frequent rearrangements during stationary growth. J Bacteriol 187(21):7176–7184

Vali H, Förster O, Amarantidis G, Petersen N (1987) Magnetotactic bacteria and their magnetofossils in sediments. Earth Planet Sci Lett 86:389–400

Microbiol Monogr (3)
D. Schüler: Magnetoreception and Magnetosomes in Bacteria
DOI 10.1007/7171_038/Published online: 16 September 2006

Ecophysiology of Magnetotactic Bacteria

Dennis A. Bazylinski[1] (✉) · Timothy J. Williams[2]

[1]School of Life Sciences, University of Nevada at Las Vegas, 4505 Maryland Parkway, Las Vegas, NV 89154-4004, USA
dennis.bazylinski@unlv.edu

[2]Department of Biochemistry, Biophysics, and Molecular Biology, Iowa State University, 1210 Molecular Biology Building, Ames, IA 50011, USA

Abstract Magnetotactic bacteria are a physiologically diverse group of prokaryotes whose main common features are the biomineralization of magnetosomes and magnetotaxis, the passive alignment and active motility along geomagnetic field lines. Magnetotactic bacteria exist in their highest numbers at or near the oxic–anoxic interfaces (OAI) of chemically stratified aquatic habitats that contain inverse concentration gradients of oxidants and reductants. Few species are in axenic culture and many have yet to be well described. The physiology of those that have been described appears to dictate their local ecology. Known Fe_3O_4-producing strains are microaerophiles that fix atmospheric nitrogen, a process mediated by the oxygen-sensitive enzyme nitrogenase. Marine Fe_3O_4-producing strains oxidize reduced sulfur species to support autotrophy through the Calvin–Benson–Bassham or the reverse tricarboxylic acid cycle. These organisms must compete for reduced sulfur species with oxygen, which chemically oxidizes these compounds, and yet the organism still requires some oxygen to respire with to catalyze these geochemical reactions. Most Fe_3O_4-producing strains utilize nitrogen oxides as alternate electron acceptors, the reductions of which are catalyzed by oxygen-sensitive enzymes. Fe_3O_4-producing magnetotactic bacteria must solve several problems. They must find a location where both oxidant (oxygen) and reductants (e.g., reduced sulfur species) are available to the cell and therefore in close proximity. They must also mediate oxygen-sensitive, ancillary biochemical reactions (e.g., nitrogen fixation) important for survival. Thus, the OAI appears to be a perfect habitat for magnetotactic bacteria to thrive since microaerobic conditions are maintained and oxidant and reductant often overlap.

1 Introduction

1.1 General Features of Magnetotactic Bacteria

The magnetotactic bacteria are defined as mainly aquatic prokaryotes whose swimming direction is influenced by the Earth's geomagnetic and other applied fields. The term "magnetotactic bacteria" has no taxonomic significance and represents a collection of morphologically, physiologically, and phylogenetically diverse prokaryotes (Bazylinski 1995; Bazylinski and Frankel 2004). Despite this diversity, these unique microorganisms have several traits in common. All known species or strains:

1. Are phylogenetically associated with the Domain Bacteria
2. Are motile by means of flagella
3. Are microaerophiles or anaerobes or both
4. Have a solely respiratory form of metabolism, with the exception of one species
5. Exhibit nitrogenase activity and presumably can fix atmospheric dinitrogen
6. Are mesophilic with respect to growth temperatures
7. All possess magnetosomes, which are intracellular membrane-bounded crystals of a magnetic mineral (Bazylinski and Frankel 2004)

The magnetosome is the most conspicuous feature of the magnetotactic bacteria and contains crystals of either the iron oxide magnetite (Fe_3O_4) or the iron sulfide greigite (Fe_3S_4). Almost all magnetotactic bacteria produce only one mineral compositional type (Bazylinski and Frankel 2004) although at least one uncultivated species is known to biomineralize both minerals

(Bazylinski et al. 1995). Some greigite-producing magnetotactic bacteria also biomineralize mackinawite (tetragonal FeS) and possibly a sphalerite-type cubic FeS, both of which are thought to be precursors to Fe_3S_4 (Pósfai et al. 1998a, 1998b).

The shape and size of the Fe_3O_4 and Fe_3S_4 crystals in magnetotactic bacteria is species- and strain-specific and thus appears to be under genetic control (Bazylinski et al. 1994). Shapes of magnetosome crystals include the equilibrium form of Fe_3O_4 and Fe_3S_4, the cubo-octahedron, and anisotropic elongated crystals that appear rectangular in projection or are tooth- or bullet-shaped (Bazylinski and Frankel 2004). The size range of mature magnetosome crystals is about 35–120 nm, which places these crystals in the single-magnetic-domain size range for both Fe_3O_4 and Fe_3S_4 (Bazylinski and Frankel 2004). This has physical significance in that by biomineralizing single-magnetic-domains, the cell has maximized the magnetic remanence per volume of the particles (discussed in detail elsewhere in this volume). The precise magnetosome crystal compositions, morphologies, and size ranges, as well as their restricted arrangement in the cell, are good indications that the biomineralization of magnetosomes is through a biologically controlled mineralization process (Bazylinski and Frankel 2003). Crystals of Fe_3O_4 and Fe_3S_4 can also be formed by microbes through biologically induced mineralization (Frankel and Bazylinski 2003) but the particles display few of the qualities exhibited by the crystals produced by the magnetotactic bacteria.

The lipid bilayer membrane surrounding the crystals, the magnetosome membrane (Gorby et al. 1988), is thought to control the size and shape of the magnetosome crystal and, in at least one species of magnetotactic bacterium, appears to originate as an invagination of the cell membrane (Komeili et al. 2005; see also chapter by Komeili, this volume). The magnetosomes are generally anchored to the cell membrane and arranged in a chain or chains within the cell; they are responsible for the cell's magnetic dipole moment and thus their magnetotactic behavior (Bazylinski and Frankel 2004). Recent evidence shows that construction of the chain motif is under genetic control (Komeili et al. 2005; Scheffel et al. 2006). Magnetotaxis is a combination of the passive alignment (due to the presence of the magnetosomes) and active swimming of cells along geomagnetic field lines (for detailed description of magnetotaxis, see also chapter by Frankel, Williams, and Bazylinski, this volume). Two types of magnetotaxis have been described: axial and polar magnetotaxis (Frankel et al. 1997). It seems clear now that both types of magnetotaxis work in conjunction with various forms of chemotaxis (e.g., aerotaxis) for bacteria to efficiently find and maintain an optimal position regarding various nutrients in aquatic habitats with vertical chemical gradients (discussed in detail elsewhere in this volume) by reducing a three-dimensional search problem to one of a single dimension (Frankel et al. 1997). However, very recent evidence suggests that the model for polar magnetotaxis may need to be modified (Simmons et al. 2006).

Although almost all cultivated strains of magnetotactic bacteria are facultatively anaerobic microaerophiles, the physiology of the strains is very different. Moreover, the physiological state of the cells (i.e., how they are grown), in some cases, has a profound affect on magnetosome formation and thus on cellular magnetism. The purpose of this chapter is to review what is known of the physiology of the known magnetotactic bacteria, taking into consideration laboratory studies and genomics and how their physiology affects magnetosome production. Genetics and genetic systems, although linked strongly to cell physiology, are discussed elsewhere in this volume.

1.2
General Ecology of Magnetotactic Bacteria

Magnetotactic bacteria are common in water columns or sediments with vertical chemical stratification. They can occur at relatively high cell numbers (e.g., 10^4 cells/ml) at the oxic–anoxic interface (OAI) and the anoxic regions of the habitat or both (Bazylinski et al. 1995; Bazylinski and Moskowitz 1997; Simmons et al. 2004; for more details see also chapter by Simmons and Edwards, this volume). In freshwater systems where the sulfate concentration is very low or zero, the OAI is generally located at the water–sediment interface or several millimeters below it. The situation in many deep sea sites is similar, although it can be markedly different in many, mostly undisturbed, marine coastal habitats. Seawater contains about 28 mM sulfate and in these latter situations, hydrogen sulfide generated by the action of anaerobic sulfate-reducing bacteria in the sediment diffuses upwards into the water column causing the OAI to occur in the water column. What results is an inverse, oxygen:sulfide concentration double gradient where O_2 diffuses downward from air at the surface and S^{2-} diffuses upwards from the anaerobic zone. Only Fe_3O_4-producing magnetotactic bacteria have been found in freshwater systems while both Fe_3O_4- and Fe_3S_4-producers have been found in marine environments. In marine systems, the Fe_3O_4-producers are found mainly at the OAI and the Fe_3S_4-producers just below the OAI where S^{2-} is present. The Fe_3S_4-producers have thus only been found in marine habitats and none have been cultivated in pure culture to date.

2
Physiology of Magnetotactic Bacteria

The isolation and axenic cultivation of magnetotactic bacteria in pure culture has proven to be very difficult. Despite the fact that these organisms were discovered more than 30 years ago (Blakemore 1975), only a small number of isolates is available (and an even smaller number of genera and species), and all biomineralize Fe_3O_4. There are currently no greigite-producing strains available in pure culture. The development of general isolation strategies is further hindered by the fact that the physiology, and thus the nutritional requirements, of the magnetotactic bacteria is (are) diverse and in many cases, particularly for those that synthesize greigite, are unknown. Thus understanding the physiology of the strains in pure culture may provide significant insight in the isolation of other related and perhaps unrelated strains.

In general, in enrichment media, the relatively slow-growing, fastidious magnetotactic bacteria are outgrown by non-magnetotactic microorganisms. Therefore, a primary strategy has been to inoculate enrichment medium with purified, magnetically-separated cells of magnetotactic bacteria. A commonly used method of obtaining purified cells of magnetotactic bacteria is through the use of a "capillary racetrack" originally devised by Wolfe et al. (1987). Briefly, sediment and/or water containing magnetotactic bacteria is placed on top of a sterile, wetted cotton plug in the wide-mouthed end of a Pasteur pipette, which is filled with membrane filter-sterilized water from the original source. The south pole of a bar magnet is placed near the sealed tip of the capillary furthest from the reservoir. Migration to and accumulation of magnetotactic cells at the end of the capillary can be observed by using a dissecting microscope with the lighting set up for dark-field. Generally, most fast-swimming cells of magnetotactic bacteria (the cocci) will have reached the tip within about 30 min while other more slow-swimming cells may take a bit longer. When enough cells have accumulated for a reasonable inoculum, the tip of the pipette is broken off and the cells are removed aseptically using a thin syringe needle. The homogenous inocula are then subsequently transferred to appropriate enrichment media.

Because all known magnetotactic bacteria are microaerophiles (high $[O_2]$ levels are inhibitory to growth and magnetite formation; see section on microaerophily, this chapter) or anaerobes or facultatively anaerobic microaerophiles, most enrichment media has been based on the formation of a semi-solid $[O_2]$-gradient or liquid anaerobic media. In general, relatively low concentrations of nutrients appear more favorable for the isolation of magnetotactic bacteria compared to rich media with higher concentrations of carbon and nitrogen sources. Although some species, including *Desulfovibrio magneticus* strain RS-1 and perhaps the greigite-producers, are obligate anaerobes, most magnetotactic bacteria tolerate short exposures to O_2 during enrichment and inoculation, making the strict exclusion of O_2 during cell manipulations unnecessary.

Magnetotactic bacteria appear to be redox-sensitive, that is, they do not grow from small inocula in growth medium without the addition of a reducing agent. Redox buffering by the addition of reducing agents such as sodium thioglycolate, ascorbic acid, or cysteine-HCl at concentrations of $0.1–0.4\ g\ L^{-1}$ or dithiothreitol at 1 mM to the medium is required for growth of these microaerophilic or anaerobic species (Bazylinski et al. 1988; Schüler et al. 1999). The inclusion of resazurin, a redox indicator that is colorless when fully reduced, is extremely helpful in the determination of whether a enrichment medium is totally reduced or whether an $[O_2]$-redox gradient has been established. Ferric citrate is the most often used iron source for growth and magnetite biomineralization, as it can be easily prepared and autoclaved together with other medium components without problems with precipitation (Schüler et al. 1999). The type of iron source is not critical, however, as long as it is kept soluble at neutral pH by the presence of chelating compounds or reducing agents in the medium. Ferrous or ferric salts at concentrations between 20–50 μM are sufficient to allow for both growth and magnetosome formation (Schüler and Baeuerlein 1996, 1998). Semi-solid $[O_2]$-redox gradient media have proven to be the most successful type used for the isolation and maintenance of various magnetotactic bacteria, as they allow for the establishment of chemical (e.g., $[O_2]$, Fe(II)) gradients in order to mimic the conditions in stratified water and sediments. The formation of semi-solid $[O_2]:[S^{2-}]$ inverse double gradient media has been used for the successful enrichment of freshwater and marine magnetotactic bacteria (Bazylinski DA, unpublished; Schüler et al. 1999). The formulation for this gradient medium, a modification of the medium originally developed by Nelson and Jannasch (1983) for the enrichment and isolation of the microaerophilic, sulfide-oxidizer *Beggiatoa*, is described in detail in Schüler et al. (1999). Growth of magnetite-producing species occurs as a sharp microaerophilic band of cells at the OAI (the pink:colorless interface) within the multiple gradients. As the number of cells in the band increases, cells deplete O_2 at the location of the band and the band of motile cells moves towards the surface.

Because many typical agars contain impurities that might be inhibitory to growth, highly purified agars, such Agar Noble (Difco) should be used in enrichment medium. Using the techniques described in the preceding paragraphs, several strains of heterotrophic, microaerophilic magnetospirilla have been isolated in media containing low concentrations (0.01%) of succinate or other organic acids as electron donor (Schüler et al. 1999). Marine magnetotactic, facultatively autotrophic, sulfide-oxidizing microaerophiles including vibrios and cocci (discussed later in this chapter), can be also be enriched for and isolated using this technique, except that natural or artificial seawater must be used as the diluent for the medium. An artificial seawater formula that has been used successfully for the isolation of marine vibrios, cocci, and spirilla can be found in Bazylinski et al. (1994). For magnetotactic bacteria from brackish environments, the salinity can be determined with a hand-held refractometer, and the seawater diluted with distilled water to that salinity. To obtain pure cultures of these strains, shake tubes of medium solidified with 10–13 g Agar Noble per liter can be used to obtain isolated colonies of cells within an $[O_2]$ gradient (the headspace of the tubes is air). For the facultatively anaerobic strains that use nitrous oxide (N_2O) as an alternate terminal electron acceptor, the air headspace of the shake tubes can be replaced and over-pressurized to 10 psi with pure N_2O. N_2O is extremely soluble at room temperature and dissolves in the solid medium even reaching the bottom of the tube. For anaerobic nitrate- or sulfate-reducers, nitrate or sulfate can be added to the medium and the air headspace of the shake tubes replaced with O_2-free N_2. Colonies of magnetotactic bacteria grown in agar shake tubes tend to dark brown to black in color due to the presence of magnetite.

2.1 *Magnetospirillum* Species

The first magnetotactic bacteria to be isolated in pure culture and cultivated was *Aquaspirillum magnetotacticum* strain MS-1 (Blakemore et al. 1979), which was later assigned to a new genus *Magnetospirillum* (Schleifer et al. 1991). The genus *Magnetospirillum*, which contains all freshwater species, belongs to the Rhodospirillaceae (α-1 group) and is nested within a cluster of α-Proteobacteria that includes the phototrophic genera *Phaeospirillum* and *Rhodospirillum*, and the non-phototrophic genus *Azospirillum* (Burgess et al. 1993). There are three well described species of *Magnetospirillum* including *M. magnetotacticum* strain MS-1 (Blakemore et al. 1979; Maratea and Blakemore 1981), *M. magneticum* strains AMB-1 and MGT-1 (Matsunaga et al. 1991; Burgess et al. 1993), and *M. gryphiswaldense* (Schleifer et al. 1991). A large number of other strains have not yet been completely characterized and named (e.g., Schüler et al. 1999). All three strains biomineralize a single chain of cubo-octahedral crystals of Fe_3O_4. In *M. magneticum*, the magnetosome membrane vesicle appears to originate from invaginations of the cell membrane and form prior to nucleation of Fe_3O_4 within the vesicle (Komeili et al. 2004, 2005; see also chapter by Komeili, this volume). All strains are motile by a single polar flagellum at both ends of the cell. Cells of *Magnetospirillum* species grown in liquid generally exhibit axial magnetotaxis while cells of newly isolated strains and those grown in semi-solid $[O_2]$-gradient medium tend to lean to having a polar preference in their swimming and thus display some polar magnetotaxis (discussed elsewhere in this volume).

Two genomes of *Magnetospirillum* species are currently available for study: the incomplete draft genome (genome size about 4.3 Mb; estimated at about 90% complete) for *M. magnetotacticum* can be found at http://genome.jgi-psf.org/draft_microbes/magma/magma.home.html and the 5.0 Mb-sized genome of *M. magneticum* strain AMB-1 has recently been completed (Matsunaga et al. 2005) and can be found at http://www.ncbi.nlm.nih.gov/entrez/query.fcgi?db=genome&cmd=Retrieve&dopt=Overview&list_uids=19021. The genome of both species appears to consist of a single, circular chromosome and plasmids were not detected. However, *M. magneticum* strain MGT-1 contains a 3.7 kb cryptic plasmid referred to as pMGT, which was characterized and used for the development of an improved expression system in strain AMB-1 through the construction of a shuttle vectors constructed by ligating pUC19 (pUMG) or a kanamycin resistance cassette (pMGTkm) into a *Bam*HI site of pMGT (Okamura et al. 2003). These recombinant plasmids were capable of replicating in both *Magnetospirillum* species and *Escherichia coli*.

2.1.1 Energy Metabolism and Electron Transport in *Magnetospirillum* Species

All *Magnetospirillum* species grow chemoorganoheterotrophically utilizing certain organic acids as sources of electrons and carbon. None utilize sugars either oxidatively or fementatively. Growth is respiratory and fermentation has not been found to occur in any *Magnetospirillum* species. Oxygen and nitrate are utilized as terminal electron acceptors by all strains and all can be considered denitrifying bacteria, reducing nitrate to the gaseous products nitrous oxide (N_2O) and N_2 (Bazylinski and Blakemore 1983a; D. Schüler, personal communication). While both *M. magneticum* and *M. gryphiswaldense* grow anaerobically with nitrate, cells of *M. magnetotacticum* appear to require some O_2 even when growing with nitrate (Bazylinski and Blakemore 1983a). The role of O_2 in this situation is unknown although cells appear to consume it while denitrifying. Molecular oxygen may simply poise the redox potential in the growth medium for growth on nitrate to initiate.

Electron transport and cytochromes have been studied in some detail in *M. magnetotacticum*. O'Brien et al. (1987) reported the presence of *a*-, a_1-, *b*-, *c*-, cd_1-, and *o*-type hemes in microaerobically denitrifying cells of *M. magnetotacticum*. More than 85% of the total cytochromes detected were of the *c*-type. The *a*- and *b*-types were mainly detected in cell membranes, whereas 70% of the *c*-type hemes were soluble. Because both a_1-type and *o*-type hemes, considered to be terminal oxidases, were synthesized by microaerobically denitrifying cells of *M. magnetotacticum*, O'Brien et al. (1987) concluded that the respiratory chain in this organism is branched. Tamegai et al. (1993) purified and characterized a novel "cytochrome a_1-like" hemoprotein that was found to be present in greater amounts in magnetic cells than non-magnetic cells. They found no evidence for the presence of a cytochrome a_1 or an *o*-type cytochrome reported by O'Brien et al. (1987) to be the terminal oxidases in the electron transport chain of *M. magnetotacticum*. The "a_1-like" hemoprotein was composed of two different subunits with molecular masses of 41 kDa (subunit I) and 17 kDa (subunit II), and exhibited very little cytochrome *c* oxidase activity. The genes encoding this unusual cytochrome were identified and sequenced (Tanimura and Fukumori 2000). Three open reading frames (ORFs) preceded by a putative ribosome-binding site were found in the sequenced region and designated *mcaII*, *mcaI*, and *hosA*. *mcaI* and *mcaII* were shown to encode subunits I and II of the "cytochrome a_1-like" hemoprotein, respectively. *hosA* showed significant sequence homology to the gene encoding for heme *o* synthase (protoheme IX farnesyltransferase), an essential enzyme for the biosynthesis of heme *o* and heme *a* (Saiki et al. 1993). Despite the fact that six histidine residues predicted to associate with prosthetic cofactors of the heme-copper oxidase superfamily are conserved in the "cytochrome a_1-like" hemoprotein, no amino acid residues proposed to participate in the O_2-reducing and the coupled proton-pumping reactions in cytochrome *c* oxidase in *Paracoccus denitrificans* (Iwata et al. 1995) are conserved in subunit I. This finding probably explains the poor cytochrome *c* oxidase activity of the protein. A new *ccb*-type cytochrome *c* oxidase (Tamegai and Fukumori 1994), a cytochrome *c*-550 that is homologous to cytochrome c_2 in some photosynthetic bacteria (Yoshimatsu et al. 1995) and a cytochrome cd_1-type nitrite reductase (Yamazaki et al. 1995) were identified and purified from *M. magnetotacticum*. The latter cytochrome is discussed in the nitrate reduction section.

2.1.1.1 Autotrophic growth of *Magnetospirillum* species

Autotrophic growth has been demonstrated for a number of marine strains of magnetotactic bacteria (Bazylinski et al. 2004; Williams et al. 2006), most appearing to use the Calvin–Benson–Bassham (CBB) cycle for fixation of CO_2 and autotrophy. One of the key enzymes of this pathway is ribulose-1,5-bisphosphate carboxylase/oxygense (RubisCO), which exists in two major forms. Form I consists of small and large subunits and form II consists of only the large subunit. Form II RubisCO genes, *cbbM*, have been found within the genomes of *M. magnetotacticum* (Bazylinski et al. 2004), *M. magneticum* (Matsunaga et al. 2005), and *M. gryphiswaldense* (D. Schüler, personal communication) although none have yet been found to grow chemolithoautotrophically or photoautotrophically, meaning that no electron donor has been found to support autotrophic growth of these strains either in the dark or under light. The amino acid sequences of cbbM proteins are generally highly conserved from those organisms that possess this protein, and cbbM from *M. magnetotacticum* and *M. magneticum* share 92.8% sequence identity, which accords well with the close phylogenetic relationship between the two strains (99.4% sequence identity for 16S rRNA genes). The amino acid sequences of cbbM in different species of magnetotactic bacteria are shown in Fig. 1. Other putative genes of the CBB pathway of autotrophy are also present in *M. magnetotacticum*, e.g., a probable phosphoribulokinase gene (*prkB*; GenBank accession no. ZP_00208632; JGI no. Magn03007198) on contig 3612 (Bazylinski et al. 2004) that overlaps *cbbM* by 129 bp. It thus seems unlikely that this bacterium cannot grow autotrophically.

Both forms of RubisCO are capable of fixing CO_2 or O_2 although they differ in the specificity factor that measures the ability of the enzyme to discriminate between the two substrates at a given $CO_2 : O_2$ ratio (Jordan and Ogren

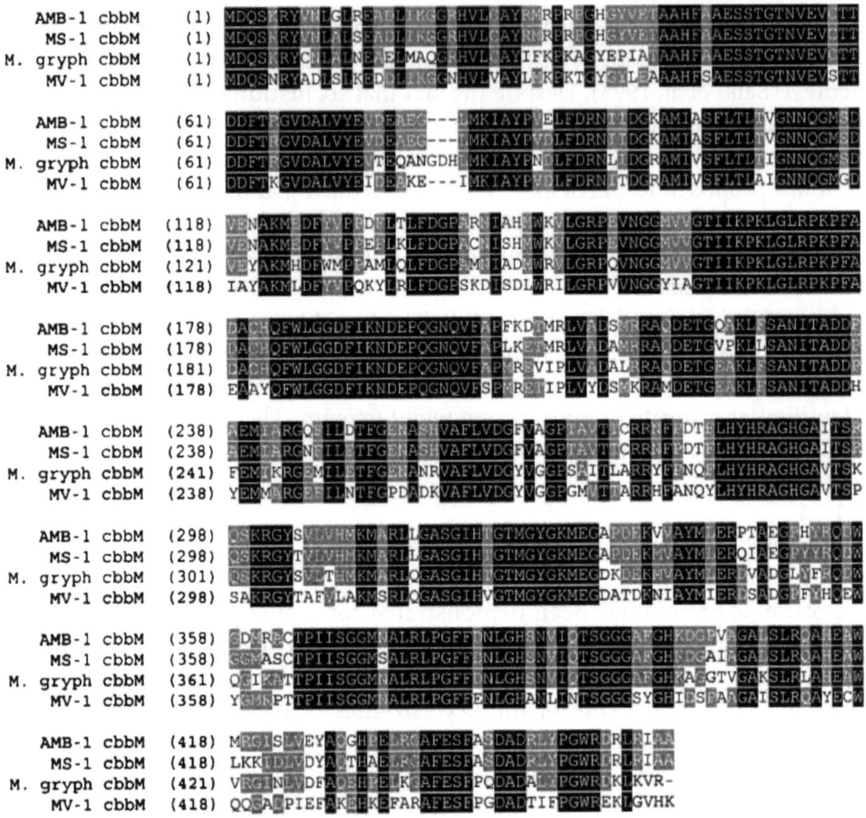

Fig. 1 Alignment of cbbM proteins from four magnetotactic strains: *Magnetospirillum magneticum* strain AMB-1, *M. magnetotacticum* MS-1, *M. gryphiswaldense*, and the marine, magnetotactic vibrio strain MV-1. Amino acid residues conserved across all four sequences are *highlighted in gray*; amino acid residues that are identical across all four sequences are *highlighted in black*

1981). Form II RubisCO, cbbM, generally has a lower specificity factor than the form I enzyme, cbbL, and requires a higher $CO_2 : O_2$ ratio to function as an efficient carboxylase (Jordan and Ogren 1981; Hernandez et al. 1996; Baker et al. 1998). All known magnetotactic bacteria, including *Magnetospirillum* species, are restricted to growth as microaerophiles and/or anaerobes (Bazylinski and Frankel 2004). Therefore, the optimal growth conditions for these organisms and/or the natural environmental conditions under which they are commonly found would have high $CO_2 : O_2$ ratios. Therefore, it appears reasonable that *Magnetospirillum* species, as well as other autotrophic magnetotactic strains, would possess and rely on a form II RubisCO enzyme for CO_2 fixation if they utilize the CBB cycle for autotrophy (Bazylinski et al. 2004).

As previously stated, some of the closest phylogenetic relatives of *Magnetospirillum* species are photosynthetic. The 16S rRNA gene of *M. magne-*

totacticum shows 95.1% sequence identity to that of *Phaeospirillum fulvum*, although there is no evidence of magnetosome production in *Phaeospirillum* (based on *P. molischianum*) (D. Schüler, personal communication). Interestingly, putative genes encoding photosynthetic reaction center proteins (e.g., *pufM*) and bacteriochlorophyll biosynthesis proteins (e.g., *bchC, bchX, bchY, bchZ*) are present in the draft genome of *M. magnetotacticum*. The precise arrangement of these putative *puf* and *bch* genes is unknown since the relevant contigs have yet to be assembled, although a putative *bchX* and *bchY* are positioned adjacently on a single contig (GenBank accession no. NZ_AAAP01001459). We could not identify any *puf* or *bch* genes in the complete genome of *M. magneticum* (Matsunaga et al. 2005) nor were they found in the genome of *M. gryphiswaldense* (D. Schüler, personal communication) and it is unlikely that such genes are contaminants in the *M. magnetotacticum* genome. It is possible that the putative *bch/puf* genes are pseudogenes that represent non-functional remnants of a photosynthetic apparatus that has been lost in other *Magnetospirillum* species.

2.1.2 Microaerophily and the Effect of O_2 and Nitrate on Fe_3O_4 Biomineralization in *Magnetospirillum* Species

Microaerophiles are defined as organisms for which oxygen is both beneficial (they respire with it as a terminal electron acceptor and thus use it for energy) and deleterious (they are poisoned by O_2 or its derivatives) (Krieg and Hoffman 1986). However, they either cannot grow or grow poorly at 21% O_2, the concentration of O_2 in air. These microorganisms prefer microaerobic conditions (< 21% O_2) for growth, the optimum $[O_2]$ for growth dependent on the bacterial species and strain. Although some microaerophiles are able to grow anaerobically using alternative terminal electron acceptors or by fermentation, their preference for low $[O_2]$, when they respire with O_2 as a terminal electron acceptor, is what distinguishes them from anaerobes that can tolerate low levels of O_2 and from aerobes or facultative anaerobes (Krieg and Hoffman 1986).

Many theories have been proposed to explain microaerophily. Various forms of oxygen that are produced in growth media and as products of O_2 respiration are known to be toxic to many bacteria (Krieg and Hoffman 1986). These include hydrogen peroxide (H_2O_2), the hydroxyl radical (OH), and the superoxide radical (O_2^-). It is thought that microaerophily in some organisms is due to the inability to produce enzymes such as catalase and superoxide dismutase that protect against these toxic forms of oxygen. The addition of these enzymes to the growth medium sometimes allows an organism to grow on the surface of agar plates under full aerobic conditions. For example, cells of *M. magnetotacticum* apparently do not produce catalase and are very sensitive to H_2O_2, and the addition of catalase to the growth medium allowed for aerobic growth of this organism although cells grown in this way did not produce magnetosomes (Blakemore 1982). Interestingly, magnetite is known to catalyze the breakdown of H_2O_2 (Blakemore 1982). Lack of cellular iron and low metabolic rates have also been used to explain microaerophily. The former explanation, now not widely accepted, was based on the fact that addition of iron salts to the growth medium enhanced aerotolerance in the known microaerophile *Campylobacter jejuni* (Hoffman et al. 1979; Krieg and Hoffman 1986). The latter explanation is also not widely accepted as a reason for microaerophily but has been used to explain how aerobic bacteria such *Azotobacter* protect O_2-sensitive nitrogenase when cells are actively fixing N_2 (Ramos and Robson 1985). It is noteworthy that almost all magnetotactic bacteria in pure culture are known to fix N_2 (Bazylinski and Frankel 2004) but are still microaerophilic when grown with ammonium ion as a nitrogen source and are not actively fixing N_2.

Regardless of the reason(s) why, all cultured magnetotactic bacteria are magnetite-producers and, with the exception of one species, *Desulfovibrio magneticus* (Sakaguchi et al. 2002), all are obligate microaerophiles when grown with O_2 as the terminal electron acceptor (Bazylinski and Frankel 2004).

Oxygen concentration and the presence of nitrogen oxides are factors that clearly influence Fe_3O_4 biomineralization in *Magnetospirillum* species. Blakemore et al. (1985) reported that microaerobic conditions (some O_2) are required for Fe_3O_4 production by cells of *M. magnetotacticum*. Cells grew in sealed, unshaken culture vessels with 0.1–21% oxygen in the headspace; maximum Fe_3O_4 production and cellular magnetism occurred at a headspace $[O_2]$ of 1%, whereas headspace $[O_2]$ values of $> 5\%$ were inhibitory. Subsequent isotope experiments showed that O_2 is not incorporated into Fe_3O_4, however, and that the oxygen in Fe_3O_4 is derived from water (Mandernack et al. 1999). Although the role of O_2 in Fe_3O_4 biomineralization is unknown, it may poise the redox potential of the growth medium at an optimal value for Fe_3O_4 synthesis. However $[O_2]$ clearly affects the synthesis of specific proteins. For example, Sakaguchi et al. (1993a) showed that the presence of O_2 in nitrate-grown cultures repressed the synthesis of a 140-kDa membrane protein in *M. magnetotacticum*. Also, Short and Blakemore (1989) showed that increasing the O_2 tension in cultures from 1% saturation to 10% caused cells to show increased activity of a manganese-type superoxide dismutase relative to that of an iron-type.

Schüler and Baeuerlein (1998) showed that Fe_3O_4 formation in *M. gryphiswaldense* is induced in non-magnetotactic cells, grown in a fermenter lacking a continuous O_2-controlling system, by a low threshold $[O_2]$ of $\sim 2–7\,\mu M$ (1.7–6.0 mbar) at 30 °C. Heyen and Schüler (2003) reported the effect of $[O_2]$ on the growth and Fe_3O_4 magnetosome synthesis of *M. gryphiswaldense*, *M. magnetotacticum*, and *M. magneticum* grown microaerobically in a continuous, O_2-controlled fermenter. For all three *Magnetospirillum* strains, Fe_3O_4 synthesis was only induced when the $[O_2]$ was below a threshold value of 20 mbar, and the optimum $[O_2]$ for Fe_3O_4 biomineralization was 0.25 mbar.

M. magneticum strain AMB-1 was originally reported to be an "aerobic" magnetotactic bacterium and the most O_2-tolerant of the cultivated magnetospirilla (Matsunaga et al. 1991). However, results from oxystat experiments described by Heyen and Schüler (2003) showed that this species is a typical microaerophile and is even less aerotolerant than *M. gryphiswaldense*. Cells of *M. magneticum* synthesize Fe_3O_4 either microaerobically or anaerobically using nitrate as the terminal electron acceptor (Matsunaga et al. 1991; Matsunaga and Tsujimura 1993).

2.1.3 Nitrogen Metabolism in *Magnetospirillum* Species

2.1.3.1 Nitrate Reduction and Denitrification in *Magnetospirillum* Species

M. magnetotacticum has been shown to be capable of assimilatory nitrate reduction (Bazylinski and Blakemore 1983a) and because other species of *Magnetospirillum* grow with nitrate as the sole nitrogen source, all are capable of this metabolic feature. As previously stated, *M. magnetotacticum*, *M. magneticum*, and *M. gryphiswaldense* all utilize nitrate as a terminal electron acceptor and are therefore also capable of dissimilatory nitrate reduction. All three strains produce N_2O and N_2 as the final products of nitrate reduction and are denitrifyers (Bazylinski and Blakemore 1983a; D. Schüler, personal communication). Cells of *M. magnetotacticum* do not accumulate nitrite during microaerobic denitrification but do accumulate N_2O transiently (Bazylinski and Blakemore 1983a). This suggests that this organism possesses a N_2O reductase although it does not appear to grow well on exogenous N_2O as a terminal electron acceptor, a phenotype similar to that of *Pseudomonas aeruginosa* (Bazylinski et al. 1986). Cells also do not grow with exogenous nitrite as a terminal electron acceptor although this may not be surprising since nitrite is known to be toxic to many bacteria, even those that denitrify.

The enzymology of denitrification in *M. magnetotacticum* has been studied in some detail. A soluble periplasmic nitrate reductase was purified from this organism that is comprised of two subunits of 86 and 17 kDa and contains molybdenum, non-heme iron, and heme *c* (Taoka et al. 2003). Molybdenum starvation of cells resulted in little periplasmic nitrate reductase activity in cell-free extracts, but the magnetosome fraction still had almost half the iron that was present in the same fraction of cells grown with molybdenum. These results indicate that nitrate reduction in this organism is not essential for Fe_3O_4 synthesis, despite the fact that nitrate appears to have a stimulatory effect on magnetosome biomineralization in growth experiments. This does not preclude the fact that a product of nitrate reduction might be the stimulatory factor although it is also possible that the presence of nitrate might poise the redox potential of the growth medium at an optimum value for Fe_3O_4 synthesis. A cytochrome cd_1-type dissimilatory nitrite reductase was also identified and purified from *M. magnetotacticum* (Yamazaki et al. 1995). The latter protein might be important in Fe_3O_4 biomineralization as it has a novel Fe(II) : nitrite oxidoreductase activity that might be linked to the oxidation of Fe(II) in the cell and, therefore, to Fe_3O_4 synthesis. Genomics also supports the presence of almost a full denitrification pathway in *M. magnetotacticum.* A periplasmic nitrate reductase *napFDAGHBC* gene locus is located on contig 3771 of the draft genome of *M. magnetotacticum* (GenBank accession no. NZ_AAAP01003771). A putative nitrous oxide (N_2O) reductase (GenBank accession no. ZP_00054320; JGI no. Magn03008954) and nitric oxide- (NO) forming nitrite reductases (GenBank accession no. ZP_00208390, JGI no. Magn03008348; and GenBank accession no. ZP_00053843, JGI no. Magn03008451) have also been identified in this draft genome. It is not known which, if either, of these latter genes encode for the cytochrome cd_1 identified by Yamazaki et al. (1995) since both these genes have heme d_1 domains. The genome of *M. magneticum* strain AMB-1 contains homologs to these genes and are designated amb1395 (GenBank accession no.BAE50199) and amb1408 (GenBank accession no. BAE50212) (Matsunaga et al. 2005). The only denitrification step in *M. magnetotacticum* not supported by enzymology or genomics is NO reduction.

2.1.3.2 N_2 fixation

All three described *Magnetospirillum* species display nitrogenase activity and grow in medium with N_2 as the sole nitrogen source in $[O_2]$-gradient medium (Bazylinski and Blakemore 1983b; Bazylinski et al. 2000) demonstrating their ability to fix atmospheric N_2 under microaerobic conditions. Genomics also support these results as many of the genes for N_2 fixation are clustered together in the genome of *M. magnetotacticum*, as shown in Fig. 2. N_2 fixation may be an important feature for survival and growth at the OAI in natural aquatic environments where fixed sources of nitrogen are limited or vigorously competed for.

2.1.4 Iron Metabolism in *Magnetospirillum* Species

Construction of the bacterial magnetosome appears to be to be a complex process that involves several discrete steps, including magnetosome vesicle formation, iron uptake by the cell, iron transport into the magnetosome vesi-

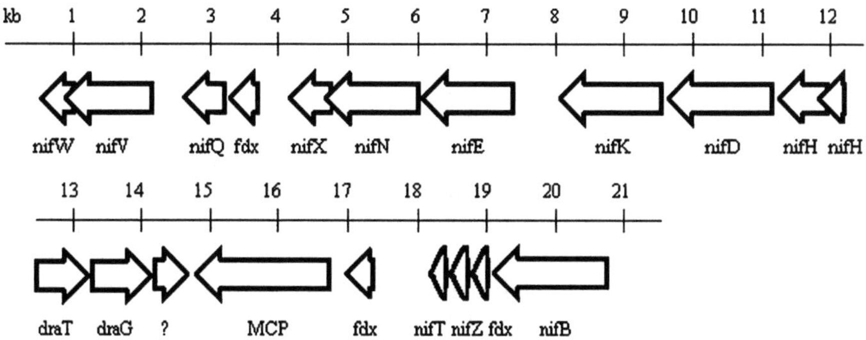

Fig. 2 Gene cluster within contig 3823 (GenBank accession no. NZ_AAAP01003823) in the draft genome of *Magnetospirillum magnetotacticum* strain MS-1 showing putative nitrogen-fixation (*nif*) genes. Each ORF is represented by an *arrow* indicating the direction of gene transcription. Putative identities of each gene are shown beneath each ORF, and are based on BLASTP searches of the translated product. All *nif* genes encode proteins that are associated with nitrogen fixation. *nifD* and *nifK* encode for α- and β-chains, respectively, of the mature molybdebum–iron nitrogenase (MoFe protein), an $\alpha_2\beta_2$ heterotetramer (Orme-Johnson 1985). There are two copies of *nifH*, which encodes for the reductase component (Fe protein) of nitrogenase (Orme-Johnson 1985); the second, smaller copy may be a pseudogene. *nifB, nifE, nifN, nifQ, nifV*, and possibly *nifX* encode for proteins involved in the synthesis of the iron–molybdenum cofactor (FeMo-co) (Orme-Johnson 1985; Shah et al. 1986, 1994, 1999; Allen et al. 1994; Mayer et al. 2002). The *nifW* product has been implicated in protecting nitrogenase from O_2 damage (Kim and Burgess 1996). The *nifZ* product has been implicated in maturation of the MoFe protein (Hu et al. 2004). The function of the *nifT* product is unclear (Simon et al. 1996). The gene labeled *?* is an unknown nitrogenase-associated protein. *draT* and *draG* encode for dinitrogenase reductase ADP-ribosyltransferase and dinitrogenase reductase activating glycohydrolase, respectively. *fdx* represents a putative ferredoxin as gene product. MCP encodes for a putative methyl-accepting chemotaxis protein

cle, and controlled Fe_3O_4 (or Fe_3S_4) biomineralization within the magnetosome vesicle. It seems now that vesicle formation, iron uptake and transport, and Fe_3O_4 mineralization steps must be temporally ordered in the magnetospirilla since there is good evidence showing that magnetosome membrane vesicles form as invaginations of the cell membrane, prior to mineral nucleation in at least one species (Komeili et al. 2004, 2005). Only iron metabolism and chemistry will be discussed here.

2.1.4.1
Iron uptake by cells of *Magnetospirillum* species

Magnetotactic bacteria can consist of > 3% iron as measured by dry weight (Blakemore 1982; D. Schüler, personal communication), which is several orders of magnitude higher than non-magnetotactic bacteria. Despite this fact, there is no evidence to suggest that they use unique iron-uptake systems.

Fe(II) is soluble up to 100 mM at neutral pH (Neilands 1984) and is presumably easily taken up by bacteria by non-specific mechanisms. However, Fe(III) is so insoluble that most microorganisms produce and rely on iron chelators, known as siderophores, which bind and solubilize Fe(III) for uptake. Siderophores are low molecular weight (< 1 kDa), specific ligands generally produced by bacteria under iron-limited conditions, usually between 0 and 1 μM (Neilands 1995), their synthesis being repressed under iron-sufficient conditions (Guerinot 1994).

Frankel et al. (1983) assumed that iron was taken up as Fe(III) by a non-specific transport system by cells of *M. magnetotacticum.* Although iron was supplied as Fe(III) quinate, the growth medium also contained chemical reducing agents (e.g., thioglycolate) that are potent enough to reduce Fe(III) to Fe(II). So, both Fe(II) and Fe(III) were present in the growth medium and it is unknown which form was taken up by the cells. Cells of *M. magnetotacticum* were reported to produce a hydroxamate siderophore when grown under high (20 μM), but not under low (5 μM), iron conditions (Paoletti and Blakemore 1986). This siderophore synthesis pattern is the reverse of what is normally observed but suggests that cells can utilize Fe(III).

Nakamura et al. (1993) reported molecular evidence for the involvement of a periplasmic-binding protein, SfuC, in the transport of iron by *M. magneticum* strain AMB-1. Siderophores were not detected in spent growth medium, although later cells of *M. magneticum* AMB-1 were found to produce both hydroxamate and phenolate siderophores (Calugay et al. 2003). Like *M. magnetotacticum*, *M. magneticum* appears to produces siderophores under growth conditions that would be considered to be iron-sufficient, if not iron-rich, for the growth of most prokaryotes. This pattern of siderophore production might be explained if iron is taken up rapidly and converted to inert Fe_3O_4, which apparently cannot be used by cells. In this situation, the concentration of iron that is available for growth would decrease rapidly and the cells would experience iron-limiting conditions, which would then stimulate siderophore production. This does not, however, explain why cells do not synthesize siderophores at low initial concentrations of iron ($\leq$ 6 μM).

Schüler and Baeuerlein (1996) described two iron uptake systems in *M. gryphiswaldense.* They showed that iron was mostly taken up as Fe(III) in an energy-dependent process. Fe(II) was also taken up by cells, but by a slow, diffusion-like process, whereas Fe(III) uptake followed Michaelis–Menten kinetics indicating that Fe(III) uptake by *M. gryphiswaldense* is a low-affinity, but high-velocity transport system. No evidence for siderophore production was found, though spent culture fluid stimulated iron uptake by iron-depleted cells. It was subsequently shown that, in *M. gryphiswaldense*, Fe(III) is taken up and rapidly converted to Fe_3O_4 without any apparent delay, indicating that there is no significant accumulation of precursors to Fe_3O_4 inside the cell, at least under optimal, microaerobic conditions for Fe_3O_4 production by *M. gryphiswaldense.*

2.1.4.2
Transport of iron into the magnetosome membrane vesicle in *Magnetospirillum*

The magnetosome membrane originates from an invagination of the cell membrane, at least in *M. magneticum* strain AMB-1 (Komeili et al. 2005). Small GTPases, such as Sar1p, are recognized to be essential for the budding reaction in the production of membrane vesicles and vesicle trafficking in eukaryotes (Kirchhausen 2000). Okamura et al. (2001) identified a 16-kDa protein that has GTPase activity, known as Mms16, in the magnetosome membrane vesicle of *M. magneticum* strain AMB-1. It was the most abundant of five proteins present in the vesicle membrane. Cells that were grown in the presence of a GTPase inhibitor showed less overall magnetism and produced fewer magnetosomes than those grown in the absence of the inhibitor. These results suggest that GTPase activity is required for magnetosome synthesis. However it now seems very unlikely that Mms16 is really a bona fide GTPase, as a protein with very high sequence similarity to Mms16 was recently shown to be involved in polyhydroxybutyrate (PHB) depolymerization in the photosynthetic bacterium *Rhodospirillum rubrum* (Handrick et al. 2004). PHB is a carbon storage product common in cells of magnetotactic bacteria either grown in culture or collected from the environment.

It is not known which redox forms of iron are transported into the magnetosome vesicle in most magnetotactic bacteria, but there is evidence that Fe(II) is transported into vesicles of *M. magneticum* AMB-1 (Nakamura et al. 1995a). Using transposon mutagenesis, Nakamura et al. (1995a; 1995b) identified a gene, *magA*, that encodes for a protein with significant sequence homology to the cation-efflux proteins, KefC, a K^+-translocating protein in *E. coli*, and NapA, a putative Na^+/H^+ antiporter from *Enterococcus hirae*. The MagA protein is present in both the cytoplasmic and magnetosome membranes of *M. magneticum*. *magA* was expressed in *E. coli* and inverted membrane vesicles prepared from these cells were shown to transport Fe(II) in an energy-dependent manner, leading to accumulation of Fe(II) in the vesicle, indicating that MagA functions as a H^+/Fe(II) antiporter in *M. magneticum*. However, *magA* expression was higher when cells of wild-type *M. magneticum* were grown under iron-limited conditions, rather than under iron-sufficient conditions where they produce more magnetosomes. Thus, despite the fact that MagA appears to be involved in iron transport, it cannot be solely responsible for magnetosome synthesis. Genes that share significant sequence homology with *magA* are present in *M. magnetotacticum* and magnetotactic coccus strain MC-1 (Grünberg et al. 2001).

MamB and MamM, two proteins abundant in the magnetosome membrane of *M. gryphiswaldense*, might also be involved in the transport of iron into magnetosome vesicles (Schübbe et al. 2003). Both proteins belong to a group of heavy-metal-ion-transporting proteins known as the cation diffusion facilitator (CDF) family (Nies and Silver 1995: Paulsen and Saier Jr 1997). A spontaneous non-magnetotactic mutant of *M. gryphiswaldense*, which lacks *mamB* and *mamM* as well as numerous other magnetosome membrane proteins and other genes, and does not biomineralize magnetosomes, was found to be deficient in iron uptake (Schübbe et al. 2003). In the yeast *Saccharomyces cerevisiae*, the overexpression of two genes that encode mitochondrial membrane proteins affects Fe(II) concentrations in the mitochondria and cytosol (Li and Kaplan 1997). These genes have been referred to as mitochondrial iron transporters (MFT; *MFT1* and *MFT2*) (Li and Kaplan 1997) but deletion of either or both genes did not affect essential Fe(II)-dependent processes in the mitochondrion, leaving the specific role of these genes unclear (Li and Kaplan 2001). Recently, iron has been shown to be bound and transported by a CDF protein in prokaryotes. The CDF protein FieF (YiiP) in *E. coli* has been shown to mediate decreased cellular accumulation of iron and to relieve iron stress (Grass et al. 2005).

2.1.4.3
Iron redox reactions in *Magnetospirillum*

Short and Blakemore (1986) showed that cells of *M. magnetotacticum*, kept under anaerobic conditions without a terminal electron acceptor, translocated protons in response to the introduction of Fe(III) suggesting that Fe(III) reduction might be linked to energy conservation in this organism. Guerin and Blakemore (1992) later reported anaerobic, Fe(III)-dependent growth of *M. magnetotacticum* in the absence of nitrate. Cells grew very slowly with poorly ordered (amorphous) Fe(III) oxides, presumably as the terminal electron acceptor under these conditions, and the growth yields were poor compared with cells that were grown on nitrate and/or O_2. However, cells grown anaerobically with Fe(III) were extremely magnetic and produced about twice as many magnetosomes compared to those grown with nitrate with 1% O_2 in the headspace (Blakemore et al. 1985). It was also suggested (Guerin and Blakemore 1992) that cells are capable of Fe(II) oxidation, which might be linked to aerobic respiratory processes, energy conservation, and Fe_3O_4 synthesis. Cell-free extracts of *M. magnetotacticum* have also been shown to exhibit Fe(III) reductase activity (Paoletti and Blakemore 1988). Noguchi et al. (1999) purified a Fe(III) reductase from this species and the enzyme appears to be loosely bound on the cytoplasmic face of the cytoplasmic membrane, has an apparent molecular weight of 36 kDa, and requires reduced nicotinamide adenine dinucleotide (NADH) and flavin mononucleotide (FMN) as an electron donor and cofactor, respectively. Enzyme activity was inhibited by zinc, which also reduced the number of magnetosomes produced by cells when included in the growth medium as $ZnSO_4$.

Blakemore (1982) attempted to determine whether the oxidation of Fe(II) to Fe_3O_4 could account for the observed growth yield of *M. magnetotacticum*:

$$3Fe(II) + 0.5O_2 + 3H_2O \rightarrow Fe_3O_4 + 6H^+$$

This reaction is energetically favorable, with a $\Delta G_{reaction}$ of – 0.85 kJ per gram of Fe_3O_4 produced under reactant concentrations for the optimal growth of the bacterium. However, based on these thermodynamic calculations, Fe(II) oxidation to Fe_3O_4 cannot provide enough energy for growth neither under environmental nor laboratory conditions (see chapter by Simmons and Edwards, this volume).

2.2
Desulfovibrio magneticus Strain RS-1

Desulfovibrio magneticus strain RS-1 is an anaerobic, sulfate-reducing, magnetotactic bacterium that was isolated from freshwater sulfidic sediments collected from the Kameno River, Wakayama Prefecture, Western Japan (Sakaguchi et al. 1996, 2002). Interestingly, magnetic separation methods were not used in the isolation of this strain (Sakaguchi et al. 1996). Cells are vibrioid and relatively large at 3–5 × 1 μm, motile by a single polar flagellum, and synthesize a chain of magnetosomes containing irregular, bullet-shaped crystals of Fe_3O_4 (Sakaguchi et al. 1993b, 2002). We recently isolated, using magnetic separation techniques, a new, second strain of this species (designated strain FH-1) from non-sulfidic water and sediment collected from a trout hatchery pond in Bozeman, Montana (USA) (Bazylinski DA, unpublished data). This latter strain has not been completely characterized but appears to be superficially similar to strain RS-1. Phylogenetically (based on 16S rDNA sequence), *D. magneticus* strain RS-1 clearly belongs to the genus *Desulfovibrio* in the family Desulfovibrionaceae and order Desulfovibrionales within the δ-Proteobacteria (Kawaguchi et al. 1995), its closest relative being *D. burkinensis* (similarity 98.7%) (Sakaguchi et al. 2002). The cellular fatty acid profile of *D. magneticus* is also consistent with members of the genus *Desulfovibrio* and the major menaquinone is MK-7(H_2) (Sakaguchi et al. 2002). The mol % G+C of the DNA is 66.

There has been some confusion as to whether this strain is actually magnetotactic or not. Sakaguchi et al. (2002) state "When pyruvate and fumarate were used, approximately 70% of cells responded to an artificial magnetic field. When sulfate was used as the electron acceptor, the majority of cells did not exhibit magnetotaxis and less than 1% were magnetotactic. Magnetotaxis in *D. magneticus* is distinctly different from *Magnetospirillum* species or magnetic cocci which swim along magnetic fields (Matsunaga et al. 1991; Frankel et al. 1997). *D. magneticus* cells swam or migrated randomly in different directions". We found that cells of both strains of *D. magneticus*, when magnetotactic at all, displayed polar magnetotaxis and accumulated at one end of the drop in the hanging drop assay (Frankel et al. 1997). While swimming towards the edge of the drop, cells appeared to be for the most part aligned along magnetic field lines.

2.2.1 Energy Metabolism in *D. magneticus* Strain RS-1

Cells of *D. magneticus* strain RS-1 are strictly anaerobic chemoorganoheterotrophs that utilize lactate, pyruvate, malate, oxaloacetate, and glycerol as electron donors and carbon sources (Sakaguchi et al. 2002). Sugars and amino acids are not utilized. Cells respire with sulfate, thiosulfate, and fumarate but not with sulfite, nitrate, elemental sulfur, Fe(III), or O_2 as terminal electron acceptors. Sulfate and thiosulfate are reduced to S^{2-}. Cells grown with a terminal electron acceptor produce *c*-type cytochromes and those grown with sulfate also produce desulfoviridin, a dissimilatory sulfite reductase common in many sulfate-reducing bacteria. Cells synthesize an average of six Fe_3O_4 crystals when pyruvate and fumarate are used as the carbon source and electron acceptor, respectively, while the vast majority of cells grown with sulfate as the terminal electron acceptor were not magnetotactic and did not produce magnetosomes. This species is the only magnetotactic bacterium known to be capable of fermentation; cells ferment pyruvate to acetate, CO_2, and H_2 in the absence of a terminal electron acceptor (Sakaguchi et al. 2002).

2.3 Magnetococcus Strain MC-1 (Magnetic Coccus #1)

Strain MC-1 represents the only magnetotactic coccus in pure culture and was isolated from brackish water collected from the OAI of the Pettaquamscutt Estuary (Rhode Island, USA). The name *Magnetococcus marinus* is being proposed for this strain (Bazylinski DA, unpublished data). Cells are roughly spherical, are 1–2 μm in diameter, and are motile by means of two bundles of flagella on one side of the cell (a flagella pattern referred to as bilophotrichous (Moench 1988)) (Frankel et al. 1997). The magnetic response of cells of strain MC-1, like that of all magnetotactic cocci, defines polar magnetotaxis (Frankel et al. 1997). Strain MC-1 and the magnetotactic cocci in general show no close affinity to any other α-Proteobacteria and appear to constitute a unique lineage that diverged early from the main branch of the α-Proteobacteria (DeLong et al. 1993: Fernández de Henestrosa et al. 2003) and are only distantly related to other magnetotactic α-Proteobacteria (e.g., *Magnetospirillum* species). Strain MC-1 produces a single chain of magnetosomes that contain elongated prismatic Fe_3O_4 crystals (Meldrum et al. 1993a). The genome of strain MC-1 appears to be a single circular chromosome about 4.5 Mb in size (Dean and Bazylinski 1999). An incomplete draft genome sequence of strain MC-1 is available for study at http://genome.jgi-psf.org/draft_microbes/magm1/magm1.home.html. The mol % G+C of the genomic DNA is about 54. Two magnetotactic coccoid strains very similar to MC-1 phenotypically and by 16S rDNA sequence, have been recently been isolated from Eel Pond (Woods Hole, MA, USA) and Chesapeake Bay (MD, USA) (Bazylinski DA, unpublished data).

Sequencing of the genome of MC-1 has revealed some interesting findings. A *lexA* gene was first identified in the published genome of MC-1 and was then cloned and the protein purified (Fernández de Henestrosa et al. 2003). Under normal conditions the SOS network, an interacting collection of genes induced as a consequence of DNA damage (Radman 1974), remains repressed by the LexA protein. LexA binds specifically to an operator region located at the promoters of the SOS genes, thereby reducing their expression to basal level (Walker 1984). Results from reverse transcription-PCR analysis revealed that the MC-1 *lexA* gene comprises a single transcriptional unit with two ORFs encoding for proteins of unknown function and one for a *rumA*-like gene, a homolog of the *E. coli umuD* gene. Mobility shift assays showed that the MC-1 LexA protein specifically binds both to its own promoter and to that of the *umuDC* operon. However, MC-1 LexA did not bind to the promoter regions of other genes, including *recA* and *uvrA*, previously reported to be regulated by LexA in bacterial species belonging to the α-Proteobacteria (Fernández de Henestrosa et al. 1998). Site-directed mutagenesis of both the *lexA* and *umuDC* operator regions demonstrated that the sequence $CCTN_{10}AGG$ is the specific target motif for the MC-1 LexA protein and represents a new LexA binding motif in prokaryotes (Fernández de Henestrosa et al. 2003).

2.3.1 Energy Metabolism and Electron Transport in Strain MC-1

Cells of strain MC-1 grow chemoheterotrophically with acetate or chemolithoautotrophically with S^{2-} or thiosulfate as the electron donor. Strain MC-1 is very fastidious and for many years growth was restricted to semi-solid $[O_2]$-gradient media. We have been recently able to grow this strain in liquid microaerobically and chemolithoautotrophically with thiosulfate and obtain a reasonable yield of cells for study (Williams et al. 2006). Attempts at growing this strain heterotrophically in liquid with acetate have not been successful. After prolonged transfer in thiosulfate-containing growth medium, the original culture of MC-1 lost the ability to utilize acetate. However, we have recently reisolated MC-1 from the Pettaquamscutt Estuary and confirmed its identity using 16S rDNA analysis, and found that this newly-isolated culture utilizes acetate (Bazylinski DA, unpublished data). Strain MC-1 appears to be an obligate microaerophile and did not grow anaerobically with nitrate, nitrite, N_2O, sulfate, thiosulfate, dimethyl sulfoxide, or trimethylamine oxide as terminal electron acceptors (Bazylinski DA, unpublished data) with either thiosulfate or acetate as the electron donor.

2.3.2 Chemolithoautotrophy in Strain MC-1

Cells of strain MC-1 oxidize S^{2-} and thiosulfate to sulfate and produce transient internal sulfur-rich globules during chemolithoautotrophic growth on these electron donors (Williams et al. 2006). It was suspected that strain MC-1 utilized the CBB cycle for autotrophy, as most aerobic chemolithoautotrophic bacteria do. Cell-free extracts of strain MC-1 did not exhibit RubisCO activity however, and the various forms of RubisCO genes were not present in the draft genome sequence of strain MC-l. We did discover ORFs in the draft genomes that have high sequence identity to genes for some key enzymes of the reverse or reductive tricarboxylic acid (rTCA) cycle, including pyruvate:ferredoxin oxidoreductase and 2-oxoglutarate:ferredoxin oxidoreductase, from autotrophic members of the Aquificales that utilize this cycle for CO_2 fixation and autotrophy. Moreover, all autotrophic organisms produce biomass characterized by discrimination against ^{13}C during CO_2 fixation. The ^{13}C content of whole cells of MC-1 relative to the ^{13}C content of the inorganic carbon source ($\Delta\delta^{13}C$) was – 11.4‰, a value consistent with operation of the rTCA cycle (Fuchs 1980; Preuss et al. 1989). Cellular fatty acids showed enrichment of ^{13}C relative to whole cells. Cell-free extracts showed activities for several key enzymes of the rTCA cycle including fumarate reductase, pyruvate:acceptor oxidoreductase, and 2-oxoglutarate:acceptor oxidoreductase. Although ATP citrate lyase (another key enzyme of the rTCA cycle) activity was not detected using commonly used assays, cell-free extracts did

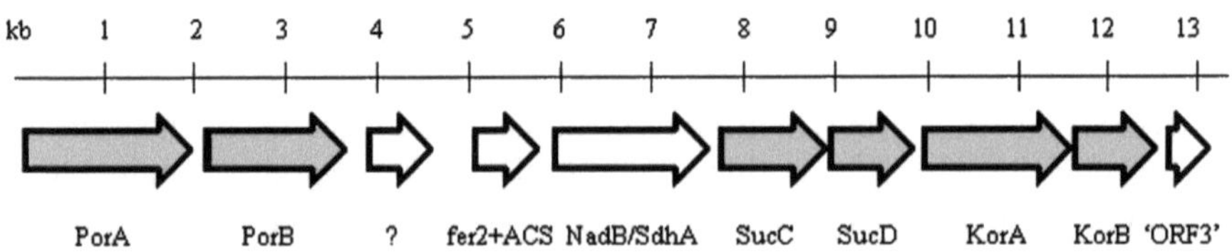

Fig. 3 Gene cluster within contig 294 (GenBank accession no. AAAN03000051) in the draft genome of strain MC-1, showing identities of ORFs that encode for putative proteins implicated in the reverse tricarboxylic acid cycle (*shaded in gray*) (Williams et al. 2006). Each ORF is represented by an *arrow* indicating the direction of gene transcription. Putative identities of proteins are based on BLASTP searches and are shown beneath each ORF. *?* indicates a protein of unknown function. PorA and PorB are the α and β subunits of pyruvate:acceptor oxidoreductase, respectively. fer2/ACS is a putative protein with an N-terminal 2Fe-2S iron–sulfur cluster binding domain (fer2) and a C-terminal acetyl-CoA synthase (ACS) domain. NadB/SdhA is a FAD-binding protein showing homology to both aspartate oxidase and succinate dehydrogenase/fumarate reductase, a flavoprotein subunit protein. SucC and SucD are the α and β subunits of succinyl-CoA synthetase, respectively. KorA and KorB are the α and β subunits of 2-oxoglutarate:acceptor oxidoreductase, respectively. "ORF3" matches an unknown putative protein that immediately follows the KorA and KorB genes in *Hydrogenobacter thermophilus* (Yun et al. 2001)

cleave citrate, the reaction being dependent upon the presence of ATP and coenzyme A. The presence of an ATP-dependent citrate-cleaving mechanism was thus inferred although the gene or genes responsible for this important step in the cycle have not been identified as yet. Several of the genes responsible for encoding enzymes involved in the rTCA cycle in strain MC-1 are located in a gene cluster within contig 294 (GenBank accession no. AAAN03000051) in the draft genome of strain MC-1 (Fig. 3). Strain MC-1 is the first known representative of the α-Proteobacteria to use the rTCA cycle for autotrophy.

2.3.3 Nitrogen Metabolism of Strain MC-1

2.3.3.1 Nitrogen oxide reduction

Strain MC-1 does not grow anaerobically with nitrite or nitrate as the terminal electron acceptor and thus is not capable of denitrification. However, a possible *napA*-like periplasmic nitrate reductase gene (GenBank accession no. ZP_00608741; JGI no. Mmc1DRAFT_0224) has been identified in the genome of MC-1, as well as a putative nitric oxide (NO) reductase (GenBank accession no. ZP_00606909; JGI no. Mmc1DRAFT_2116). Interestingly, *napA* is part of *napFDAGHBC* gene locus located on contig 282 (GenBank accession no. AAAN03000076) of the draft genome. It is possible that these ORFs are remnants of a denitrification pathway that once was present in this organism.

2.3.3.2 N_2 Fixation in Strain MC-1

Cells of strain MC-1 grow well in [O_2]-gradient media containing either thiosulfate or acetate as the electron donor and no fixed nitrogen source. Cells exhibit nitrogenase activity as evidenced by their ability to reduce acetylene to ethylene in these cultures (Bazylinski DA, unpublished data). As with *M. magnetotacticum*, genome data supports the ability of strain MC-1 to fix N_2 in that many of the key genes for N_2 fixation, including the structural genes for nitrogenase (*nifH*, *nifD*, and *nifK*), are clustered together as shown in Fig. 4.

2.4 Magnetotactic Marine Vibrio Strain MV-1 (Magnetic Vibrio #1)

Strain MV-1 is a small ($1-3 \times 0.3-0.4\,\mu m$) marine vibrio that is comma-shaped or helical in morphology and has a single polar flagellum (Bazylinski 1990). Phylogenetically, strains like MV-1 represent a new genus within the

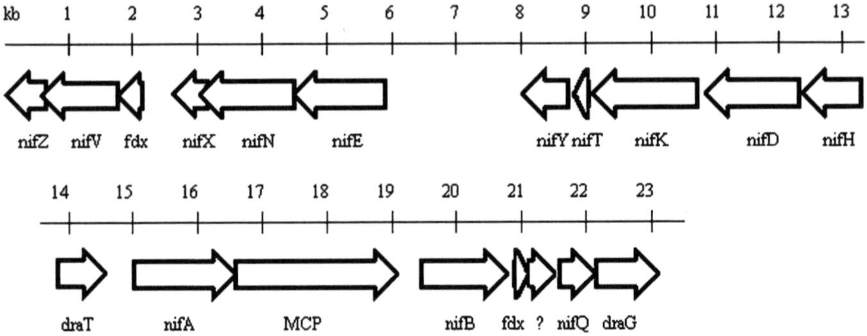

Fig. 4 Gene cluster within contig 354 (GenBank accession no. AAAN03000001) in the draft genome of strain MC-1 showing putative nitrogen-fixation (*nif*) genes. Each ORF is represented by an *arrow* indicating the direction of gene transcription. Identities of each gene are shown beneath each ORF, and are based on BLASTP searches of the translated product. Putative identities of specific genes are given in the caption of Fig. 2

α-Proteobacteria, possibly within the Rhodospirillales, and have no known close phylogenetic relatives (DeLong et al. 1993). Cells swim in both directions similar to those of *Magnetospirillum* species but many cells appear to accumulate at the edge of the drop in the magnetic hanging drop assay suggesting that this organism may show aspects of both polar and axial magnetotaxis (Bazylinski DA, unpublished data). Strain MV-1 is a facultatively anaerobic microaerophile (Bazylinski et al. 1988) that produces a single chain of elongated Fe_3O_4-containing magnetosomes oriented along the long axis of the cell (Bazylinski et al. 1988), which sometimes have large gaps between them (Bazylinski et al. 1995). The morphology of the Fe_3O_4 crystals has been described as truncated hexa-octahedrons (Thomas-Keprta et al. 2001). The genome of strain MV-1 appears to consist of a single circular chromosome whose size is estimated at about 3.7 Mb (Dean and Bazylinski 1999). Strains of magnetotactic vibrios have been isolated from a number of marine sites (e.g., strain MV-2 isolated from water collected from the OAI of the Pettaquamscutt Estuary (Dean and Bazylinski 1999)) appear to be very similar to strain MV-1 both phenotypically and by 16S rDNA sequence analysis (Bazylinski DA and Williams TJ, unpublished data) indicating that this species may be widespread in marine, chemically stratified aquatic habitats.

2.4.1 Energy Metabolism and Electron Transport in Strain MV-1

Cells of strain MV-1 and MV-2 grow chemoorganoheterotrophically as well as chemolithoautotrophically and chemoorganoautotrophically (Bazylinski et al. 2004). Cells utilize certain organic acids and amino acids as electron donors for chemoorganoheterotrophic growth (Bazylinski et al. 1988). Sugars are not utilized. Chemoorganoheterotrophic growth of the strains occurs both microaerobically with O_2 or anaerobically with N_2O as terminal electron acceptors, respectively (Bazylinski et al. 1988). However, N_2O evidently cannot be used as an electron acceptor when some electron donors are used during autotrophic growth (Bazylinski et al. 2004).

Autotrophic growth of strains MV-1 and MV-2 is through the CBB pathway. Cell-free extracts of both strains showed RubisCO activity and a form II RubisCO gene (*cbbM*) but no form I (*cbbL*) was found in strain MV-1 (Fig. 1) (Bazylinski et al. 2004). A gene encoding for cbbQ, a putative posttranslational activator of RubisCO, was also identified just downstream from *cbbM* in strain MV-1 (Bazylinski et al. 2004).

Cells of strain MV-1 and MV-2 are able to utilize S^{2-}, thiosulfate, and possibly soluble Fe(II) but not tetrathionate or Fe(II) as FeS or $FeCO_3$ (siderite) microaerobically as electron donors to support autotrophy (Bazylinski et al. 2004; Bazylinski DA, unpublished data). N_2O as a terminal electron acceptor supports growth on thiosulfate but not on S^{2-}. Cells also microaerobically oxidize the C_1 compound formate but not methanol to CO_2, supporting chemoorganoautotrophic growth. Because formate is more oxidized than formaldehyde, these organisms probably do not use the ribulose monophosphate pathway or the serine pathway for C assimilation (typical of type I and type II methanotrophs, respectively) and rely on RubisCO to obtain cell C. Cell-free extracts of formate-grown cells of strain MV-1 show RubisCO activity, supporting this idea (Bazylinski et al. 2004). Thus these strains are facultative methylotrophs. This situation is not unique and there are other methylotrophic bacteria that oxidize C_1 compounds chemoorganoautotrophically using CO_2 as the major source of cell carbon, relying on RubisCO and the CBB pathway for CO_2 fixation and autotrophy. For example, cells of *Paracoccus denitrificans* grow aerobically on methanol in this way (Cox and Quayle 1975; Shively et al. 1978). Strains MV-1 and MV-2 initially grew as a biofilm on the surface of liquid growth medium containing formate and O_2 (N_2O did not support growth on formate), generating the first report of a biofilm produced by a magnetotactic bacterium (Bazylinski et al. 2004).

2.4.2
Iron Metabolism in Strain MV-1

Strain MV-1 appears to differ greatly from *Magnetospirillum* species with regard to the cells' magnetotactic response to iron concentration in the growth medium. Non-magnetotactic cells of *M. magnetotacticum* can be generated by omitting the major source of iron (20 μM Fe(III) quinate) (Blakemore et al. 1979). Growth yields from cultures with and without Fe(II) quinate are comparable. A similar situation exists for *M. gryphiswaldense* (Schüler and Baeuerlein 1996). Cells of MV-1 continue to biomineralize magnetite and remain magnetotactic when grown anaerobically with N_2O even when the medium is iron limiting (when the major source of iron is omitted from the growth medium) (Dubbels et al. 2004). Cells in these growth experiments (medium contained about 3 μM iron from mineral solution) restricted their own growth and the growth yield, based on cell numbers and total protein, was about 25% of that obtained when the major source of iron (25 μM $FeSO_4$) was included. The number of magnetosomes per cell was about half of those produced under iron sufficient conditions. This suggests that at least under anaerobic conditions magnetosome synthesis is obligatory in strain MV-1. This experiment has not been performed under microaerobic conditions.

2.4.2.1
Iron uptake and iron redox reactions in strain MV-1

Stable, spontaneous, non-magnetotactic mutants are regularly observed when cells of MV-1 are cultured on solid media incubated under anaerobic or microaerobic conditions (Dubbels et al. 2004). Similar mutants have been described for *M. gryphiswaldense* (Schübbe et al. 2003). Randomly amplified polymorphic DNA analysis showed that these mutants are not genetically identical. Cellular iron content of one non-magnetotactic mutant strain, designated MV-1nm1, grown anaerobically, was ~ 20- to 80-fold less than the iron content of wild-type (wt) MV-1 for the same iron concentrations, indicating that MV-1nm1 is deficient in some form of iron uptake. MVnm-1 is a genomic deletion mutant (Dean and Bazylinski 1999) but also has two point mutations (transversions) in a gene called *chpA* (discussed below).

Both wt MV-1 and MV-1nm1 appear to synthesize a hydroxamate type of siderophore but differed in their patterns of putative siderophore production (Dubbels et al. 2004). Wt MV-1 produces the highest levels of siderophore, based on the CAS assay, between initial media iron concentrations of about 8–28 μM, a pattern similar to that of several *Magnetospirillum* species (Calugay et al. 2003; Paoletti and Blakemore 1986). Siderophore production by MVnm-1 was more typical of non-magnetotactic bacteria and the highest levels of siderophores were produced at iron concentrations of ≤ 8 μM. The siderophore produced by the MV strains appeared to be of a hydroxamate-type but appeared to be unstable at the pH of the CAS assay (Schwyn and Neilands 1987). These results did not appear to explain the phenotypic difference between MV-1 and MVnm-1.

Comparative protein profiles of the two strains showed that MV-1nm1 did not synthesize several proteins produced by wt MV-1. One of these proteins (ChpA, copper-handling protein) was purified and characterized. The protein, a homodimer with an apparent subunit mass of about 19 kDa, was iron regulated, and periplasmic. There are homologous genes to *chpA* in many other, mostly pathogenic, species of prokaryotes as well as in

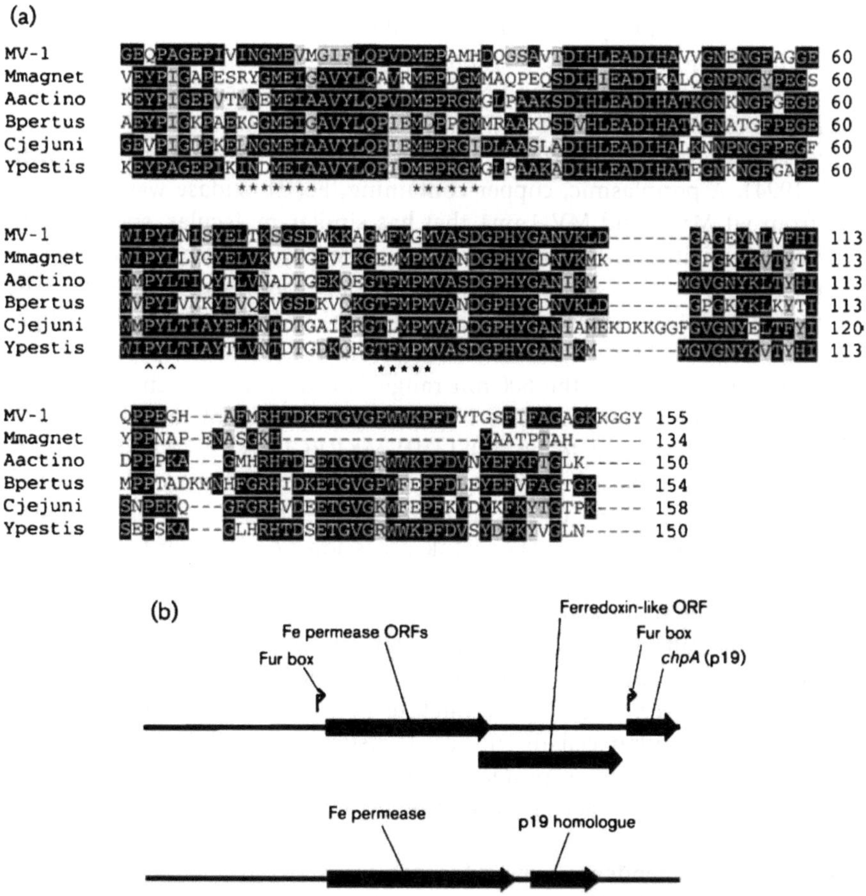

Fig. 5 **a** Alignment of derived amino acid sequences of p19 homologs. *White lettering on black background* indicates identical amino acids; *black lettering on gray background* indicates a conservative amino acid substitution; * indicates MXM/MX2M copper transporter motifs and an MX3M motif; ^ indicates the location of a proline hinge motif. Aactino, *Actinobacillus actinomycetemcomitans*; Bpertus, *Bortedella pertussis*; Cjejuni, *Campylobacter jejuni*; MV-1, wt MV-1, Mmagnet, *Magnetospirillum magnetotacticum*; Ypestis, *Yersinia pestis*: **b** Genetic organization of *chpA* operons. Putative *chpA* operon from strain MV-1 (*upper*), representation of putative *chpA* operons from *C. jejuni*, *Y. pestis*, and *Treponema pallidum* (*lower*). (Figure adapted from Dubbels et al. 2004)

another magnetotactic bacterium, *M. magnetotacticum* (GenBank accession no. ZP_00054566) (Fig. 5a). Two potential "copper-handling" motifs (MXM/MX2M), as well as a MX3M motif, are present in its amino acid sequence, together with the native protein-bound copper in a 1 : 1 ratio. These features are characteristic of copper transport proteins, the best characterized of which is Ctr1p, an integral plasma membrane component of a copper-dependent, high-affinity iron transport system in the yeast *S. cerevisiae* (Dancis et al. 1994a, b; Koch et al. 1997). Ctr1p is responsible for the transport of copper across the plasma membrane for eventual incorporation into the multi-copper iron oxidase Fet3p, which is involved and required in the secretory pathway for high-affinity iron transport in this organism (Askwith et al. 1994). A periplasmic, copper-containing, Fe(II) oxidase was also purified from wt MV-1 and MV-1nm1 that has similar molecular, spectral, and enzymatic activities as Fet3p. This enzyme, like ChpA, was regulated by media iron concentration and contained four copper atoms per molecule. The multi-copper Fe(II) oxidase is neutral with a pI of 7.0. The enzyme is light blue in color and displays absorbance maxima at 280 and 660 nm. The blue color and absorbance in the 600 nm range indicates the presence of a type I Cu(II) site in the enzyme. The N-terminal sequence of the MV-1 Fe(II) oxidase was determined to be ADSLDDVMKRRGLTEKDVLAAAKTYVPXG and there is an apparent homolog to the gene encoding for this protein within the draft genome of *M. magnetotacticum* that has been annotated as a putative N_2O reductase (GenBank accession no. ZP_00054320; JGI no. Magn03008954). This suggests that the copper-containing Fe(II) oxidase in strain MV-1 might act as a N_2O reductase (strain MV-1 respires with N_2O anaerobically), although the purified protein showed no N_2O reductase activity (Dubbels BL and Bazylinski DA, unpublished data). This may not be surprising, however, since N_2O reductase is a labile enzyme whose activity is often lost upon cell disruption (Kristjansson and Hollocher 1981), even when the enzyme is protected from oxygen (Kristjansson and Hollocher 1980). Most characterized N_2O reductases contain copper (Charnock et al. 2000).

The copper-dependent pathway for iron transport in *S. cerevisiae* also includes Ftr1p, a high-affinity plasma membrane iron permease that forms a complex with Fet3p (Stearman et al. 1996). The structural gene encoding for ChpA and two other putative genes upstream of *chpA* were cloned and sequenced. These putative genes encode for a protein similar to the iron permease, Ftr1, from the yeast *S. cerevisiae*, and a ferredoxin-like protein of unknown function (Fig. 5b).

Dubbels et al. (2004) hypothesized that ChpA, the iron permease, and the Fe(II) oxidase might have analogous functions for the three components of the *S. cerevisiae* copper-dependent high-affinity iron uptake system (Ctr1, Ftr1, and Fet3, respectively), and thus that strain MV-1 may have a similar iron uptake system. Given its cellular location and the fact that it binds copper, ChpA might have a similar copper transport function as Ctr1 in *S. cerevisiae*, providing copper to an Fe(II) oxidase for the formation of an active oxidase–permease complex for the transport of iron(III) across the cell membrane. Moreover, because *M. magnetotacticum* appeared to have homologs to ChpA and the Fe(II) oxidase, this copper-dependent transport system might be operational in other magnetotactic bacteria as well. However, the Fe(II) oxidase purified from both wt MV-1 and MV-1nm1 displayed comparable Fe(II) oxidase activities using O_2 as the electron acceptor, indicating that ChpA does not supply the multi-copper Fe(II) oxidase with copper.

A specific function for ChpA in strain MV-1 was not found (Dubbels et al. 2004) and the protein has only been studied in one organism, *Campylobacter jejuni* (van Vliet et al. 1998), despite the fact that *chpA* gene homologs are found in a large number of pathogenic bacteria. The gene encoding for the ChpA homolog in *C. jejuni* is under transcriptional control of the Fur protein and is thus regulated by iron concentration. However, the function of the ChpA homolog has not been determined in this organism and a knock-out mutant of the gene shows no obvious phenotypic change from the wild-type (CW Penn, personal communication). The ChpA gene homolog in *C. jejuni* is repressed at 40 μM iron (van Vliet et al. 1998) while the expression of *chpA* is not repressed even at 100 μM iron in strain MV-1. If ChpA is involved in an iron uptake system in magnetotactic bacteria and its expression is not repressed at high iron concentrations, the result might be the uptake of extremely high amounts of iron, which could be toxic. Forming Fe_3O_4 may be a way of detoxifying accumulating free iron ions since it is relatively inert.

Fe(III) reductase activity, an activity possibly linked to energy conservation and Fe_3O_4 biomineralization (Frankel et al. 1983) and demonstrated in *Magnetospirillum* species, was found in the cytoplasm of strain MV-1 (Dubbels et al. 2004).

2.4.3 Nitrogen Metabolism of Strain MV-1

2.4.3.1 Nitrogen oxide reduction

Cells of MV-1 reduce N_2O to N_2 and initially appeared not to have any of the other nitrogen oxide reductases involved in denitrification (Bazylinski 1990). However, recent studies show that strain MV-1 can grow anaerobically with nitrate and nitrite as terminal electron acceptors if the concentrations of these compounds are kept relatively low (Schübbe S and Bazylinski DA, unpublished data). When cells are actively reducing N_2O during growth, the growth medium, which contains the redox indicator resazurin, turns pink. Resazurin is colorless when reduced under anaerobic conditions and pink when oxidized. This observation suggests that cells either raise the redox potential high enough during growth for the resazurin to become oxidized, for example by oxidizing reduced Fe(II) (Fe(III) will cause resazurin to turn pink under anaerobic conditions) or utilize reduced resazurin as an electron donor to N_2O reductase reductase. Acetylene, a known inhibitor of N_2O reductase (Bazylinski et al. 1986), inhibits growth of MV-1 on N_2O and the medium does not turn pink (Bazylinski DA, unpublished data). To our knowledge, resazurin is not known to act as an electron donor in nitrogen oxide reduction. Resazurin is not required for anaerobic growth of MV-1 on N_2O, however (Dubbels BL and Bazylinski DA, unpublished data). Cells of MV-1 biomineralize more Fe_3O_4 crystals per cell when grown anaerobically with N_2O than when grown microaerobically with O_2 in $[O_2]$-gradient cultures (Bazylinski et al. 1988).

2.4.3.2
N_2 Fixation in Strain MV-1

Cells of strain MV-1 as well as MV-2, a closely related strain (DeLong et al. 1993), show nitrogenase activity as evidenced by the reduction of acetylene to ethylene under conditions where a fixed nitrogen source is missing from the growth medium both under chemoorganoheterotrophic and chemolithoautotrophic conditions (Bazylinski DA, unpublished data).

2.5
Magnetotactic Marine Spirillum Strain MMS-1 (Marine Magnetotactic Spirillum #1)

Very little characterization has been performed with this strain as it is very difficult to grow. It was formerly known as MV-4 until it was found to actually be a spirillum and not a vibrio and was isolated from mud and water collected from School Street marsh (Woods Hole, MA, USA). Cells of this marine strain are small (1–3 μm long) and range in morphology from being bean-shaped to partially helical to fully helical (Bazylinski and Frankel 2000). A second strain of this species has recently been isolated (MMS-2) (Bazylinski DA, unpublished data). Like strain MV-1, the spirillar strains MMS-1 and MMS-2 represent a new genus within the α-Proteobacteria and have no close phylogenetic relatives although they appear to be more closely related to strain MV-1 than to the freshwater *Magnetospirillum* species (Williams TJ and Bazylinski DA, unpublished data). Cells of strain MMS-1 biomineralize a single chain of elongated, prismatic crystals of Fe_3O_4 in their magnetosomes (Meldrum et al. 1993b). As in most magnetotactic bacteria, the chain lies along the long axis of the cell. Cells of both MMS-1 and MMS-2 display a strong polar magnetotactic response in the hanging drop assay (Frankel RB and Bazylinski DA, unpublished data). Cells appear to be obligately microaerophilic.

2.5.1
Energy Conservation, Metabolism, and Electron Transport in Strain MMS-1

Like those of strain MV-1, cells of strain MMS-1 grow chemoorganoheterotrophically as well as chemolithoautotrophically (Bazylinski DA, unpublished data). Cells utilize certain organic acids as electron donors for chemoorganoheterotrophic growth. Sugars are not utilized. O_2 appears to be the only terminal electron acceptor for growth. Chemolithoautotrophic growth is supported by the oxidation of thiosulfate but not sulfide in $[O_2]/[S^{2-}]$ inverse gradient cultures (Bazylinski DA, unpublished data). Cells grown chemolithoautotrophically with thiosulfate produce intracellular sulfur-rich globules and polyphosphate. No other electron donors have been found to support autotrophic growth of strain MMS-1 and the pathway of autotrophy in this organism has not yet been determined.

2.5.2
Nitrogen Metabolism in Strain MMS-1

2.5.2.1
N_2 Fixation in Strain MMS-1

Cells of strain MMS-1, like those of MV-1, show nitrogenase activity as evidenced by the reduction of acetylene to ethylene under conditions where a fixed nitrogen source is missing from the growth medium both under chemoorganoheterotrophic and chemolithoautotrophic conditions (Bazylinski DA, unpublished data).

2.6
Fe_3S_4-Producing Magnetotactic Bacteria

Little is known about the physiology of the Fe_3S_4-producing magnetotactic bacteria since there is not a single organism of this type in pure culture for study. Two species, the many-celled magnetotactic prokaryote (MMP) and a large rod, have been partially characterized morphologically and phylogenetically. The MMP is a large, multicellular bacterium that consists of about 10–30 cells arranged in a roughly spherical manner, and is motile as an entire unit but not as separate cells (Rogers et al. 1990). Each cell is multiflagellated on one side. Phylogenetically the MMP is affiliated with the sulfate-reducing bacteria in the δ-Proteobacteria, suggesting that it also is a sulfate-reducing bacterium (DeLong et al. 1993). A rod-shaped Fe_3S_4-producer was found to be affiliated with γ-Proteobacteria, suggesting a metabolism based on iron rather than sulfur (Simmons et al. 2004).

3
Ecophysiology of Magnetotactic Bacteria: Conclusions

The accumulated results described in this chapter illustrate the importance of magnetotactic bacteria in the biogeochemical cycling of many key elements including iron, sulfur, nitrogen, and carbon. Some of the reactions mediated by magnetotactic bacteria described herein include: S^{2-}, thiosulfate, and Fe(II) oxidation; nitrogen oxide, sulfate, and Fe(III) reduction; N_2 and CO_2 fixation; and Fe_3O_4 and Fe_3S_4 biomineralization. Many of these reactions are important physiologically to the cell although some have not been shown to be as yet. Because the magnetotactic bacteria are mainly present at the OAI in chemically stratified aquatic habitats, it seems likely that many of these biogeochemical transformations occur here. The OAI has long been known as an eminent site for many chemically and microbially mediated redox reactions. It is clear that many of the physiological and biogeochemical reactions mediated by magnetotactic bacteria described in this chapter are sensitive to but require O_2 in some way. These include S^{2-} oxidation, N_2 fixation fixation, Fe(II) oxidation, sulfate reduction, and CO_2 fixation. In some cases, there is a strong competition between the organism and O_2 for the reduced substrates (e.g., S^{2-} and Fe(II)) yet the organism must have some O_2 to respire. Thus, the organism must have access to both O_2 and the reduced substrates although a low $[O_2]$ is crucial. In other cases, the substrate is not oxidized by O_2 but the enzyme involved in the reaction is O_2-labile (e.g., nitrogenase). However, even in this situation, the organism would likely require some O_2 as oxidant since N_2 and CO_2 fixation (during autotrophy) are energetically expensive and not all magnetotactic bacteria can use terminal electron acceptors other than O_2. Again, a low $[O_2]$ appears to be crucial. In the case of the reduction of nitrogen oxides, which are generally inhibited by O_2, nitrate concentration peaks overlap those of O_2 in many aquatic habitats where nitrate is present. If the $[O_2]$ is low enough, simultaneous use of O_2 and nitrate as terminal electron acceptors may occur and thus cells maximize the amount of energy conserved by using both terminal electron acceptors. Given the above, it seems rather obvious why cells of magnetotactic bacteria appear to thrive at the OAI and why they might expend some energy synthesizing magnetosomes to more efficiently locate and maintain position there.

Many non-magnetotactic bacteria are primarily located at the OAI that rely solely on chemotaxis and not magnetotaxis. Non-magnetotactic mutants of magnetotactic bacterial strains that do not synthesize magnetosomes also form bands of cells in $[O_2]$-gradient cultures at the OAI (Bazylinski DA, unpublished data). Moreover, cells of the magnetotactic coccoid strain MC-1 form microaerobic bands of cells in $[O_2]$-gradient cultures even in a Mu-metal can where the geomagnetic field is eliminated. Disrupted bands of MC-1 in $[O_2]$-gradient cultures reform faster if the culture is left in the geomagnetic field; however, this is only on the order of tens of minutes (Bazylinski DA, unpublished data). This time difference may not be significant in the life of a magnetotactic bacterium in the environment. Finally, the finding of large numbers of south-seeking polar magnetotactic bacteria in the northern hemisphere raises additional serious questions regarding the biological value of magnetotaxis (Simmons et al. 2006). All together, these results might suggest that magnetotaxis might not function as is currently believed and that there is a physiological reason for cells to expend a significant amount of energy taking up large amounts of iron and biomineralizing magnetosomes. However, this physiological reason, if one exists, is unknown at the present time.

Acknowledgements We thank K.J. Edwards, R.B. Frankel, and S.L. Simmons for collaboration and stimulating discussion. We are particularly grateful to D. Schüler for sharing unpublished information. Research in our laboratory is supported by US National Science Foundation Grant EAR-0311950.

References

Allen RM, Chatterjee R, Madden MS, Ludden PW, Shah VK (1994) Biosynthesis of the iron-molybdenum cofactor of nitrogenase. Crit Rev Biotechnol 14:225–249

Askwith C, Eide D, Van Ho A, Bernard PS, Li L, Davis-Kaplan S, Sipe DM, Kaplan J (1994) The *FET3* gene of *S. cerevisiae* encodes a multicopper oxidase required for ferrous iron uptake. Cell 76:403–410

Baker SH, Jin S, Aldrich HC, Howard GT, Shively JM (1998) Insertion mutation of the form I *cbbL* gene encoding ribulose bisphosphate carboxylase/oxygenase (RuBisCO) in *Thiobacillus neapolitanus* results in expression of form II RuBisCO, loss of carboxysomes, and an increased CO_2 requirement for growth. J Bacteriol 180:4133–4139

Bazylinski DA (1990) Anaerobic production of single-domain magnetite by the marine, magnetotactic bacterium, strain MV-1. In: Frankel RB, Blakemore RP (eds) Iron biominerals. Plenum, New York, pp 69–77

Bazylinski DA (1995) Structure and function of the bacterial magnetosome. ASM News 61:337–343

Bazylinski DA, Blakemore RP (1983a) Denitrification and assimilatory nitrate reduction in *Aquaspirillum magnetotacticum*. Appl Environ Microbiol 46:1118–1124

Bazylinski DA, Blakemore RP (1983b) Nitrogen fixation (acetylene reduction) in *Aquaspirillum magnetotacticum*. Curr Microbiol 9:305–308

Bazylinski DA, Frankel, RB (2000) Biologically controlled mineralization of magnetic iron minerals by magnetotactic bacteria. In: Lovley DR (ed) Environmental microbe-metal interactions. ASM, Washington DC, pp 109–144

Bazylinski DA, Frankel RB (2003) Biologically controlled mineralization in prokaryotes. Rev Mineral Geochem 54:217–247

Bazylinski DA, Frankel RB (2004) Magnetosome formation in prokaryotes. Nature Rev Microbiol 2:217–230

Bazylinski DA, Garratt-Reed AJ, Frankel RB (1994) Electron microscopic studies of magnetosomes in magnetotactic bacteria. Microsc Res Tech 27:389–401

Bazylinski DA, Moskowitz BM (1997) Microbial biomineralization of magnetic iron minerals: microbiology, magnetism, and environmental significance. Rev Mineral Geochem 35:181–223

Bazylinski DA, Dean AJ, Schüler D, Phillips EJP, Lovley DR (2000) N_2-dependent growth and nitrogenase activity in the metal-metabolizing bacteria, *Geobacter* and *Magnetospirillum* species. Environ Microbiol 2:266–273

Bazylinski DA, Dean AJ, Williams TJ, Kimble-Long L, Middleton SL, Dubbels BL (2004) Chemolithoautotrophy in the marine, magnetotactic bacterial strains MV-1 and MV-2. Arch Microbiol 182:373–387

Bazylinski DA, Frankel RB, Heywood BR, Mann S, King JW, Donaghay PL, Hanson AK (1995) Controlled biomineralization of magnetite (Fe_3O_4) and greigite (Fe_3S_4) in a magnetotactic bacterium. Appl Environ Microbiol 61:3232–3239

Bazylinski DA, Frankel RB, Jannasch HW (1988) Anaerobic magnetite production by a marine magnetotactic bacterium. Nature 334:518–519

Bazylinski DA, SooHoo CK, Hollocher TC (1986) Growth of *Pseudomonas aeruginosa* on nitrous oxide. Appl Environ Microbiol 51:1239–1246

Blakemore RP (1975) Magnetotactic bacteria. Science 190:377–379

Blakemore RP (1982) Magnetotactic bacteria. Annu Rev Microbiol 36:217–238

Blakemore RP, Maratea D, Wolfe RS (1979) Isolation and pure culture of a freshwater magnetic spirillum in chemically defined medium. J Bacteriol 140:720–729

Blakemore RP, Short K, Bazylinski DA, Rosenblatt C, Frankel RB (1985) Microaerobic conditions are required for magnetite formation within *Aquaspirillum magnetotacticum*. Geomicrobiol J 4:53–71

Burgess JG, Kawaguchi R, Sakaguchi T, Thornhill RH, Matsunaga T (1993) Evolutionary relationships among *Magnetospirillum* strains inferred from phylogenetic analysis of 16S rDNA sequences. J Bacteriol 175:6689–6694

Calugay RJ, Miyashita H, Okamura Y, Matsunaga T (2003) Siderophore production by the magnetic bacterium *Magnetospirillum magneticum* AMB-1. FEMS Microbiol Lett 218:371–375

Charnock JM, Dreusch A, Körner H, Neese F, Nelson J, Kannt A, Michel H, Garner CD, Kroneck PMH, Zumft WG (2000) Structural investigations of the Cu_A centre of nitrous oxide reductase from *Pseudomonas stutzeri* by site-directed mutagenesis and X-ray absorption spectroscopy. Eur J Biochem 267:1368–1381

Cox RB, Quayle JR (1975) The autotrophic growth of *Micrococcus denitrificans* on methanol. Biochem J 150:569–571

Dancis A, Haile D, Yuan,DS, Klausner RD (1994a) The *Saccharomyces cerevisiae* copper transport protein (Ctr1p). Biochemical characterization, regulation by copper, and physiologic role in copper uptake. J Biol Chem 269:25660–25667

Dancis A, Yuan DS, Haile D, Askwith C, Eide D, Moehle C, Kaplan J, Klausner RD (1994b) Molecular characterization of a copper transport protein in *S. cerevisiae*: an unexpected role for copper in iron transport. Cell 76:393–402

Dean AJ, Bazylinski DA (1999) Genome analysis of several magnetotactic bacterial strains using pulsed-field gel electrophoresis. Curr Microbiol 39:219–225

DeLong EF, Frankel RB, Bazylinski DA (1993) Multiple evolutionary origins of magnetotaxis in bacteria. Science 259:803–806

Dubbels BL, DiSpirito AA, Morton JD, Semrau JD, Neto JN, Bazylinski DA (2004) Evidence for a copper-dependent iron transport system in the marine, magnetotactic bacterium strain MV-1. Microbiol 150:2931–2945

Fernández de Henestrosa AR, Cuñé J, Mazón G, Dubbels BL, Bazylinski DA, Barbé J (2003) Characterization of a new LexA binding motif in the marine magnetotactic bacterium strain MC-1. J Bacteriol 185:4471–4482

Fernández de Henestrosa AR, Rivera E, Tapias A, Barbé J (1998) Identification of the *Rhodobacter sphaeroides* SOS box. Mol Microbiol 28:991–1003

Frankel RB, Bazylinski DA (2003) Biologically induced mineralization by bacteria. Rev Mineral Geochem 54:95–114

Frankel RB, Bazylinski DA, Johnson M, Taylor BL (1997) Magneto-aerotaxis in marine, coccoid bacteria. Biophys J 73:994–1000

Frankel RB, Papaefthymiou GC, Blakemore RP, O'Brien W (1983) Fe_3O_4 precipitation in magnetotactic bacteria. Biochim Biophys Acta 763:147–159

Fuchs G (1980) Alternate pathways of autotrophic CO_2 fixation. In: Schlegel HG, Bowien B (eds) Autotrophic bacteria. Springer, Berlin Heidelberg New York, pp 365–382

Gorby YA, Beveridge TJ, Blakemore RP (1988) Characterization of the bacterial magnetosome membrane. J Bacteriol 170:834–841

Grass G, Otto M, Fricke B, Haney CJ, Rensing C, Nies DH, Munkelt D (2005) FieF (YiiP) from *Escherichia coli* mediates decreased cellular accumulation of iron and relieves iron stress. Arch Microbiol 183:9–18

Grünberg K, Wawer C, Tebo BM, Schüler D (2001) A large gene cluster encoding several magnetosome proteins is conserved in different species of magnetotactic bacteria. Appl Environ Microbiol 67:4573–4582

Guerin WF, Blakemore RP (1992) Redox cycling of iron supports growth and magnetite synthesis by *Aquaspirillum magnetotacticum*. Appl Environ Microbiol 58:1102–1109

Guerinot ML (1994) Microbial iron transport. Annu Rev Microbiol 48:743–772

Handrick R, Reinhardt S, Schultheiss D, Reichart T, Schüler D, Jendrossek V, Jendrossek D (2004) Unraveling the function of the *Rhodospirillum rubrum* activator of polyhydroxybutyrate (PHB) degradation: the activator is a PHB-granule-bound protein (phasin). J Bacteriol 186:2466–2475

Hernandez JM, Baker SH, Lorbach SC, Shively JM, Tabita FR (1996) Deduced amino acid sequence, functional expression, and unique enzymatic properties of the form I and form II ribulose bisphosphate carboxylase/oxygenase from the chemoautotrophic bacterium *Thiobacillus denitrificans*. J Bacteriol 178:347–356

Heyen U, Schüler D (2003) Growth and magnetosome formation by microaerophilic *Magnetospirillum* strains in an oxygen-controlled fermentor. Appl Microbiol Biotechnol 61:536–544

Hoffman PS, Krieg NR, Smibert RM (1979) Studies of the microaerophilic nature of *Campylobacter fetus* subsp, *jejuni*. I. Physiological aspects of enhanced aerotolerance. Can J Microbiol 25:1–7

Hu Y, Fay AW, Dos Santos PC, Naderi F, Ribbe MW (2004) Characterization of *Azotobacter vinelandii nifZ* deletion strains. Indication of stepwise MoFe protein assembly. J Biol Chem 279:54963–54971

Iwata S, Ostermeier C, Ludwig B, Michel H (1995) Structure at 2.8 Å resolution of cytochrome *c* oxidase from *Paracoccus denitrificans*. Nature 376:660–669

Jordan DB, Ogren WL (1981) Species variation in the specificity of ribulose bisphosphate carboxylase/oxygenase. Nature 291:513–515

Kawaguchi R, Burgess JG, Sakaguchi T, Takeyama H, Thornhill RH, Matsunaga T (1995) Phylogenetic analysis of a novel sulfate-reducing magnetic bacterium, RS-1, demonstrates its membership of the δ-proteobacteria. FEMS Microbiol Lett 126:277–282

Kim S, Burgess BK (1996) Evidence for the direct interaction of the *nifW* gene product with the MoFe protein. J Biol Chem 271:9764–9770

Kirchhausen, T. (2000) Three ways to make a vesicle. Nature Rev Mol Cell Biol 1:187–198

Koch KA, Pena MM, Thiele DJ (1997) Copper-binding motifs in catalysis, transport, detoxification and signaling. Chem Biol 4:549–560

Komeili A, Li Z, Newman DK, Jensen GJ (2005) Magnetosomes are invaginations organized by the actin-like protein MamK. Science 311:242–245

Komeili A, Vali H, Beveridge TJ, Newman DK (2004) Magnetosome vesicles are present prior to magnetite formation and MamA is required for their activation. Proc Natl Acad Sci USA 101:3839–3844

Krieg NR, Hoffman PS (1986) Microaerophily and oxygen toxicity. Annu Rev Microbiol 40:107–130

Kristjansson JK, Hollocher TC (1980) First practical assay for soluble nitrous oxide reductase of denitrifying bacteria and a partial kinetic characterization. J Biol Chem 255:704–707

Kristjansson JK, Hollocher TC (1981) Partial purification and characterization of nitrous oxide reductase from *Paracoccus dentrificans*. Curr Microbiol 6:247–251

Li L, Kaplan J (1997) Characterization of two homologous yeast genes that encode mitochondrial iron transporters. J Biol Chem 272:28485–28493

Li L, Kaplan J (2001) The yeast gene *MSC2*, a member of the cation diffusion facilitator family, affects the cellular distribution of zinc. J Biol Chem 276:5036–5043

Mandernack KW, Bazylinski DA, Shanks WC, Bullen TD (1999) Oxygen and isotope studies of magnetite produced by magnetotactic bacteria. Science 285:1892–1896

Maratea D, Blakemore RP (1981) *Aquaspirillum magnetotacticum* sp. nov., a magnetic spirillum. Int J Syst Bacteriol 31:452–455

Matsunaga T, Okamura Y, Fukuda Y, Wahyudi AT, Murase Y, Takeyama H (2005) Complete genome sequence of the facultative anaerobic magnetotactic bacterium *Magnetospirillum* sp. strain AMB-1. DNA Res 12:157–166

Matsunaga T, Sakaguchi T, Tadokoro F (1991) Magnetite formation by a magnetic bacterium capable of growing aerobically. Appl Microbiol Biotechnol 35:651–655

Matsunaga T, Tsujimura N (1993) Respiratory inhibitors of a magnetic bacterium *Magnetospirillum* sp. AMB-1 capable of growing aerobically. Appl Microbiol Biotechnol 39:368–371

Mayer SM, Gormal CA, Smith BE, Lawson DM (2002) Crystallographic analysis of the MoFe protein of nitrogenase from a *nifV* mutant of *Klebsiella pneumoniae* identifies citrate as a ligand to the molybdenum of iron molybdenum cofactor (FeMoco). J Biol Chem 277:35263–35266

Meldrum FC, Heywood BR, Mann S, Frankel RB, Bazylinski DA (1993a) Electron microscopy study of magnetosomes in a cultures coccoid magnetotactic bacterium. Proc R Soc London B 251:231–236

Meldrum FC, Heywood BR, Mann S, Frankel RB, Bazylinski DA (1993b) Electron microscopy study of magnetosomes in two cultured vibrioid magnetotactic bacteria. Proc R Soc London B 251:237–242

Moench TT (1988) *Biliphococcus magnetotacticus* gen. nov. sp. nov., a motile, magnetic coccus. Antonie van Leeuwenhoek 54:483–496

Nakamura C, Sakaguchi T, Kudo S, Burgess JG, Sode K, Matsunaga T (1993) Characterization of iron uptake in the magnetic bacterium *Aquaspirillum* sp. AMB-1. Appl Biochem Biotechnol 39/40:169–176

Nakamura, C, Burgess JG, Sode K, Matsunaga T (1995a) An iron-regulated gene, *magA*, encoding an iron transport protein of *Magnetospirillum* AMB-1. J Biol Chem 270:28392–28396

Nakamura C, Kikuchi T, Burgess JG, Matsunaga T (1995b) Iron-regulated expression and membrane localization of the MagA protein in *Magnetospirillum* sp. strain AMB-1. J Biochem 118:23–27

Neilands JB (1984) A brief history of iron metabolism. Biol Metals 4:1–6

Neilands JB (1995) Siderophores: structure and function of microbial iron transport compounds. J Biol Chem 270:26723–26726

Nelson DC, Jannasch HW (1983) Chemoautotrophic growth of a marine *Beggiatoa* in sulfide-gradient cultures. Arch Microbiol 136:262–269

Nies DH, Silver S (1995) Ion efflux systems involved in bacterial metal resistances. J Ind Microbiol 14:186–199

Noguchi Y, Fujiwara T, Yoshimatsu K, Fukumori Y (1999) Iron reductase for magnetite synthesis in the magnetotactic bacterium *Magnetospirillum magnetotacticum*. J Bacteriol 181:2142–2147

O'Brien W, Paoletti LC, Blakemore, RP (1987) Spectral analysis of cytochromes in *Aquaspirillum magnetotacticum*. Curr Microbiol 15:121–127

Okamura Y, Takeyama H, Matsunaga T (2001) A magnetosome-specific GTPase from the magnetic bacterium *Magnetospirillum magneticum* AMB-1. J Biol Chem 276:48183–48188

Okamura Y, Takeyama H, Sekine T, Sakaguchi T, Wahyudi AT, Sato R, Kamiya S, Matsunaga T (2003) Design and application of a new cryptic-plasmid-based shuttle vector for *Magnetospirillum magneticum*. Appl Environ Microbiol 69:4274–4277

Orme-Johnson WH (1985) Molecular basis of biological nitrogen fixation. Annu Rev Biophys Biophys Chem 14:419–459

Paoletti LC, Blakemore RP (1986) Hydroxamate production by *Aquaspirillum magnetotacticum*. J Bacteriol 167:73–76

Paoletti LC, Blakemore RP (1988) Iron reduction by *Aquaspirillum magnetotacticum*. Curr Microbiol 17:339–342

Paulsen IT, Saier Jr MH (1997) A novel family of ubiquitous heavy metal ion transport proteins. J Membr Biol 156:99–103

Pósfai M, Buseck PR, Bazylinski DA, Frankel RB (1998a) Reaction sequence of iron sulfides in bacteria and their use as biomarkers. Science 280:880–883

Pósfai M, Buseck PR, Bazylinski DA, Frankel RB (1998b) Iron sulfides from magnetotactic bacteria: structure, compositions, and phase transitions. Am Mineral 83:1469–1481

Preuss A., Schauder R, Fuchs G, Stichler W (1989) Carbon isotope fractionation by autotrophic bacteria with three different CO_2 fixation pathways. Z Naturforsch Sect C 44:397–402

Radman M (1974) Phenomenology of an inducible mutagenic DNA repair pathway in *Escherichia coli*: SOS repair hypothesis. In: Prakash L, Sherman F, Miller M, Lawrence CW, Tabor HW (eds) Molecular and environmental aspects of mutagenesis. Charles C. Thomas, Springfield IL, pp 128–142

Ramos JL, Robson RL (1985) Isolation and properties of mutants of *Azotobacter chroococcum* defective in aerobic nitrogen fixation. J Gen Microbiol 131:1449–1458

Rogers FG, Blakemore RP, Blakemore NA, Frankel RB, Bazylinski DA, Maratea D, Rogers C (1990) Intercellular structure in a many-celled magnetotactic procaryote. Arch Microbiol 154:18–22

Sakaguchi H, Hagiwara H, Fukumori Y, Tamaura Y, Funaki M, Hirose S (1993a) Oxygen concentration-dependent induction of a 140-kDa protein in magnetic bacterium *Magnetospirillum magnetotacticum* MS-1. FEMS Microbiol Lett 107:169–174

Sakaguchi T, Arakaki A, Matsunaga T (2002) *Desulfovibrio magneticus* sp. nov., a novel sulfate-reducing bacterium that produces intracellular single-domain-sized magnetite particles. Int J Syst Evol Microbiol 52:215–221

Sakaguchi T, Burgess JG, Matunaga T (1993b) Magnetite formation by a sulphate-reducing bacterium. Nature 365:47–49

Sakaguchi T, Tsujimura N, Matsunaga T (1996) A novel method for isolation of magnetic bacteria without magnetic collection using magnetotaxis. J Microbiol Methods 26:139–145

Saiki K, Mogi T, Ogura K, Anraku Y (1993) *In vitro* heme O synthesis by the *cyoE* gene product from *Escherichia coli.* J Biol Chem 268:26041–26044

Scheffel A, Gruska M, Faive D, Linaroudisn A, Plitzko JM, Schüler D (2006) An acidic protein aligns magnetosomes along a filamentous structure in magnetotactic bacteria. Nature 441:248

Schleifer K-H, Schüler D, Spring S, Weizenegger M, Amann R, Ludwig W, Kohler M (1991) The genus *Magnetospirillum* gen. nov., description of *Magnetospirillum gryphiswaldense* sp. nov. and transfer of *Aquaspirillum magnetotacticum* to *Magnetospirillum magnetotacticum* comb. nov. Syst Appl Microbiol 14:379–385

Schübbe S, Kube M, Scheffel A, Wawer C, Heyen U, Meyerdierks A, Madkour MH, Mayer F, Reinhardt R, Schüler D (2003) Characterization of a spontaneous nonmagnetic mutant of *Magnetospirillum gryphiswaldense* reveals a large deletion comprising a putative magnetosome island. J Bacteriol 185:5779–5790

Schüler D, Baeuerlein E (1996) Iron-limited growth and kinetics of iron uptake in *Magnetospirillum gryphiswaldense.* Arch Microbiol 166:301–307

Schüler D, Baeuerlein E (1998) Dynamics of iron uptake and Fe_3O_4 mineralization during aerobic and microaerobic growth of *Magnetospirillum gryphiswaldense.* J Bacteriol 180:159–162

Schüler D, Spring S, Bazylinski DA (1999) Improved technique for the isolation of magnetotactic spirilla from freshwater sediment and their phylogenetic characterization. Syst Appl Microbiol 22:466–471

Schwyn B, Neilands JB (1987) Universal chemical assay for the detection and determination of siderophores. Anal Biochem 160:47–56

Shah VK, Allen JR, Spangler NJ, Ludden PW (1994) In vitro synthesis of the iron-molybdenum cofactor of nitrogenase. Purification and characterization of NifB cofactor, the product of NIFB protein. J Biol Chem 269:1154–1158

Shah VK, Imperial J, Ugalde RA, Ludden PW, Brill WJ (1986) In vitro synthesis of the iron-molybdenum cofactor of nitrogenase. Proc Natl Acad Sci USA 83:1636–1640

Shah VK, Rangaraj P, Chatterjee R, Allen RM, Roll JT, Roberts GP, Ludden PW (1999) Requirement of NifX and other *nif* proteins for in vitro biosynthesis of the iron-molybdenum cofactor of nitrogenase. J Bacteriol 181:2797–2801

Shively JM, Saluja A, McFadden BA (1978) Ribulose-bisphophate carboxylase from methanol-grown *Paracoccus denitrificans.* J Bacteriol 134:1123–1132

Short KA, Blakemore RP (1986) Iron respiration-driven proton translocation in aerobic bacteria. J Bacteriol 167:729–731

Short KA, Blakemore RP (1989) Periplasmic superoxide dismutases in *Aquaspirillum magnetotacticum.* Arch Microbiol 152:342–346

Simmons SL, Bazylinski DA, Edwards KJ (2006) South-seeking magnetotactic bacteria in the northern hemisphere. Science 311:371–374

Simmons SL, Sievert SM, Frankel RB, Bazylinski DA, Edwards KJ (2004) Spatiotemporal distribution of marine magnetotactic bacteria in a seasonally stratified coastal pond. Appl Environ Microbiol 70:6230–6239

Simon HM, Homer MJ, Roberts GP (1996) Perturbation of *nifT* expression in *Klebsiella pneumoniae* has limited effect on nitrogen fixation. J Bacteriol 178:2975–2977

Stearman R, Yuan DS, Yamaguchi-Iwai Y, Klausner RD, Dancis A (1996) A permease-oxidase complex involved in high-affinity iron uptake in yeast. Science 271:1552–1557

Tamegai H, Fukumori Y (1994) Purification, and some molecular and enzymatic features of a novel *ccb*-type cytochrome *c* oxidase from a microaerobic denitrifier, *Magnetospirillum magnetotacticum.* FEBS Lett 347:22–26

Tamegai H, Yamanaka T, Fukumori Y (1993) Purification and properties of a "cytochrome a_1"-like hemoprotein from a magnetotactic bacterium, *Aquaspirillum magnetotacticum*. Biochim Biophys Acta 1158:237–243

Tanimura Y, Fukumori Y (2000) Heme-copper oxidase family structure of *Magnetospirillum magnetotacticum* "cytochrome a_1-like" hemoprotein without cytochrome *c* oxidase activity. J Inorg Biochem 82:73–78

Taoka A, Yoshimatsu K, Kanemori M, Fukumori Y (2003) Nitrate reductase from the magnetotactic bacterium *Magnetospirillum magnetotacticum* MS-1: purification and sequence analysis. Can J Microbiol 49:197–206

Thomas-Keprta KL, Clemett SJ, Bazylinski DA, Kirschvink JL, McKay DS, Wentworth SJ, Vali H, Gibson EK Jr, McKay MF, Romanek CS (2001) Truncated hexa-octahedral magnetite crystals in ALH84001: presumptive biosignatures. Proc Natl Acad Sci USA 98:2164–2169

van Vliet AH, Wooldridge KG, Ketley JM (1998) Iron-responsive gene regulation in a *Campylobacter jejuni* fur mutant. J Bacteriol 180:5291–5298

Walker GC (1984) Mutagenesis and inducible responses to deoxyribonucleic acid damage in *Escherichia coli*. Microbiol Rev 48:60–93

Williams TJ, Zhang CL, Scott JH, Bazylinski DA (2006) Evidence for autotrophy via the reverse tricarboxylic acid cycle in the marine magnetotactic coccus strain MC-1. Appl Environ Microbiol 72:1322–1329

Wolfe RS, Thauer RK, Pfennig N (1987) A capillary racetrack method for isolation of magnetotactic bacteria. FEMS Microbiol Lett 45:31–36

Yamazaki T, Oyanagi H, Fujiwara T, Fukumori,Y (1995) Nitrite reductase from the magnetotactic bacterium *Magnetospirillum magnetotacticum*; a novel cytochrome cd_1 with Fe(II):nitrite oxidoreductase activity. Eur J Biochem 233:665–671

Yoshimatsu K, Fujiwara T, Fukumori Y (1995) Purification, primary structure, and evolution of cytochrome *c*-550 from the magnetic bacterium, *Magnetospirillum magnetotacticum*. Arch Microbiol 163:400–406

Yun NR, Arai H, Ishii M, Igarashi Y (2001) The genes for anabolic 2-oxoglutarate: ferredoxin oxidoreductase from *Hydrogenobacter thermophilus* TK-6. Biochem Biophys Res Commun 282:589–594

Microbiol Monogr (3)
D. Schüler: Magnetoreception and Magnetosomes in Bacteria
DOI 10.1007/7171_039/Published online: 7 October 2006

Geobiology of Magnetotactic Bacteria

Sheri L. Simmons[1,2] · Katrina J. Edwards[1,3] (✉)

[1]Geomicrobiology Group, Department of Marine Chemistry and Geochemistry, Woods Hole Oceanographic Institution, Woods Hole, MA 02543, USA

[2]*Present address:*
Department of Earth and Planetary Science, University of California Berkeley, Berkeley, CA 94720-4767, USA

[3]*Present address:*
Geomicrobiology Group, Department of Biological Sciences, University of Southern California, 3616 Trousdale Parkway, Los Angeles, CA 90089-0371, USA
kje@usc.edu

Abstract MTB population dynamics and their contribution to iron cycling in modern chemically stratified environments have not been previously evaluated in a systematic fashion. Using the tools of modern molecular microbial ecology and mineralogy, it is possible to determine the population dynamics of magnetite and greigite producing MTB with respect to environmental geochemistry. This integrated approach can help provide an understanding of the contribution of MTB to iron and sulfur cycling and export to sediments. Additionally, understanding the coupled biology and geochemistry of MTB in modern environments can help interpret the significance of magnetofossils present in the rock record.

1 Introduction

The ecology of magnetotactic bacteria (MTB) and their potentially substantial role in iron cycling are largely unexplored questions amidst the extensive literature on MTB behavior, phylogeny, and magnetofossil morphology. Quantitative microbial ecology seeks, in part, to define how the population density of different species varies over both time and space in conjunction with relevant environmental parameters. This requires the development and application of tools for species-level identification and enumeration of uncultivated organisms, and for appropriate geochemical measurements to be made in order to place population structure into environmental context. Molecular tools have been applied to individual species of MTB (e.g. DeLong et al. 1993; Spring et al. 1992, 1993, 1994, 1998), but have not been used to track the population dynamics of MTB communities over time. Such an integrated approach is required to understand both the impacts of MTB on their environment and the biosignatures they leave in the sedimentary and rock records. MTB communities in chemically stratified coastal ponds provide ideal model systems for this approach. They are analogues of larger, less accessible stratified marine basins (e.g. the Black Sea and the Cariaco Basin) where the contribution of MTB to iron flux could be quantitatively significant, but are more compact and dynamic systems that are easier to sample and monitor for focused study.

The widespread prevalence of such chemically stratified marine environments (water columns and sediments) suggests that marine MTB are globally more abundant than freshwater MTB, the focus of the limited previous work on MTB ecology in natural environments (e.g. Spring et al. 1993). Mostly qualitative observations of marine MTB show that they are abundant in the suboxic to anoxic zones of salt marsh sediments and semi-anaerobic basins worldwide (Farina et al. 1990; Mann et al. 1990a; Mann et al. 1990b; Bazylinski and Frankel 1992; Bazylinski et al. 1995; Spring and Bazylinski 2000; Simmons et al. 2004) (Simmons et al. 2006, manuscript in preparation) and occur in deep-sea sediments (Stolz et al. 1986; Petermann and Bleil 1993). Marine MTB can reach population densities of 10^4 cells ml^{-1} (Simmons et al. 2006, 2006a, personal communication). The large amounts of Fe concentrated by MTB in intracellular magnetosomes (10^{-13} to 10^{-15} g Fe per cell), considering their estimated population density, indicate that nanomolar concentrations of Fe may be sequestered in biomass (Stolz 1992; Simmons et al. 2006, personal communication). These observations suggest a significant MTB contribution to iron cycling in the environment that has not been quantitatively addressed.

The potential contribution of MTB to sulfur cycling and deposition is also significant. Nearly all cultivated marine magnetite-producing MTB are chemolithoautotrophic S-oxidizers (Spring and Bazylinski 2000), and many store intermediates (S^0) intracellularly for use as electron donors (Spring et al. 1993; Cox et al. 2002; Bazylinski et al. 2004). Greigite-producing MTB, which have not yet been isolated in axenic culture, sequester large amounts of Fe and S in the greigite mineral (Fe_3S_4) by an unknown mechanism. On the basis of the identification of one greigite-producer in the *Deltaproteobacteria* (DeLong et al. 1993), these have been suggested to function as sulfate reducers, but there is no data to date about their metabolic function.

The general characteristics of environments where MTB are found are fairly well defined by numerous observations (Blakemore 1982; Mann et al. 1990a). These include a neutral to slightly alkaline pH, suboxic to anoxic conditions, the presence of dissolved iron, and the presence of sulfide for greigite-producing MTB. The minimum concentration of dissolved iron required for the presence of MTB is not known. Observations in Salt Pond, a small stratified marine basin, suggest that MTB are most abundant in slightly reducing conditions (Simmons and Edwards 2006). MTB also appear to be most abundant in organic-rich environments such as salt marshes (Blakemore 1975; Simmons and Edwards 2006), the water column of stratified eutrophic freshwater (Kim et al. 2005) and salt ponds (e.g. Simmons et al. 2004), and freshwater sediments (Spring et al. 1993; Cox et al. 2002). Heterotrophic consumption of organic matter in these environments leads to the depletion of oxygen and the establishment of oxic-anoxic gradients favorable to the growth of MTB. At least some MTB are autotrophs (Bazylinski et al. 2004; Williams et al. 2006), but it is possible that others make use of high organic content by anaerobic heterotrophy (sulfate or iron reduction).

A complete description of the geobiology of MTB would necessarily include a comprehensive description of the microbial ecology of MTB and the governing factors controlling their population dynamics, as well as a quantitative understanding of the biogeochemical role they play in Fe and S cycling. A more complete understanding of the geobiology of MTB would enable the development of, for example, a predictive model for the amount of magnetite and greigite deposition by MTB given certain environmental parameters. Such a model has two potential benefits: it would enable us to quantify the impact of MTB on iron and sulfur cycling in modern environments, as well as better interpret magnetofossils as paleoenvironmental indicators. Clearly, the best way to develop this model is to study modern MTB populations in situ. The key parameters to measure are the relative abundances of magnetite- and greigite-producing MTB, their turnover time, and cellular iron and sulfur content with respect to environmental chemistry. The standard tools of modern molecular ecology can be used to identify individual species of MTB and track their population dynamics over time. In combination with analysis of MTB mineral content and post-depositional modification, it will be possible to obtain estimates of how fluctuations in MTB populations regulate the amount of iron and sulfur deposited as magnetofossils in sediments.

1.1 Magnetofossils as Paleoindicators

Fossilized magnetosomes produced by MTB have been previously suggested as useful paleoenvironmental indicators (Stolz 1992), both in the sedimentary and rock records. Magnetosomes have characteristic morphologies that are distinct from inorganic minerals, and are important carriers of remanent magnetism in the sedimentary rock record (Petersen and von Dobeneck 1986; Stolz 1992). Magnetite magnetofossils have been identified in rocks as old as 700 Mya (Chang 1989). Fossil magnetite from MTB appears to be well preserved in suboxic freshwater sedimentary environments (Snowball et al. 2002). Magnetite dissolves easily in highly sulfidic environments, but a fraction can be preserved by formation of a pyrite coating (Canfield and Berner 1987). Greigite is also abundant and well-preserved in the rock record (Roberts et al. 1996; Pósfai et al. 2001; Kao et al. 2004). The distinctive morphology of magnetofossil greigite has not yet been used to distinguish it from inorganic greigite, although biogenic greigite crystals have been shown to display different size distributions than inorganic greigite (Pósfai et al. 2001).

It is much less clear, however, exactly what the presence of fossilized magnetite and/or greigite reveals about the original depositional environment. Magnetite formed by MTB was once thought to indicate the presence of oxygen, due to the microaerophilic habitat of most of these species, but subsequent observations showed that MTB can also produce magnetite under anaerobic conditions (Bazylinski et al. 1988; Sakaguchi et al. 2002) and are found in anaerobic zones in nature (Simmons et al. 2006, personal communication). Greigite-producing MTB require a sulfur source to synthesize these biominerals, most likely sulfide in the sulfidic habitats they reside in. Some greigite-producers appear to require an oxygen-sulfide gradient, while others are found at low abundance in highly sulfidic waters (up to 1 mM sulfide) (Simmons et al. 2006, personal communication). These observations indicate that the presence of particular types of magnetofossils in the rock record could be a strong indicator of the redox environment existing at the time of deposition. The key factor limiting the use of magnetofossils as reliable paleo-indicators, however, is the lack of quantitative constraints on the redox environments favoring magnetite vs. greigite production by MTB.

Studies of magnetosomes preserved in lacustrine and marine sediments show broad correlations between the concentration and/or type of magnetosome and bulk chemical properties of the sediment, suggesting that environmental chemistry regulates both the total number of MTB and the relative abundance of different species. A positive linear relationship between total organic carbon and sediment magnetism, dominated by single domain magnetite and greigite of probable bacterial origin (Snowball et al. 1999, 2002), appeared in varved lake cores from Swedish freshwater lakes covering 9000 years of sediment deposition. A similar positive relationship between organic carbon content and magnetosome concentration was found in recent sediments from freshwater Lake Ely (Kim et al. 2005). The magnetic properties of magnetosomes were correlated with redox potential in sediment cores from a Swiss freshwater lake (Egli 2004a). Two groups of magnetosomes with distinct morphology and magnetic properties were present, which were interpreted as originating from two or more species of MTB (Egli 2004a; Egli 2004b). The first group of magnetosomes, characterized by greater morphological diversity, was associated with more oxygenated sediments. The second group was characterized by magnetosomes with less morphological diversity that were either magnetite with an elongated morphology or greigite. The organic carbon content of Pacific deep-sea sediments was also strongly correlated with magnetosome crystal morphology (Yamazaki and Kawahata 1998), again suggesting a habitat preference by distinct species of MTB with species-specific magnetosome morphologies.

These observations hint at a more precise governance of the population dynamics of individual MTB species by environmental chemistry. Study of modern stratified environments where MTB are abundant can provide some of the information required to interpret these magnetofossils, by linking the quantitative population dynamics of magnetite- and greigite- producing MTB to environmental geochemistry and mineralogy. Modern stratified marine environments are good analogues for the ancient oceans, now widely thought (Anbar and Knoll 2002; Shen et al. 2003; Arnold et al. 2004; Rouxel et al. 2005) to have contained oxic surface and iron-rich, sulfidic deep waters, as originally proposed by (Canfield 1998). Moreover, stratified water columns with euxinic bottom waters provide expanded chemical gradients that are easily sampled for simultaneous microbiological and chemical analysis. The lessons learned from MTB dynamics in small ponds could have widespread implications for larger such systems in the modern Earth environment, as well as for the ancient stratified oceans.

2 The Ecology of MTB in Stratified Marine Environments

2.1 Salt Marshes and Salt Ponds

Magnetotactic bacteria were first reported by Richard Blakemore from sediments of Eel Pond in Woods Hole, MA (Blakemore 1975), and since then, the salt marshes and salt ponds of New England have played a key role in our understanding of MTB ecology. A wide variety of MTB are abundant in salt marshes, particularly in the surface layers of sulfidic sediments where they co-occur with photo- and chemo-synthetic sulfide oxidizing bacteria (unpublished observations). The many-celled magnetotactic prokaryote (MMP), a greigite-producer, was identified as a *Deltaproteobacterium* with samples from Sippewissett salt marsh (Falmouth MA, DeLong et al. 1993). Large numbers of morphologically diverse, phylogenetically unidentified greigite rods were found in the Neponset, MA salt marsh (Heywood et al. 1990; Bazylinski and Frankel 1992). The chemical gradients in salt marsh sediments are steep, however, and it is difficult to match chemistry and microbiology on such small scales.

Glacially formed kettle ponds and basins such as Salt Pond and Siders Pond (Falmouth, MA) and the Lower Basin of the Pettaquamscutt River (Narragansett, RI) develop seasonal or permanent chemical gradients between oxic surface waters and euxinic bottom waters, due to a combination of restricted water exchange, thermal heating of surface waters, and high organic matter input. The chemical and biological gradients in salt ponds are expanded versions of the surface layers of salt marsh sediments. They are compact, easily sampled analogues of large stratified basins like the Black Sea, and are the only natural environments where the distribution of MTB has been studied with respect to chemical gradients (Stolz 1992; Bazylinski et al. 1995; Simmons et al. 2004; Simmons et al. 2006, personal communication).

Microcosms have also been used to examine the distribution of freshwater MTB with respect to chemical gradients (Flies et al. 2005a; Flies et al. 2005b). In these studies, sediments from various nutrient-rich freshwater environments were incubated at room temperature in bottles or small aquaria (0.1–5 l). MTB were sampled and counted in conjunction with detailed measurement of several chemical parameters. MTB were most abundant in suboxic conditions immediately below the oxic-anoxic transition zone (about 1% of total cell numbers), consistent with MTB distributions in stratified marine environments (Simmons et al. 2004) (Simmons et al. 2006, personal communication). The main drawback of MTB microcosms is that ecological succession in such a restricted environment inevitably leads to the dominance of one organism (typically a magnetite-producing coccus), eliminating much of the natural diversity that can be readily observed in environmental samples. Additionally, sulfide tends to accumulate to high levels in marine microcosms, poisoning some sulfide-sensitive MTB (unpublished observations). Microcosms provide valuable information due to the high resolution of sampling possible, but cannot replace detailed studies of natural MTB communities.

A stylized chemical profile for a stratified marine basin is shown in Fig. 1, based on measurements in Salt Pond, MA (Simmons et al. 2004) (Simmons et al. 2006, personal communication). Oxygen declines gradually as depth increases, due to aerobic heterotrophic consumption of organic matter. A chemocline of varying width develops where oxygen reaches zero, produced by the continuously diffusing supply of sulfide generated by sulfate-reducing bacteria in anaerobic bottom waters. Intermediate sulfur compounds (S^0, thiosulfate, and polysulfides) are often found at the upper

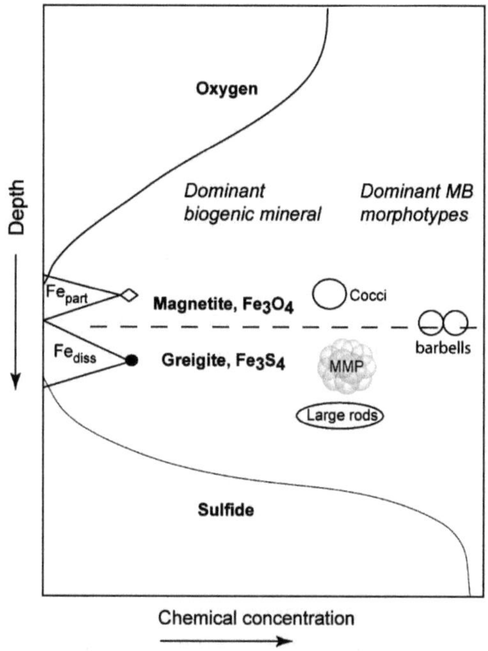

Fig. 1 Schematic of MB distribution observed in Salt Pond, showing approximate locations of magnetite and greigite producing MB relative to a typical stratified water column chemical profile. Fe_{part} = total particulate iron and Fe_{diss} = total dissolved iron. Reprinted from Simmons et al. (2004) with slight modifications

boundary of the sulfide zone (Jørgensen et al. 1979; Jørgensen et al. 1991; Ramsing et al. 1996). A peak in dissolved Fe(II) is present at the base of the chemocline, where sulfide concentrations are insufficient to sequester it as solid Fe-sulfides. A peak of particulate Fe is often found immediately above the chemocline, due to the upward flux of Fe(II) into oxygenated waters (Murray et al. 1995). The presence of oxidants and reductants at the chemocline, as well as a physical density gradient between warm surface waters and cool bottom waters, results in elevated bacterial counts and dense plates of anoxic phototrophs where light is available (e.g. Tonolla et al. 2000). MTB and magnetic protists (Bazylinski et al. 2000) are most abundant within and below the chemocline (Bazylinski et al. 1995; Simmons et al. 2004) (Simmons et al. 2006, personal communication).

2.2
Methods for Studying MTB Population Dynamics

Microscopic observations of MTB in the environment generally show a high morphological diversity (Stolz 1992; Bazylinski et al. 1995; Kim et al. 2005), particularly of greigite-producing MTB, that as yet is not completely accounted for in studies of the diversity of their 16*s* ribosomal RNA genes. Most published MTB sequences are from magnetite-producing freshwater cocci in the *Alphaproteobacteria* (Spring and Bazylinski 2000). Only two known greigite-producing MTB have been phylogenetically identified to the species level: the many-celled magnetotactic prokaryote (MMP) in the *Deltaproteobacteria* (DeLong et al. 1993) and a large rod in the *Gammaproteobacteria* (Simmons et al. 2006, personal communication). The overall lack of phylogenetic knowledge hinders the development of molecular probes that are required for accurate enumeration for different species of MTB in the environment, and consequently, a better understanding of the population dynamics of MTB communities. Here we briefly review previous work on the molecular identification of MTB, current methods that can be applied to newly discovered MTB, and their application to enumerating MTB in the environment.

Most previous observations of uncultivated MTB have been based on light microscopy. It is straightforward to collect sediments or waters known to contain MTB, place a drop of water on a microscope coverslip, put a bar magnet next to the drop, and observe MTB accumulating on the north or south side of the drop. For inverted microscopes, the drop can be viewed directly; for upright microscopes, the widely used "hanging drop" method is necessary. In this method, a small rubber O-ring is placed on a coverslip and a drop of water put inside. The O-ring is covered with an additional coverslip and the sandwich is gently inverted. The disadvantage of direct observation for quantitative cell counts is that it relies on cells being abundant enough in the natural sample to observe without prior magnetic enrichment. In our experience, natural waters often contain low abundances of MTB, which can make accurate direct counts difficult. Past direct counts in the literature also did not take into account the presence of MTB with south polarity in the Northern Hemisphere (Simmons et al. 2006b), as it was assumed that all Northern Hemisphere MTB possessed north polarity. The degree of undercount depends on the environment; in our recent experience, the majority of marine MTB usually appear to show north polarity. Even in samples and particular taxa that predominately exhibit north polarity, however, it is common to observe a significant fraction (> 10%) of the population displaying south-seeking polarity. We have also observed up to 100% south polarity in some natural samples (Simmons et al. 2006b). We suggest, therefore, that routine observations include counts for both north- and south-seeking polarities. However, it is important to note that MTB with especially slow or axial swimming behavior that do not accumulate rapidly at a drop edge can also be undercounted by the direct count method.

Molecular identification methods make no prior assumptions about the abundance or swimming behavior of MTB. Full-cycle 16*s* ribosomal RNA analysis, a widely used technique in microbial ecology for the identification and quantification of micro-organisms (reviewed by Amann and Ludwig 2000), can easily be applied to MTB. This method requires separation of MTB from the bulk matrix, PCR and sequencing of 16*s* ribosomal DNA genes, design of oligonucleotide probes based on those sequences, and subsequent verification of probes on environmental samples. Once a probe is verified to correspond to a particular species of MTB, it can then be applied to water column or sediment samples to count cells of that species. There are multiple methods available for physically isolating MTB from other microbes. These include the classic "racetrack" technique (Wolfe et al. 1987; Schüler et al. 1999), in which a glass Pasteur pipette is flame-sealed at the tip and a sterile cotton plug is placed at the base of the wide end. Sterile, chemically reduced medium or environmental water is injected through the plug into the tip, while water containing MTB is placed in the wide end. A magnet is placed near the tip, and over a short period (10–30 min) MTB migrate through the cotton plug into the tip, which can be broken off to collect cells. A dissecting scope can be used to observe cells accumulating in a black pellet at the tip. It is recommended not to incubate the racetrack for excessive periods, as other small, fast-swimming non-magnetotactic bacteria can swim through the cotton plug. Flies et al. (2005a) described a novel racetrack-like method for MTB isolation using a separation chamber constructed on a microscope slide with cellulose strips separating slurry containing MTB from sterile water under a coverslip. As in the large-scale racetrack, application of a magnetic field causes MTB to swim out of the slurry into sterile water. Using an inverted microscope, it is also possible to directly isolate particular morphotypes of MTB using a small drop of water placed on a microscope slide in conjunction with a small bar magnet. MTB accumulating on either side of the drop can be gently removed with a small-volume (5–10 μl) glass capillary pipettor (Simmons et al. 2004). This method has the advantage that particular morphotypes of MTB can be observed microscopically prior to isolation, and samples collected in this fashion can be used directly for PCR or fixed on slides for in situ hybridization. Flow cytometry has also been used on chemically fixed samples containing MTB to separate them based on size and magnetosome content (Spring et al. 1993; Wallner et al. 1997).

Once samples are isolated, small aliquots can be used directly in PCR with universal bacterial primers without further DNA extraction. A subsample of cells should also be preserved for in situ hybridization (ISH) by fixation with 2–4% paraformaldehyde. Because often only small numbers of MTB are available, it is best to immobilize live cells directly on a gelatin- or agarose-spotted welled slide, followed by immersion of the slide in paraformaldehyde and subsequent washing (see Simmons et al. 2006b for details). In the ideal case, sequencing of the PCR product reveals one or two clusters of closely related sequences that can be used for probe design (Simmons et al. 2006, personal communication). If a wide variety of unrelated sequences are obtained, it is better (and less expensive) to optimize the physical isolation step and repeat PCR rather than designing many probes. Alternatively, denaturing gradient gel electrophoresis (DGGE) can be used prior to sequencing to identify the most abundant sequence. Flies et al. (2005a) used this method, followed by gel extraction and sequencing, to identify a freshwater coccus in MTB microcosms. The drawbacks of DGGE are that it is primarily used with relatively short amplicons (Muyzer et al. 1993), limiting the amount of available phylogenetic information available for probe design, and requires a high abundance of the target population.

Once sequences are obtained, tools in the free software program ARB (Ludwig et al. 2004) are useful for sequence alignment, tree-building, and probe design. A variety of detection methods can be used with labeled oligonucleotide probes, with the goal of connecting a particular sequence to a particular morphotype of MTB. Probes should be tested on a suite of positive and negative control cells, either from cultures or manufactured using the clone-FISH method (Schramm et al. 2002). The most common technique is direct labeling of a probe with a fluorescent dye, usually fluorescein or Cy3. Spring et al. (1993) used flow cytometry, PCR, sequencing, and detection with this method to identify and enumerate cells of "*Magnetobacterium bavaricum*" in eutrophic freshwater sediment. Spring et al. (1992, 1994) also used fluorescently labeled probes to identify uncultivated freshwater magnetotactic cocci. Secondary detection methods such as catalyzed reporter deposition-in situ hybridization (CARD-FISH; Pernthaler et al. 2002) and gold-antibody conjugation can be used to increase the sensitivity of ISH. We used CARD-FISH with fluorescent tyramides to identify two uncultivated marine MTB (Simmons et al. 2006b) (Simmons et al. 2006, personal communication), based on morphology and relative abundance in fixed enriched samples (Fig. 2).

The gold-labeling method has the advantage of combining TEM visualization with phylogenetic identification. Spring et al. (1998) designed polyribonucleotide probes labeled with fluorescein or dioxygenin to target MTB with unusually large magnetosomes. Anti-FITC and anti-dioxygenin antibodies labeled with nanometer-size gold particles visible in TEM were applied to thin sections of resin-embedded cells. This is probably the only method that can definitively connect a particular phylogenetic probe with cells that are unquestionably magnetotactic, but it is technically challenging and time-consuming. It may be sufficient in many cases to apply fluorescent probes to highly enriched samples of a particular morphotype and make arguments based on the relative abundance of probe-positive cells in enriched and non-enriched samples, as well as any unique morphological characteristics.

Once the phylogenetic identity of cells is known, FISH probes can be used to count cells in the environment. Using FISH probes, Spring et al. (1993) found that "*Magnetobacterium bavaricum*" occurred in the microaerobic zone of freshwater sediments at 7×10^5 cells cm^3 and constituted approximately 30% of the microbial biovolume in that layer. FISH is not necessarily the best technique for counting extremely low-abundance microbes in the environment, however, as it requires a minimum number of cells per

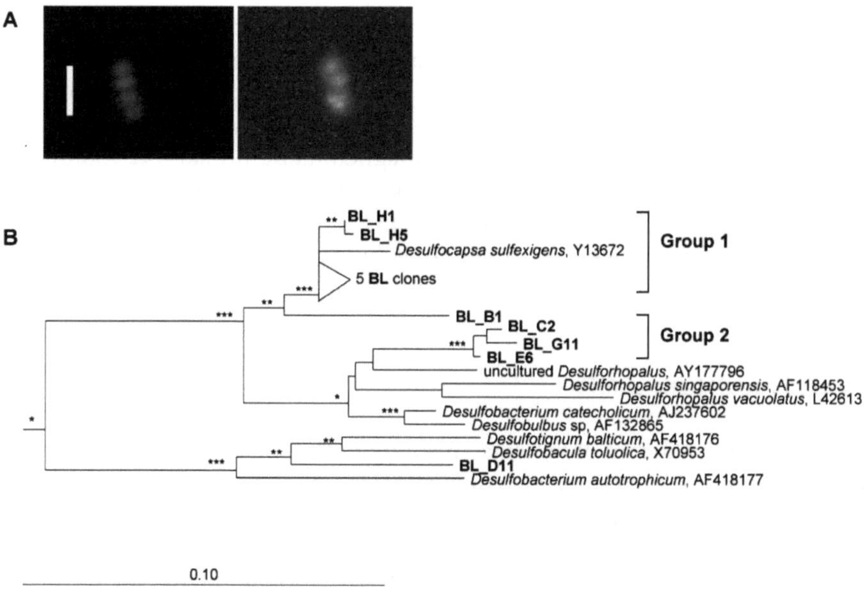

Fig. 2 The use of 16*s* rDNA sequencing and in situ hybridization to identify individual species of magnetotactic bacteria, in this case a barbell-shaped organism exhibiting south polarity (Simmons et al. 2006b). **A** Greyscale image of the barbell hybridizing with a HRP-labeled probe (visualized with Cy3-tyramides) targeting Group 2 sequences (related to *Desulforhopalus* spp.). *Scale bar* represents 5 microns. Modified from Figure 2 in Simmons et al. (2006b) and reprinted with permission. **B** Two clusters of 16*s* rDNA sequences (marked with **BL***) were obtained from samples highly enriched in the barbell bacterium. The tree was constructed using maximum likelihood analysis and 1000 bootstrap replicates in Phylip. Nodes with greater than 90% bootstrap support are marked with ***, 60–90% with **, and 50–60% with *

filter or slide and is time-consuming to apply to multiple species in multiple samples. Quantitative PCR (qPCR) is a more sensitive, high-throughput method to count multiple species in a microbial community over a time series. Species-specific primers are designed to amplify short segments of the 16*s* rRNA gene, and amplification is quantitatively tracked by the incorporation of a DNA-binding dye (SybrGreen) or a fluorescently labeled probe (TaqMan) (see Nolan 2004 for an introduction). The SybrGreen method is widely used in marine microbial ecology. We have developed a SybrGreen assay for several species of MTB in Salt Pond (Simmons et al. 2006, personal communication), and strongly recommend qPCR as a method for describing MTB communities.

Key genes involved in magnetosome biomineralization (Schüler 2004; Komeili et al. 2005; Scheffel et al. 2005) are also potential targets for quantitative PCR to enumerate MTB in the environment. At present, this is limited by the very few sequences of such genes available, all of them from magnetite-producers in the *Alphaproteobacteria*, and many with no identified function. Degenerate primers designed from the amino acid sequences of proteins identified as biomineralization-related (Grünberg et al. 2001) were not successful in amplifying potential biomineralization genes from purified DNA of the greigite-producing many-celled magnetotactic prokaryote (Simmons and Edwards, 2006, personal communication). Much more work is needed to determine what, if any, relationship exists between these genes and the biomineralization pathway in greigite-producers before they can be used as functional markers for MTB in the environment.

2.3 Layered Populations of MTB

Most MTB are gradient organisms, meaning both that they derive energy for growth from the proximity of reductants and oxidants at a chemical interface, and occur in thin layers at particular locations relative to this interface. We frequently observe blooms of different species of MTB in Salt Pond as the chemocline expands during summer stratification, implying the transient presence of optimal conditions for their growth. It is often possible to infer possible metabolic pathways based on the depth profile of the organism with respect to water chemistry, particularly reduced sulfur and iron compounds. Qualitative observations in the Pettaquamscutt (Stolz 1992) and stratified Lake Ely (Kim et al. 2005) both showed a transition in MTB morphotypes across the redoxcline. Bazylinski et al. (1995) showed that this transition corresponds to a boundary between magnetite and greigite producing MTB, with magnetite-producers occurring higher in the water column than greigite-producers. A similar transition occurs in Salt Pond (Simmons et al. 2004) (Simmons et al. 2006, personal communication) and presumably in other stratified marine systems (Fig. 1). An organism capable of co-precipitating magnetite and greigite was found in high numbers in the Pettaquamscutt (Bazylinski et al. 1993), but due to the lack of phylogenetic identification we do not know if it occurs elsewhere. Thus far, there have been no other reports of simultaneous magnetite and greigite precipitation in a single bacterium.

Most observations of MTB in stratified systems have been qualitative. In an attempt to understand the quantitative contributions of MTB to iron and sulfur cycling, we developed and applied a quantitative PCR assay for four major groups of MTB in Salt Pond (Simmons et al. 2006, personal communication) (see below for description of methods). Each group was previously known to produce magnetite or greigite. We tracked population densities and distributions across the chemocline in Salt Pond during summer 2003 (Simmons et al. 2006, personal communication). The results broadly confirm previous qualitative observations (Fig. 1). Magnetite-producing MTB are most abundant at the top of the chemocline. The dominant magnetite-producers in Salt Pond and the Pettaquamscutt are uncultivated cocci closely related to the isolate MC-1, a microaerophilic sulfide-oxidizer with the capacity for autotrophy (Williams et al. 2006). They peaked directly above the peak in ferrous iron, a typical distributional pattern for sulfide-oxidizers. We have observed 10^5 cocci ml^{-1} in the Pettaquamscutt (unpublished data) and 10^3 cocci ml^{-1} in Salt Pond (Simmons et al. 2006, personal communication). Smaller populations of magnetite-producing vibrios have been observed but not counted at similar depths in the Pettaquamscutt (Stolz 1992), and occasionally in Salt Pond.

We discovered, identified, and quantified a new MTB with a unique barbell-shaped morphology and unknown mineral content in Salt Pond (Fig. 2). This organism is the first known MTB to exhibit south polarity (Simmons et al. 2006b), though the significance of this behavior, if any, is not known for its ecological niche. It appeared in higher numbers than all other MTB groups and in the late season comprised 2–10% of total *Bacteria* (Simmons et al. 2006, personal communication). Cell counts of the barbell varied inversely with that of the magnetite-producing cocci, and it typically occurred slightly lower in the water column. The distributional patterns of the barbell, along with the metabolic capabilities of co-existing species, suggest that the barbell is either a sulfate reducer with a low tolerance for sulfide or is capable of growth on intermediate sulfur compounds.

Known greigite-producing MTB tended to occur at lower cell densities than the magnetite-producers, though the reasons are unknown. None of these organisms have yet been cultivated in the laboratory. The many-celled magnetotactic prokaryote (MMP) typically occurred at around 100 aggregates ml^{-1} as determined by qPCR in 2003 (Simmons et al. 2006, personal communication), but we have observed higher abundances during blooms in other years. Its depth distribution was not as well-defined with qPCR as the other groups due to a higher detection limit with the MMP primer set, which occured due to the high diversity within this group (Simmons and Edwards 2006). Interestingly, the MMP comprised $1.9 \pm 1.4\%$ of all bacteria in Salt Pond sediments during early stratification in June, suggesting that MTB survive the oxic mixed water column during winter, most likely by retreating to sediments. A large greigite-producing rod distantly related to *Thiomicrospira crunogena* is found from the base of the chemocline into the hypolimnion, where sulfide concentrations do not change much during seasonal stratification (Simmons et al. 2006, personal communication). The lack of bloom dynamics as revealed by qPCR and numerous qualitative observations and the low cell abundances (100 cells ml^{-1} on average) suggest a slow growth rate. Its distribution also indicates that it is not a gradient organism like the other three groups of MTB described here. Higher concentrations of unidentified large rods (10^5 cells ml^{-1}) were observed in sulfidic waters of the Pettaquamscutt (Bazylinski et al. 1995), suggesting that rod-shaped greigite producers can occasionally reach higher concentrations.

The phylogenetic diversity of the greigite producers has barely been tapped, and our observations are almost certainly an underestimate of their total concentration. Further work is needed to develop suites of qPCR primers for more groups of magnetite- and greigite-producers to achieve more accurate counts of their true abundance and distribution.

2.4 Protist Grazing of MTB

Grazing of bacteria by heterotrophic nanoflagellates (HNF) and ciliates can be an important regulator of bacterial populations in aquatic environments, although this coupling is less strong under eutrophic conditions (Pernthaler 2005). Abundances of HNF and bacteria show typical predator/prey dynamics in experimental microcosms, with maximum populations of HNF appearing shortly after maximum populations of bacteria. These dynamics also appear to be applicable to blooms of individual species in natural waters (Pernthaler 2005). We and others (Bazylinski et al. 2000) have observed magnetically responsive protists in Salt Pond on numerous occasions (Fig. 3). These consist of flagellates, biflagellates, dinoflagellates, and ciliates (Bazylinski et al. 2000). TEM analysis showed that the protist cells contained magnetosomes with morphologies and dimensions identical to those of magnetite-producing MTB (Bazylinski et al. 2000), suggesting that they accumulate within protists as a result of consumption and digestion of MTB. The magnetosomes are also occasionally visible on a light microscope (Fig. 3).

The magnetically responsive protists appear to be most abundant in the early and late season of stratification in Salt Pond, when the chemocline narrows and MTB are concentrated within a smaller volume. We have observed up to $2.9 \pm 0.59 \times 10^3$ magnetotactic protists ml^{-1} of diverse morphologies in the chemocline of Salt Pond during late September (unpublished data). About half of these showed north polarity and the other half south polarity. They exhibited a slow looping swimming motion near the drop edge, unlike the swimming behavior of individual MTB.

The possible correlation between MTB blooms and protist abundance suggests that protists could play a role in mediating the flux of magnetosomes out of the chemocline. Protists could increase Fe flux in three ways: by packaging large quantities of magnetosomes into a single cell, which sinks faster than individual magnetosomes by Stokes' law, consuming MTB and subsequently egesting magnetosomes, and stimulating the growth of MTB. The observations of (Bazylinski et al. 2000) show that protists can accumulate large numbers of magnetosomes. During blooms of protists, we have observed that direct feeding on MTB and subsequent egestion of magnetosomes appears qualitatively to be an important mechanism. This behavior is consistent with experiments on cultivated protists, which have shown that different species egest non-optimal food particles with a passage time ranging from a few seconds when prey abundance is high to several minutes when prey abundance is low (Boenigk et al. 2001). If magnetosome egestion rates are

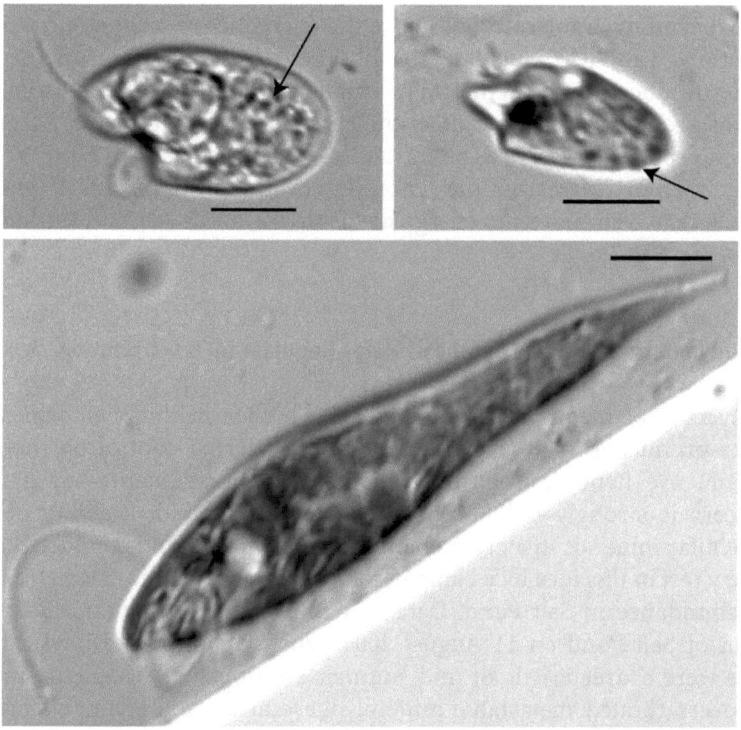

Fig. 3 Three morphologies of magnetotactic protists observed in the chemocline of Salt Pond, September 2004. *Arrows* indicate possible magnetosomes. All *scale bars* represent 10 microns

higher during blooms of MTB, protist grazing could be an important driver of Fe export from the system.

These possible connections between protists, MTB, and Fe flux are poorly understood. Coupled measurements of protist and MTB abundance in the environment are needed to determine the degree to which MTB populations are regulated by grazing. Culture-based experiments are needed to indicate whether grazing stimulates MTB population growth, the rates of MTB consumption, as well as whether protists mostly internalize or expel magnetosomes from digested MTB.

3
Thermodynamic Constraints on MTB

A consideration of the energetics of intracellular mineral synthesis by MTB can inform ecological observations. Our chemical data from Salt Pond, collected in conjunction with observations of MTB, provides a basis for some simple thermodynamic calculations on the energetics of magnetotaxis. These calculations can help elucidate the observed layering of magnetite- and greigite-producers and whether MTB can derive sufficient energy from magnetite or greigite synthesis to meet their metabolic requirements. Magnetite has been proposed to serve as an intracellular electrochemical "battery" powered by MTB cycling across the redoxcline (Vali and Kirschvink 1990; Kopp et al. 2004). We consider the energy available from this mechanism for MTB observed in Salt Pond.

3.1 Thermodynamics of Mineral Synthesis Helps Regulate MTB Distribution

The layering of magnetite-producing MTB higher in the chemocline than greigite-producing MTB (Fig. 1) suggests an external control on their distribution. We hypothesized that the presence of magnetite- or greigite-producers is strongly influenced by the thermodynamic stability of their intracellular minerals in a given chemical environment, and conducted a preliminary test of this idea by a comparison of thermodynamic calculations and MTB abundance in Salt Pond. Data was collected from the stratified water column of Salt Pond on 11 August 2004, when chemical and physical conditions were characteristic of mid-summer stratification. Magnetite was the most oversaturated metastable mineral phase at the top of the chemocline (Canovas 2006), where sulfide was not detectable. The corresponding stable phase is hematite (Fe_2O_3). Magnetite-producing cocci were the most abundant MTB at $1.4 \pm 0.24 \times 10^4$ cells ml^{-1}, while greigite-producing MMP were only present at about 100 cells ml^{-1}. Just 15 cm deeper in the water column, greigite was the most thermodynamically stable intermediate mineral phase due to the abrupt appearance of dissolved sulfide (Canovas 2006). Pyrite is the corresponding stable mineral phase. The abundance of greigite-producing MMP rose to $8.7 \pm 0.71 \times 10^3$ aggregates ml^{-1}, while the abundance of the cocci fell an order of magnitude to $2.5 \pm 0.42 \times 10^3$ cells ml^{-1}.

At this location in the top of the sulfide zone, pyrite formation is favored by the simultaneous presence of reduced sulfur species and a weak oxidant such as S^0, polysulfides, trace dissolved oxygen, Fe^{3+}, or organics (Wilkin and Barnes 1997b). Most pyrite formation in stratified euxinic water columns occurs in this narrow zone directly below the oxic-anoxic transition (Lyons 1997; Wilkin and Barnes 1997b), while pyrite is generally supersaturated in deeper, more sulfidic waters. The very low abundance of greigite-producing MTB in these highly sulfidic regions (Simmons et al. 2006, personal communication) suggests that conditions in the top of the sulfide zone are more favorable for their growth. Further determinations of cell abundances, chemical measurements, and thermodynamic modeling are required to substantiate these initial observations, which suggest that MTB generally seek to minimize the redox differential between their intracellular minerals and the external environment.

3.2 The Energetic Yield of Mineral Synthesis

Blakemore (1982) calculated that the magnetite synthesis reaction $3Fe^{2+} + 0.5O_2 + 3H_2O \rightarrow Fe_3O_4 + 6H^+$ is energetically favorable, with a $\Delta G_{\text{reaction}}$ of -0.85 kJ per gram of magnetite produced under reactant concentrations which maximized the growth of *Aquaspirillum magnetotacticum* (now called *Magnetospirillum magnetotacticum* (Schleifer et al. 1991)). He then estimated that this reaction could only account for less than 1% of the observed growth yield. This relatively simple thermodynamic calculation showed that the amount of energy produced by one possible reaction pathway for magnetite synthesis was insufficient for biomass synthesis by two orders of magnitude. Since Blakemore's calculation was based on growth yields in culture, however, its applicability to MTB in the environment is not immediately apparent.

We used concomitant estimates of MTB density, dissolved iron, and oxygen concentrations in Salt Pond collected over three summers (Simmons et al. 2004) (Simmons et al. 2006, personal communication), as well as newer thermodynamic data (Robie and Hemmingway 1995), to calculate the ΔG for magnetite synthesis in the environment. Using the same equation as above, the maximum ΔG at depths where magnetite-producing MTB were observed was -0.17 kJ per gram of magnetite. For comparison, the iron and oxygen concentrations used by Blakemore give a ΔG of -0.20 kJ gram^{-1} at temperatures typical of Salt Pond. Magnetite synthesis in the environment is less energetically favorable than under *M. magnetotacticum* culture conditions, primarily due to about ten-fold lower concentrations of dissolved iron in the environment compared with optimized culture conditions.

Using the concept of Gibbs energy dissipation (Heijnen 1993), it is possible to calculate the amount of biomass that could theoretically be produced from the energetic yield of magnetite synthesis. Approximately 2900 kJ are required to produce one mole of biomass carbon for iron-oxidizing microorganisms (Heijnen 1993). Aquatic bacteria contain about 5–20 fg C per cell (Whitman et al. 1998), or 1.7×10^{-15} mol C per cell at the upper end of the range. It therefore requires 4.8×10^{-12} kJ to manufacture the carbon biomass of one cell using ferrous iron oxidation. Magnetotactic bacteria contain 10^{-13} to 10^{-14} g of Fe per cell sequestered in magnetite (Chang et al. 1987; Suzuki et al. 2006) (Simmons et al. 2006, personal communication). Multiplied by a $\Delta G_{reaction}$ of -0.2 kJ per gram of magnetite, the value for optimal growth conditions for MTB, the synthesis of this amount of magnetite yields from 2×10^{-14} to 2×10^{-15} kJ. These values are two to three orders of magnitude lower than the amount of energy required for biosynthesis of a single cell. From these simple calculations, it is obvious that magnetite synthesis cannot provide enough energy for growth—the same conclusion reached by Blakemore in his culture experiments.

Inorganic low temperature greigite (Fe_3S_4) synthesis proceeds through the solid state conversion of mackinawite (FeS) to greigite (and eventually pyrite), via loss of Fe^{2+} from the crystal structure (Wilkin and Barnes 1996; Lennie et al. 1997). The presence of oxidants is required, either sulfur species of intermediate oxidation state (Benning et al. 2000) or other non-sulfur oxidants such as Fe^{3+} or bicarbonate (Wilkin and Barnes 1996). Greigite formation by MTB appears to follow this pathway (Pósfai et al. 1998). The Fe^{2+} lost from mackinawite in this process is oxidized to form amorphous Fe-oxyhydroxides (Lennie et al. 1997), which have been observed to surround greigite magnetosomes in one study (Farina et al. 1990) but not in another (Pósfai et al. 1998). A possible balanced equation for greigite formation from mackinawite via this "iron loss" pathway is: $4FeS_{(s)} + 0.5O_2 + 2H^+ \rightarrow Fe_3S_{4(s)} + Fe^{2+} + H_2O$ (Wilkin and Barnes 1997a). ΔG_r^0 for this reaction with oxygen as the oxidant is -130.2 kJ mol^{-1}, or -0.5 kJ per gram of greigite, with ΔG_f^0 values for greigite and mackinawite from Benning et al. (2000) and all other values from Stumm and Morgan (1996). Other oxidants are likely substituted for oxygen under the mildly reducing conditions where greigite-producing MTB are typically found, reducing the potential energetic yield from this reaction. As with magnetite formation, the energy yielded by greigite synthesis appears insufficient for primary biomass production. The apparently higher energetic yield from greigite synthesis, coupled to the overall higher intracellular greigite content in these bacteria (Simmons et al. 2006, personal communication) compared with magnetite-producers, indicates that it could potentially supplement energetic needs if coupled with an electron transport system. There is no biological mechanism known for harnessing energy from mineral synthesis.

3.3
The Magnetite/Maghemite Battery Hypothesis

Magnetite in MTB has been proposed to serve as an "electrochemical storage battery" (Vali and Kirschvink 1990; Kopp et al. 2004): cells migrating across a redoxcline could access additional energy for growth by oxidizing magnetite to maghemite in oxidizing conditions and reversing the reaction in reducing conditions. The transformation of magnetite ($Fe^{2+}Fe^{3+}{}_2O_4$) to maghemite ($Fe^{3+}{}_2O_3$) releases an atom of ferrous iron while maintaining the same crystal structure and magnetic properties. The abiotic reaction kinetics of this transformation in aqueous solution are very slow on the time scale of a bacterium (months to years; (Murad and Schwertmann 1993)), so it is a good candidate for active catalysis by the cell. Kopp et al. (2004) observed that cells of *Magnetospirillum magnetotacticum* in the high-oxygen zone of oxygen gradient media contained more oxidized magnetosomes than cells in the reduced zone, but it is unknown whether MTB are actively mediating this process, or whether they are actually cycling back and forth across a redoxcline on relevant timescales.

A plausible set of reactions for each "half-cell" of the battery is:

Oxic $4Fe_3O_{4(s)} + O_{2(aq)} \rightarrow 6Fe_2O_{3(s)}$

$\Delta G^0 = -333\ \mathrm{kJ\ mol^{-1}}$

Reducing $12Fe_2O_{3(s)} + HS^- \rightarrow 8Fe_3O_{4(s)} + SO_4^{2-} + H^+$

$\Delta G^0 = -123.5\ \mathrm{kJ\ mol^{-1}}$.

These assume complete conversion of magnetite to maghemite in the oxic zone with no formation of other ferric oxyhydroxides (Tang et al. 2003). The conversion is unlikely to be complete (Cornell and Schwertmann 1996), in which case this equation gives a maximum value. Sulfide is the most abundant reductant in the reducing zone; at circumneutral pH about half is HS^- and the other half H_2S. It is unlikely that this reaction proceeds directly as written because the reaction of maghemite and sulfide would be likely to produce pyrite. Another possibility is that cells use H_2 as a reductant for maghemite, similar to known iron reducers in the *Geobacteraceae* (Lovley 2003). The ΔG_f for maghemite is from Majzlan et al. (2003), the ΔG_f for magnetite is from Robie and Hemmingway (1995), and all other values are from Stumm and Morgan (1996).

We calculated the expected ΔG_r for each reaction using measured oxygen, sulfide, and pH values from Salt Pond during its 2003 summer stratification. The per-cell ΔG_r available from the aerobic reaction was about 1.3×10^{-14} kJ for oxygen concentrations typical of depths where magnetite-producers are found. The ΔG_r available per cell for the reducing reaction (normalized to 4×10^{-14} g magnetite/cell) was $3.9 - 4.2 \times 10^{-15}$ kJ, with the smaller values corresponding to lower sulfide concentrations found at the top of the sulfide zone. Since magnetite-producing MTB are primarily located at the top of the chemocline, if they are cycling back and forth over a 10 cm scale they are not likely to encounter sulfide concentrations higher than 0.1 mM. The maximum total energy available from a single cycle is about 1.7×10^{-14} kJ. To produce enough energy to synthesize a cell's worth of biomass as calculated above, a magnetotactic bacterium would have to cycle through the redoxcline 282 times.

If cells were continually cycling up and down, we might expect to see a diffuse band of MTB in Salt Pond bounded by oxygen concentrations at the top and sulfide at the bottom. Instead, we see fairly distinct peaks in MTB abundance with a resolution of 10–15 cm (Simmons et al. 2006, personal communication). This sampling resolution is not high enough to conclusively resolve the question, however. The number of cycles required suggests that the cell might use this method for maintenance energy, but is unlikely to rely on it exclusively to fill its energy requirements. Culture-based measurements of the magnetite-producers MV-1 and MV-2 indicate that those strains primarily oxidize reduced sulfur compounds in gradient media to obtain energy for growth (Bazylinski et al. 2004). The observation that a non-magnetotactic mutant strain of MV-1 showed higher growth rates than wild-type MV-1 under Fe concentrations typical of Salt Pond (Dubbels et al. 2004) also suggests that the “battery” is not a primary energy source in the environment. Since MV-1 was grown in anaerobic conditions, however, there is the possibility that results would have differed in gradient media. The presence of greigite-producing MTB in highly sulfidic waters, away from a redox gradient, also argues against the universality of this mechanism among MTB. It seems unlikely to be a major driving force for the evolution and maintenance of magnetotaxis.

4 The Influence of MTB on Reduced Fe Cycling

4.1 The Contribution of MTB to Reactive Fe Flux

The abundance of MTB in chemically stratified marine environments with high concentrations of dissolved iron indicates that they could substantially contribute to the flux of reduced Fe to sediments. The presence of MTB in stratified ponds like Salt Pond suggests that they could also be present in much larger but chemically similar stratified environments, such as the Cariaco Basin and the Black Sea. Since detailed geochemical data on reduced iron fluxes is available for these environments, it is possible to constrain the total potential contribution of MTB to Fe flux with measurements from Salt Pond and the Pettaquamscutt. A fairly limited set of parameters is required: cell density, Fe content per cell, chemocline width, and population turnover times. The first three parameters were measured directly in Salt Pond, and can be approximated for other environments based on published data. The population turnover time of MTB is not known, but an estimate of 2 days is reasonable based on observations of marine heterotrophic bacteria (Noble and Fuhrman 2000). With this information we can calculate the contribution of MTB to the total flux of reduced Fe out of euxinic basins (μg Fe cm^{-2} yr^{-1}), as follows: (the depth-integrated concentration of MTB) × (Fe content per cell) × (chemocline width) × (number of MTB populations per year).

Our counts of MTB, as well as numerous other observations, provide a bound on the abundance of MTB in stratified marine environments. The highest MTB concentrations observed in stratified marine basins are approximately 10^5 cells ml^{-1} (Bazylinski et al. 1995) (Simmons et al. 2006, personal communication). MTB in marine sediments in New England average 10^3–10^4 cells ml^{-1} (Blakemore 1982), and integration of MTB cell counts with depth in Salt Pond results in an average MTB concentration of 10^2–10^4 cells ml^{-1} at and below the chemocline. Estimates for Fe content per cell are in the range of 10^{-13} to 10^{-14} g (Chang et al. 1987) (Simmons et al. 2006, personal communication). It is possible to estimate this value for novel MTB by

TEM analysis of magnetosome size and numbers per cell. Chemocline width is environment-dependent and simple to measure using analytical techniques for determining oxygen, sulfide, and dissolved iron.

If we estimate that MTB constitute 1% of the microbial biomass in the Black Sea chemocline, a reasonable number given their proportion of total bacteria in Salt Pond (Simmons et al. 2006, personal communication), precipitation of Fe in intracellular minerals by MTB accounts for 0.5 to 5% of the Fe-sulfide flux in the deep basin. The total flux of reduced Fe in the form of Fe-sulfides to sediments of the Black Sea deep euxinic basin is 38 ± 28 μg Fe $cm^{-2}\,yr^{-1}$ (Anderson and Raiswell 2004), almost all of which forms in a narrow zone directly below the oxic-anoxic interface (Lyons 1997), the location where MTB would be most likely to occur. Further work is needed to determine the abundance and distribution of MTB in the Black Sea chemocline. This estimate is supported by observed concentrations of MTB (Bazylinski et al. 1995) and measured Fe-sulfide flux (Wilkin and Barnes 1997b) in the stratified Pettaquamscutt estuary. MTB constitute an estimated 1.3–13% of the total Fe-sulfide flux out of the chemocline in the Pettaquamscutt, consistent with our estimates for the Black Sea. These numbers suggest that MTB could be major unrecognized contributors to geochemically important fluxes of reduced Fe.

4.2 Integrated Studies

The combination of traditional molecular microbial ecological techniques and geochemical measurements provides a powerful tool for determining the contribution of MTB to iron cycling in the environment, as well as the energetic factors dictating their distribution. We now have a fairly good understanding of the type of work necessary to fill in the many gaps in our knowledge of MTB. It is important to track the changing abundances of MTB over time in a single environment to obtain an integrated picture of population dynamics, as MTB can occur in sudden blooms and in narrow layers driven by the formation of particular favorable chemical conditions.

Ideally, the study of MTB in a novel stratified environment would proceed as follows. First, light microscopy can be used to identify the major species of MTB present. These can be isolated for molecular identification as described above, as well as placed on TEM grids for elemental analysis of their mineral contents. The goal should be to link molecular phylogenetic probes with organisms known to produce magnetite or greigite. Next, samples should be collected over a time series using a fine-scale sampler, such as the inverted cone described by Jørgensen et al. (1979) or a multi-syringe instrument (Tonolla et al. 2000). Fine-scale chemical sampling should be conducted in conjunction with sample collection, perhaps with voltammetric microelectrodes to obtain information on intermediate sulfur species

(Luther III et al. 2001). Fluorescently labeled probes or qPCR primers can be applied to the samples to count magnetite- and greigite-producing MTB. Sediment traps could be placed to obtain information on both the total flux of particulate Fe and the percentage of that comprised by magnetosomes, perhaps using the sediment-trap design recommended by Kim et al. (2005). This information can be compared with the estimated flux due to MTB obtained from cell counts and Fe content per cell.

With sufficient observations of this type, it may be possible to develop a predictive model for the amount of magnetite and greigite deposition by MTB given certain environmental parameters. Such a model would also greatly aid in the development of magnetofossils as paleoenvironmental indicators. To date, little is known about the diversity, ecology, and geochemistry of MTB communities in the environment. The tools we have described already exist; all that is needed is their application to this fascinating group of organisms in new environments.

References

Amann R, Ludwig W (2000) Ribosomal RNA-targeted nucleic acid probes for studies in microbial ecology. FEMS Microbiol Rev 24:555–565

Anbar A, Knoll A (2002) Proterozoic ocean chemistry and evolution: a bioinorganic bridge? Science 297:1137–1142

Anderson TF, Raiswell R (2004) Sources and mechanisms for the enrichment of highly reactive iron in euxinic Black Sea sediments. Am J Sci 304:203–233

Arnold GL, Anbar AD, Barling J, Lyons TW (2004) Molybdenum Isotope Evidence for Widespread Anoxia in Mid-Proterozoic Oceans. Science 304:87–90

Bazylinski DA, Dean AJ, Williams TJ, Long LK, Middleton SL, Dubbels BL (2004) Chemolithoautotrophy in the marine, magnetotactic bacterial strains MV-1 and MV-2. Arch Microbiol 182:373–387

Bazylinski DA, Frankel RB (1992) Production of iron sulfide minerals by magnetotactic bacteria in sulfidic environments. In: Skinner HC, Fitzpatrick RW (eds) Biomineralization Processes, Iron, Manganese. Catena, pp 147–159

Bazylinski DA et al. (1995) Controlled biomineralization of magnetite (Fe_3O_4) and gregite (Fe_3S_4) in a magnetotactic bacterium. Appl Environ Microbiol 61:3232–3239

Bazylinski DA, Frankel RB, Jannasch HW (1988) Anaerobic magnetite production by a marine magnetotactic bacterium. Nature 334:518–519

Bazylinski DA, Heywood BR, Mann S, Frankel RB (1993) Fe_3O_4 and Fe_3S_4 in a bacterium. Nature 366:218

Bazylinski DA, Schlezinger DR, Howes BH, Frankel RB, Epstein SS (2000) Occurrence and distribution of diverse populations of magnetic protists in a chemically stratified coastal salt pond. Chem Geol 169:319–328

Benning LG, Wilkin RT, Barnes HL (2000) Reaction pathways in the Fe-S system below 100 degrees C. Chem Geol 167:25–51

Blakemore RP (1975) Magnetotactic bacteria. Science 190:377–379

Blakemore RP (1982) Magnetotactic bacteria. Ann Rev Microbiol 36:217–238

Boenigk J, Matz C, Jurgens K, Arndt H (2001) Confusing Selective Feeding with Differential Digestion in Bacterivorous Nanoflagellates. J Eukaryot Microbiol 48:425–432

Canfield DE (1998) A new model for Proterozoic ocean chemistry. Nature 396:450–453

Canfield DE, Berner RA (1987) Dissolution and pyritization of magnetite in anoxic marine sediments. Geochem Cosmochim Acta 51:645–659

Canovas P (2006) The redox and iron-sulfide geochemistry of Salt Pond and the thermodynamic constraints on native marine magnetotactic bacteria. Master's thesis, Department of Marine Chemistry and Geochemistry. Massachusetts Institute of Technology-Woods Hole Oceanographic Institution Joint Program, Woods Hole, MA, 145 p

Chang S-BR (1989) Magnetofossils, the magnetization of sediments, and the evolution of magnetite biomineralization. Ann Rev Earth Planet Sci 17:169–195

Chang S-BR, Kirschvink JL, Stolz JF (1987) Biogenic magnetite as a primary remanence carrier in limestone deposits. Phys Earth Planet In 46:289–303

Cornell RM, Schwertmann U (1996) The iron oxides. Wiley, Weinheim

Cox BL et al. (2002) Organization and elemental analysis of P-, S-, and Fe-rich inclusions in a population of freshwater magnetococci. Geomicrobiol J 19:387–406

DeLong EF, Frankel RB, Bazylinski DA (1993) Multiple evolutionary origins of magnetotaxis in bacteria. Science 259:803–806

Dubbels BL, DiSpirito AA, Morton JD, Semrau JD, Neto JNE, Bazylinski DA (2004) Evidence for a copper-dependent iron transport system in the marine, magnetotactic bacterium strain MV-1. Microbiology 150:2931–2945

Egli R (2004a) Characterization of individual rock magnetic components by analysis of remanence curves, 1. Unmixing natural sediments. Stud Geophys Geod 48:391–446

Egli R (2004b) Characterization of individual rock magnetic components by analysis of remanence curves. 3. Bacterial magnetite and natural processes in lakes. Phys Chem Earth, Parts A/B/C 29:869–884

Farina M, Esquivel D, Lins de Barros H (1990) Magnetic iron-sulphur crystals from a magnetotactic microorganism. Nature 383:256–258

Flies CB, Jonkers HM, de Beer D, Bosselmann K, Bottcher ME, Schüler D (2005a) Diversity and vertical distribution of magnetotactic bacteria along chemical gradients in freshwater microcosms. FEMS Microbiol Ecol 52:185–195

Flies CB, Peplies J, Schüler D (2005b) Combined Approach for Characterization of Uncultivated Magnetotactic Bacteria from Various Aquatic Environments. Appl Environ Microbiol 71:2723–2731

Grünberg K, Wawer C, Tebo BM, Schüler D (2001) A Large Gene Cluster Encoding Several Magnetosome Proteins Is Conserved in Different Species of Magnetotactic Bacteria. Appl Environ Microbiol 67:4573–4582

Heijnen JJ (1993) In search of a thermodynamic description of biomass yields for chemotrophic growth of microorganisms. Biotechnol Bioeng 39:833–858

Heywood BR, Bazylinski DA, Garrett-Reed AJ, Mann S, Frankel RB (1990) Controlled biosynthesis of greigite (Fe3S4) in magnetotactic bacteria. Naturwissenschaften 77:536-538

Jørgensen BB, Fossing H, Wirsen CO, Jannasch HW (1991) Sulfide oxidation in the Black Sea chemocline. Deep-Sea Res A: Oceanograph Res Pap 38:S1083–S1103

Jørgensen BB, Kuenen JG, Cohen Y (1979) Microbial transformations of sulphur compounds in a stratified lake (Solar Lake, Sinai). Limnol Oceanogr 24:799–822

Kao S-J, Horng C-S, Roberts AP, Liu K-K (2004) Carbon-sulfur-iron relationships in sedimentary rocks from southwestern Taiwan: influence of geochemical environment on greigite and pyrrhotite formation. Chem Geol 203:153–168

Kim BY, Kodama KP, Moeller RE (2005) Bacterial magnetite produced in water column dominates lake sediment mineral magnetism: Lake Ely, USA. Geophys J Int 163: 26–37

Komeili A, Li Z, Newman DK, Jensen GJ (2005) Magnetosomes Are Cell Membrane Invaginations Organized by the Actin-Like Protein MamK. Science 311:242–245
Kopp RE, Nash CZ, Kirschvink JL, Leadbetter JR (2004) A possible magnetite/maghemite electrochemical battery in the magnetotactic bacteria. Eos Trans. AGU 85:Fall Meet. Suppl., GP34A-06
Lennie AR, Redfern SAT, Champness PE, Stoddart CP, Schofield PF, Vaughan DJ (1997) Transformation of mackinawite to greigite; an in situ X-ray powder diffraction and transmission electron microscope study. Am Mineral 82:302–309
Lovley DR (2003) Dissimilatory Fe(III)- and Mn(IV)-reducing prokaryotes. In: Dworkin M (ed) The Prokaryotes: An Evolving Electronic Resource for the Microbiological Community, electronic release 3.4. Springer, Berlin Heidelberg New York
Ludwig W et al. (2004) ARB: a software environment for sequence data. Nucl Acids Res 32:1363–1371
Luther III G et al. (2001) Sulfur speciation monitored in situ with solid state gold amalgam voltammetric microelectrodes: polysulfides as a special case in sediments, microbial mats, and hydrothermal vent waters. J Environ Monit 3:61–66
Lyons TW (1997) Sulfur isotopic trends and pathways of iron sulfide formation in upper Holocene sediments of the anoxic Black Sea. Geochem Cosmochim Acta 61: 3367–3382
Majzlan J, Grevel KD, Navrotsky A (2003) Thermodynamics of Fe oxides: Part II. Enthalpies of formation and relative stability of goethite (alpha-FeOOH), lepidocrocite (gamma-FeOOH), and maghemite (gamma-Fe2O3). Am Mineral 88:855–859
Mann S, Sparks NHC, Board RG (1990a) Magnetotactic bacteria: Microbiology, biomineralization, paleomagnetism and biotechnology. Adv Microbial Physiol 31:125–181
Mann S, Sparks NHC, Frankel RB, Bazylinski DA, Jannasch HW (1990b) Biomineralization of ferrimagnetic greigite (Fe_3S_4) and iron pyrite (FeS_2) in a magnetotactic bacterium. Nature 343:258–261
Murad E, Schwertmann U (1993) Temporal Stability of a Fine-Grained Magnetite. Clay Clay Miner 41:111–113
Murray J, Codispoti L, Friedrich G (1995) Oxidation-reduction environments: the suboxic zone in the Black Sea. In: Huang C et al. (ed) Aquatic Chemistry. American Chemical Society, pp 157–176
Muyzer G, De Waal E, Uitterlinden A (1993) Profiling of complex microbial populations by denaturing gradient gel electrophoresis analysis of polymerase chain reaction-amplified genes coding for 16s rRNA. Appl Env Microbiol 59:695–700
Noble RT, Fuhrman JA (2000) Rapid Virus Production and Removal as Measured with Fluorescently Labeled Viruses as Tracers. Appl Environ Microbiol 66:3790–3797
Nolan T (2004) Getting Started- The Basics of Setting up a qPCR Assay. In: Bustin SA (ed) A–Z of Quantitative PCR. International University Line, La Jolla, CA, pp 529–543
Pernthaler A, Pernthaler J, Amann R (2002) Fluorescence In Situ Hybridization and Catalyzed Reporter Deposition for the Identification of Marine Bacteria. Appl Environ Microbiol 68:3094–3101
Pernthaler J (2005) Predation on prokaryotes in the water column and its ecological implications. Nat Rev Micro 3:537–546
Petermann H, Bleil U (1993) Detection of live magnetotactic bacteria in deep-sea sediments. Earth Planet Sci Lett 117:223–228
Petersen N, von Dobeneck T (1986) Fossil bacterial magnetite in deep-sea sediments from the South Atlantic Ocean. Nature 320:611–615
Pósfai M, Buseck PR, Bazylinski DA, Frankel RB (1998) Reaction sequence of iron sulfide minerals in bacteria and their use as biomarkers. Science 280:880–883

Pósfai M et al. (2001) Crystal-size distributions and possible biogenic origin of Fe sulfides. Eur J Mineral 13:691–703

Ramsing N, Fossing H, Ferdelman T, Andersen F, Thamdrup B (1996) Distribution of bacterial populations in a stratified fjord (Mariager Fjord, Denmark) quantified by in situ hybridization and related to chemical gradients in the water column. Appl Environ Microbiol 62:1391–1404

Roberts AP, Reynolds RL, Verosub KL, Adam DP (1996) Environmental magnetic implications of greigite (Fe3S4) formation in a 3 my lake sediment record from Butte Valley, northern California. Geophys Res Lett 23:2859–2862

Robie RA, Hemmingway BS (1995) Thermodynamic Properties of Minerals and Related Substances at 298.15 K and 1 Bar (105 Pascals) Pressure and at Higher Temperatures. In: US Geological Survey, Washington, DC

Rouxel OJ, Bekker A, Edwards KJ (2005) Iron isotope constraints on the archean and paleoproterozoic ocean redox state. Science 307:1088–1091

Sakaguchi T, Arakaki A, Matsunaga T (2002) Desulfovibrio magneticus sp. nov., a novel sulfate-reducing bacterium that produces intracellular single-domain-sized magnetite particles. Int J Syst Evol Microbiol 52:215–221

Scheffel A, Gruska M, Faivre D, Linaroudis A, Plitzko JM, Schuler D (2005) An acidic protein aligns magnetosomes along a filamentous structure in magnetotactic bacteria. Nature 440:110–114

Schleifer K-H et al. (1991) The genus Magnetospirillum gen. nov.: description of Magnetospirillum gryphiswaldense sp. nov. and transfer of Aquaspirillum magnetotacticum to Magnetospirillum magnetotacticum comb. nov. Syst Appl Microbiol 14:379–385

Schramm A, Fuchs BM, Nielsen JL, Tonolla M, Stahl DA (2002) Fluorescence in situ hybridization of 16S rRNA gene clones (Clone-FISH) for probe validation and screening of clone libraries. Environ Microbiol 4:713–720

Schüler D (2004) Molecular analysis of a subcellular compartment: the magnetosome membrane in Magnetospirillum gryphiswaldense. Arch Microbiol 181:1–7

Schüler D, Spring S, Bazylinski DA (1999) Improved technique for the isolation of magnetotactic spirilla from a freshwater sediment and their phylogenetic characterization. Syst Appl Microbiol 22:466–471

Shen Y, Knoll AH, Walter MR (2003) Evidence for low sulphate and anoxia in a mid-Proterozoic marine basin. Nature 423:632–635

Simmons SL, Bazylinski DA, Edwards KJ (2006b) South-seeking magnetotactic bacteria in the Northern Hemisphere. Science 311:371–374

Simmons SL, Edwards KJ (2006) Unexpected diversity in populations of the many-celled magnetotactic prokaryote. Env Microbiol, in press

Simmons SL, Sievert SM, Frankel RB, Bazylinski DA, Edwards KJ (2004) Spatiotemporal Distribution of Marine Magnetotactic Bacteria in a Seasonally Stratified Coastal Salt Pond. Appl Environ Microbiol 70:6230–6239

Snowball IF, Sandgren P, Petterson G (1999) The mineral magnetic properties of an annually laminated Holocene lake-sediment sequence in northern Sweden. Holocene 9:353–362

Snowball IF, Zillen L, Sandgren P (2002) Bacterial magnetite in Swedish varved lake-sediments: a potential bio-marker of environmental change. Quatern Int 88:13–19

Spring S, Amann R, Ludwig W, Schleifer K, van Gemerden H, Petersen N (1993) Dominating role of an unusual magnetotactic bacterium in the microaerobic zone of a freshwater sediment. Appl Environ Microbiol 59:2397–2403

Spring S, Amann R, Ludwig W, Schleifer K-H, Petersen N (1992) Phylogenetic diversity and identification of nonculturable magnetotactic bacteria. System. Appl Microbiol 15:116–122

Spring S et al. (1994) Phylogenetic analysis of uncultured magnetotactic bacteria from the alpha-subclass of Proteobacteria. Syst Appl Microbiol 17:501–508

Spring S, Bazylinski D (2000) Magnetotactic Bacteria. In: Dworkin M (ed) The Prokaryotes: An Evolving Electronic Resource for the Microbiological Community, The Prokaryotes: An Evolving Electronic Resource for the Microbiological Community, release 3.4 edn. Springer, Berlin Heidelberg New York

Spring S et al. (1998) Phylogenetic affiliation and ultrastructure of uncultured magnetic bacteria with unusually large magnetosomes. Arch Microbiol 169:136–147

Stolz JF (1992) Magnetotactic bacteria: Biomineralization, ecology, sediment magnetism, environmental indicator. In: Skinner HC, Fitzpatrick RW (eds) Biomineralization Processes of Iron and Manganese: Modern and Ancient Environments. Catena Verlag, Cremlingen-Destedt, pp 133–145

Stolz JF, Chang S-BR, Kirschvink JL (1986) Magnetotactic bacteria and single-domain magnetite in hemipelagic sediments. Nature 321:849–851

Stumm W, Morgan JJ (1996) Aquatic Chemistry, 3rd edn. Wiley, New York

Suzuki T, Okamura Y, Calugay RJ, Takeyama H, Matsunaga T (2006) Global Gene Expression Analysis of Iron-Inducible Genes in Magnetospirillum magneticum AMB-1. J Bacteriol 188:2275–2279

Tang J, Myers M, Bosnick KA, Brus LE (2003) Magnetite Fe_3O_4 nanocrystals: Spectroscopic observation of aqueous oxidation kinetics. J Phys Chem B 107:7501–7506

Tonolla M, Demarta A, Peduzzi S, Hahn D, Peduzzi R (2000) In situ analysis of sulfate-reducing bacteria related to Desulfocapsa thiozymogenes in the chemocline of meromictic Lake Cadagno (Switzerland). Appl Environ Microbiol 66:820–824

Vali H, Kirschvink JL (1990) Observations of magnetosome organization, surface structure, and iron biomineralization of undescribed magnetotactic bacteria: evolutionary speculations. In: Frankel RB, Blakemore RP (eds) Iron Biominerals. Plenum Press, New York, pp 97–115

Wallner G, Fuchs B, Spring S, Beisker W, Amann R (1997) Flow sorting of microorganisms for molecular analysis. Appl Environ Microbiol 63:4223–4231

Whitman WB, Coleman DC, Wiebe WJ (1998) Prokaryotes: The unseen majority. PNAS 95:6578–6583

Wilkin RT, Barnes HL (1996) Pyrite formation by reactions of iron monosulfides with dissolved inorganic and organic sulfur species. Geochim Cosmochim Acta 60:4167–4179

Wilkin RT, Barnes HL (1997a) Formation processes of framboidal pyrite. Geochim Cosmochim Acta 61:323–339

Wilkin RT, Barnes HL (1997b) Pyrite formation in an anoxic estuarine basin. Am J Sci 297:620–650

Williams TJ, Zhang CL, Scott JH, Bazylinski DA (2006) Evidence for autotrophy via the reverse tricarboxylic acid cycle in the marine magnetotactic coccus strain MC-1. Appl Environ Microbiol 72:1322–1329

Wolfe R, Thauer R, Pfennig N (1987) A "capillary racetrack" method for isolation of magnetotactic bacteria. FEMS Microbiol Ecol 45:31–35

Yamazaki T, Kawahata H (1998) Organic carbon flux controls the morphology of magnetofossils in marine sediments. Geology 26:1064–1066

Microbiol Monogr (3)
D. Schüler: Magnetoreception and Magnetosomes in Bacteria
DOI 10.1007/7171_040/Published online: 30 September 2006

Structure, Behavior, Ecology and Diversity of Multicellular Magnetotactic Prokaryotes

Carolina N. Keim[1] · Juliana Lopes Martins[1] · Henrique Lins de Barros[2] · Ulysses Lins[1] · Marcos Farina[3] (✉)

[1]Instituto de Microbiologia Professor Paulo de Góes, Universidade Federal do Rio de Janeiro, Centro de Ciências da Saúde, Bloco I, 21941-590 Rio de Janeiro, Brazil

[2]Centro Brasileiro de Pesquisas Físicas, Rua Xavier Sigaud 150, Urca, 22290-180 Rio de Janeiro, Brazil

[3]Instituto de Ciências Biomédicas, Universidade Federal do Rio de Janeiro, Centro de Ciências da Saúde, Bloco F, sala F2-027, 21941-590 Rio de Janeiro, Brazil
mfarina@anato.ufrj.br

Abstract Multicellular magnetotactic prokaryotes (MMPs) show a spherical morphology and are composed of 15–45 cells organized around an internal acellular compartment. Each cell presents a pyramidal shape with the apex of the pyramid facing this compartment The base, where the flagella are attached, faces the environment. MMPs display either a straight or a helical trajectory, and the sense of rotation of the trajectory (clockwise) is the same as the rotation of the microorganism's body during swimming. This is different to what would be expected if the flagella formed a bundle. The fact that MMPs present non-uniform velocities during "ping-pong movement" or "escape motility" further confirms the need for complex coordination of the action of flagella. The organisms express an unusual life cycle. Each organism grows by enlarging the cell volume, not the cell number; then, the number of cells doubles, the organism elongates, then it becomes eight-shaped, and finally splits into two equal spherical organisms. Most multicellular magnetotactic prokaryotes produce greigite (Fe_3S_4) magnetosomes, whereas recent observations show that these organisms can also biomineralize magnetite (Fe_3O_4). All data available on MMPs indicate that they constitute an important model for studies on multicellularity, biomineralization, and evolution in prokaryotes.

1 Introduction

In 1975 Richard Blakemore reported the astonishing discovery of magnetotactic bacteria (MB). These microorganisms contained chains of nanometer-sized iron-rich crystals and were capable of aligning and migrating along the lines of a magnetic field (Blakemore 1975). It was hypothesized that MB use the interaction with the geomagnetic field to swim to the bottom of marine and freshwater environments, where they could find microenvironments with lower levels of oxygen. This suggestion was strengthened after the finding that bacteria collected in either the northern or southern hemispheres (Frankel et al. 1979; Blakemore et al. 1980; Kirschvink 1980) swim to the bottom, towards the sediments. This behavior was called magnetotaxis (the passive orientation along the lines of a magnetic field, and active migration because of the flagella). Later, Frankel et al. (1997) described the behavior of a marine magnetotactic coccus that used magnetotaxis in conjunction with aerotaxis (magneto-aerotaxis), "to more efficiently migrate to and maintain position at their preferred oxygen concentration".

Several studies characterized the crystalline inclusions and the bacteria that produce them (for a review, see Bazylinski and Frankel, 2004). Initially, crystals were identified as magnetite by different methods and crystal morphologies were shown to be species- or strain-specific by high resolution transmission electron microscopy. However, in 1990, the crystalline inclusions found in multicellular magnetotactic prokaryotes were shown to contain iron sulfides (Farina et al. 1990; Mann et al. 1990). This finding opened new questions about the biomineralization phenomenon in prokaryotes.

2
Similar Microorganisms, Different Designations

In 1983, a spherical, large, south-seeking magnetotactic microorganism was reported in the coastal metropolitan lagoon "Rodrigo de Freitas", in Rio de Janeiro city (Fig. 1) (Farina et al. 1983). Their crystalline inclusions were not studied in detail at that time and were assumed to contain magnetite. Actually, the authors stated that a fraction of the crystalline inclusions observed by transmission electron microscopy (TEM) were composed of magnetite. The inclusions were estimated to contain about 10% of magnetite (Farina et al. 1983) based on TEM data and calculation of the magnetic moment of a population of microorganisms, obtained from the trajectories under an applied magnetic field of known intensity that was reversed (the U-turn method). Later, the microorganism was called magnetotactic multicellular aggregate (MMA) (Farina et al. 1990). This microorganism was different from the MB previously described, in several aspects. It was larger (5–10 μm in diameter) and composed of several cells enveloped by double membranes (Farina et al. 1983, 1990; Lins de Barros et al. 1990a).

Rodgers et al. found a similar microorganism in marine and brackish coastal sites in New England and called it many-celled magnetotactic prokaryote (Rodgers et al. 1990a) or multicellular magnetotactic prokaryote (Rodgers et al. 1990b), both denominations shortened to MMP. The microorganism presented a "rosette or mulberry-like" morphology, was approximately 12.5 μm in diameter, showed a swimming speed ranging from 67 to 175 $\mu m\, s^{-1}$ and was similar in general morphology to the one described by Farina et al. (1983, 1990). It was argued that its complex motility (never observed when the individual cell components of the MMP were seen separated from the body) implied coordination of flagellar activity, and thus communication between cells of the whole organism. Based on the putative membrane adhesion structure observed and on the microorganism's behavior, the authors proposed that the MMP was in fact a multicellular prokaryote (Rodgers et al. 1990a,b). When exposing MMPs to a 60 Hz AC magnetic field, the microorganisms could be remagnetized (many originally north-seeking MMPs could migrate as south-seeking microorganisms), indicating that the intact organism probably has an axis of motility (Rodgers et al. 1990a,b), similar to unicellular MB (Blakemore 1982).

Keim et al. (2004a) described a morphologically similar microorganism in one of the largest hypersaline lagoons in the world, the Araruama Lagoon in Rio de Janeiro State, Brazil (Fig. 1). Because the designation "magnetotactic multicellular aggregate" or MMA became inadequate after the results shown in the paper, and the dgnation MMP referred to a 16S rDNA study (DeLong et al. 1993), the denomination magnetotactic multicellular organism (MMO) was proposed for these microorganisms. Here we will use the designation MMP for all of them.

Table 1 Comparison of MMP main characteristics and the environments where they are found

Size (µm)	Cell number	Cell size	Speed (µm/s)	Type of magnetic crystal	Crystal size (nm)	Site of collection	Salinity (‰)	pH	Refs.
5–9	15–25	1.5–2.5 × 1–2 µm	30 and 100	Fe sulfide	75–100	RFL	~ 8*		Farina et al. 1983, 1990; Lins de Barros et al. 1990
2.2–4.6		$0.33 \pm 0.2\ \mu m^3$		Fe sulfide *		RFL	~ 8*		Lins and Farina, 1999
4–12	10–30			Fe sulfide	75 (mean)	SP, WH, NRM, PRWR	12–32		Bazylinski et al. 1990
3–8	7–20			Fe sulfide	75 (mean)	Marine and brakish sites	12–31		Mann et al. 1990
3–12.5	10–30	0.8–1.4 × 0.6–0.8 µm	105 (mean)	Fe sulfide	85–90	NE	12–32	7.3–6.5	Rodgers et al. 1990a
	15–20			Fe sulfide	30–120	SP, PRWR, SSNP	~ 29	~ 7.4	Pósfai et al. 1998a,b
				Fe sulfide	74 ± 18	PRWR, SSNP			Pósfai et al. 2001
						SP	18–28	7–9	Simmons et al. 2004
5.5–9.5	17.4 ± 3.59		90 ± 20	Fe sulfide		Ara	74.55	~ 7.4	Keim et al. 2004a,b; Almeida et al. 2005
			~ 100			WH			Greenberg et al. 2005
4–7				Fe sulfide	45–55	Enc	~ 50		Martins et al. submitted
				Fe oxide+ Fe sulfide		Rob	~ 50	~ 7.0*	Martins et al. submitted
5.3 ± 0.6*				Fe oxide+ Fe sulfide	104 ± 29 70 ± 8	Ita	32–35*	~ 7.0*	Keim et al. 2003; Lins et al. submitted

Table 1 (continued)

Abbreviations:	
Ara	Araruama Lagoon, a large lagoon at the coast of Rio de Janeiro State, Brazil
Enc	Encantada Lagoon, a small coastal lagoon at Restinga de Jurubatiba National Park, Rio de Janeiro State, Brazil
Ita	Itaipu Lagoon, a coastal lagoon in the metropolitan area of Niterói, Rio de Janeiro State, Brazil
NE	Brackish coastal sites on New England coast
NRM	Salt marshes in Neponset River Marsh, Boston, MA, USA
PRWR	Shallow salt-marsh pools in the Parker River Wildlife Refuge, Rowley, MA, USA
RFL	Rodrigo de Freitas Lagoon, a coastal lagoon in Rio de Janeiro city, Brazil
Rob	Robalo Lagoon, a coastal lagoo at Restinga de Jurubatiba National Park, RJ, Brazil
SP	Salt Pond, Woods Hole, MA, USA
SSNP	Sweet Springs Nature Preserve, Morro Bay, CA, USA
WH	Salt marshes in Woods Hole, MA, USA
WNB	Salt marsh pond near Wood Neck beach in Woods Hole, MA
*	Personal observation

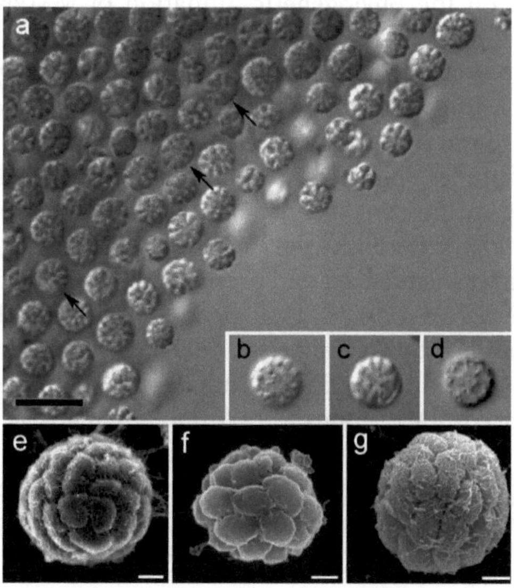

Fig. 1 Diferential interference contrast light microscopy (DIC) and scanning electron microscopy (SEM) of MMPs from: Araruama (**a**), Robalo (**b**), Rodrigo de Freitas (**c**), and Itaipu (**d**) Lagoons (DIC); and Araruama (**e**), Rodrigo de Freitas (**f**), and Encantada (**g**) Lagoons (SEM). *Scale bar* in (**a**) corresponds to 15 μm for (**a**) and 10 μm for (**b**), (**c**), and (**d**). *White bars* = 1 μm for (**e**), 1.5 μm for (**f**), and 3 μm for (**g**). Note the morphological similarity and the spherical shape of the microorganisms. The lack of flagella in (**f**) is due to the sample preparation procedure. The images in (**a**) and (**e**) are reprinted from Keim et al. 2004a with permission from Elsevier

All MMPs seem to share main particularities such as: ~ 3–12 μm in diameter (Farina et al. 1983, 1990; Bazylinski et al. 1990; Lins de Barros et al. 1990a; Mann et al. 1990; Rodgers et al. 1990a,b; Lins and Farina 1999; Keim et al. 2004a,b, Table 1), the fact that they contain several prokaryotic cells apparently organized in a helical symmetry (Farina et al. 1983, 1990; Lins de Barros et al. 1990a; Lins and Farina 1999; Keim et al. 2004b) with all cells facing both the environment and an internal acellular compartment (Keim et al. 2004a), the cells present a glycocalix (Farina et al. 1983; Keim et al. 2004a), several flagella facing the environment (Keim et al. 2004a; Rodgers et al. 1990a; Silva et al., in press), the individual cells are never motile when separated from the whole microorganism (Farina et al. 1990; Lins and Farina 1999; Keim et al. 2004a), and the organism probably has an unique life cycle (Keim et al. 2004b).

These microorganisms have not been cultured until now, and thus all studies were done with samples collected directly from the environment.

2.1 Phylogenetic Studies

Few studies exist on the phylogenetic affiliation of MMPs. The 16S rDNA of MMPs, collected in coastal sites in New England (USA), were sequenced and found to be related to the sulfate-reducing bacteria group within the δ-Proteobacteria, *Desulfosarcina variabilis* being the most closely related cultivated organism found in the databases (DeLong et al. 1993). The 16S rDNA of two different samples collected in 1990 and 1991 showed more than 99% similarity, leading to the conclusion that MMPs comprised a single species (Bazylinski et al. 1993). Recently, part of the 16S rDNA gene of MMPs from Araruama Lagoon was sequenced and presented 87% similarity with the same region of the 16S rDNA gene of MMP 1991 (Keim et al. 2004b), one of the two sequences obtained by DeLong et al. (1993). Several δ-Proteobacteria sequences were identified in a MMP-rich sample, but no sequences identical to previously sequenced MMP clones were found (GenBank accession numbers AY589482, AY589485, and AY589481) (Simmons et al. 2004). All these preliminary results suggest that MMP may represent different species, but further work is necessary to evaluate de phylogenetic diversity within the group.

3 Ecology of the Multicellular Magnetotactic Prokaryotes

MMPs have been observed in different regions around the globe (Bazylinski et al. 1990, 1993; DeLong et al. 1993; Farina et al. 1983, 1990; Flies et al. 2005; Greenberg et al. 2005; Keim et al. 2003, 2004a,b; Lins de Barros et al. 1990a; Lins and Farina 1999, 2001; Lins et al. 1992, 2000, submitted; Martins et al., submitted; Penninga et al. 1995; Pósfai et al. 1998a,b; Rodgers et al. 1990a,b; Silva et al., in press; Simmons et al. 2004, 2006; Winklhofer et al., in press). Table1 summarizes different characteristics of these microorganisms including the type of magnetic mineral precipitated in their magnetosomes. They occur in anaerobic, brackish to hypersaline, sulfide-rich environments (Bazylinski et al. 1990; Farina et al. 1990; Lins de Barros et al. 1990a; Rodgers et al. 1990a,b; Pósfai et al. 1998b; Lins and Farina, 1999; Keim et al. 2004a; Simmons et al. 2004), in the water column of stratified lagoons (Bazylinski et al. 1990; Rodgers et al. 1990a,b; Pósfai et al. 1998b; Simmons et al. 2004) or in the sediment of aquatic environments with non-stratified water columns (Farina et al. 1990; Lins de Barros et al. 1990a,b; Keim et al. 2004a).

Each type of magnetotactic bacterium seems to be adapted to different chemical and physical gradients in the stratified microenvironments where they thrive (Simmons et al. 2004). Greigite-producing magnetotactic bacteria (including the MMP) are found in the anaerobic zone, just below the oxic-anoxic transition zone (Bazylinski et al. 1995; Pósfai et al. 1998b), whereas unicellular magnetite-producing magnetotactic bacteria are usually found at the oxycline (Bazylinski et al. 1995; Simmons et al. 2004).

Simmons et al. (2004) reported results of studies on the distribution of different magnetotactic bacteria with respect to the physical and chemical condition of the environment, at Salt Pond (Falmouth, Massachussetts, USA). They studied the water column (between June and August, 2002) and found that the magnetite-producing bacteria were most abundant at the top of the oxycline, whereas MMP occurred at the bottom of the oxycline, or slightly below it, in regions of low sulfide. In this environment, the MMPs were spatially separated from the magnetotactic cocci.

Because MMPs were found in a relatively narrow window in the chemocline, below the oxycline and at relatively low sulfide concentrations, and the fact that an order of magnitude shift in their population was found within 15 cm, suggest that the MMP is a gradient microorganism, but that its location in the chemical gradient is not directed by the oxygen concentration. Instead, they could be chemotactic to iron or a sulfur compound, since Fe(III) concentrations peaked at the depth where most MMPs were found (Simmons et al. 2004).

3.1 Distribution of MMPs in Coastal Lagoons of Rio de Janeiro State

We have been studying the presence of MMPs in several coastal lagoons near Rio de Janeiro city, located in the South Atlantic Geomagnetic Anomaly (low field intensity of about 0.23 Gauss). The geomagnetic field points up with a typical inclination angle of about 25°.

Araruama Lagoon ($22°50'21''S, 42°13'44''W$) is the largest hypersaline coastal lagoon in Rio de Janeiro State (Kjerfve et al. 1996). Encantada Lagoon is a small coastal lagoon located at the Restinga de Jurubatiba National Park (22° – 22°30′S and 41°15′ – 42°W). Both lagoons are hypersaline environments, with salinity around 50‰, but have a great diversity in water physico-chemical characteristics (Kjerfve et al. 1996; Enrich-Prast et al. 2003). We have also been studying MB from Itaipu Lagoon, a brackish coastal lagoon located in Niterói, north-east of Rio de Janeiro, Brazil. Table 1 shows some characteristics of these environments.

MMPs can be obtained from samples of water and sediment after magnetic concentration, as described by Lins et al. (2003). Briefly, samples are put in a special chamber and exposed to a properly aligned magnetic field using a home-made coil connected to a DC power supply that aligns bacteria to swim towards a capillary end, from where the microorganisms are collected.

The only magnetotactic microorganisms found at Araruama Lagoon were MMPs (Figs. 1a and 2). We investigated the distribution of MMPs in microcosms consisting of Araruama Lagoon sediment and water stored in 10 L containers. MMPs were directly counted under the microscope after magnetic exposure of a fixed volume of sediment. MMPs were detectable even 42 days after sampling. Monitoring five microcosms, we observed different density population behaviors: the rate that MMP population decreased was variable, and in some cases their number increased (Fig. 2a). This occurs probably because of the biotic and abiotic variations among the microcosms.

To analyze the vertical distribution of MMPs within the microcosms, the top (1 cm) layer of the sediment was repeatedly removed (up to 5 cm) from a sample core, immediately diluted with sterile lagoon water, and placed onto

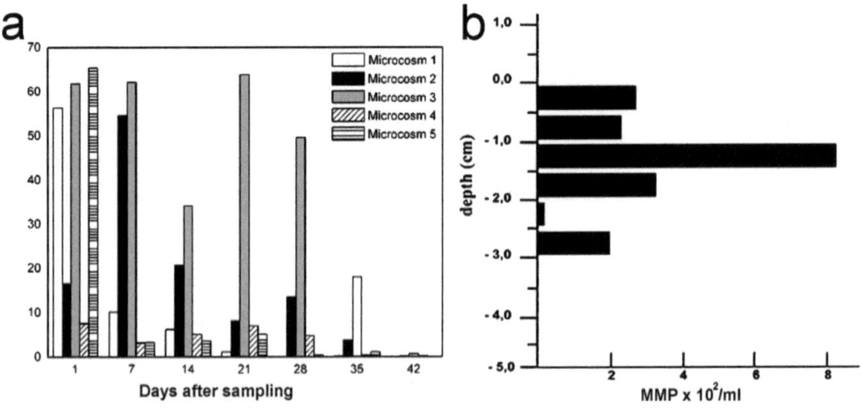

Fig. 2 Distribution of MMPs in microcosms consisting of Araruama Lagoon sediment and water stored in 10 L containers. **a** Number of MMPs in five different microcosms with time, evaluated from small core samples. **b** Distribution of MMPs in sediment cores from microcosms after two weeks. The *bars* show the concentration of MMPs at different depths in the same microcosm, as number of MMPs/mL of sediment

a microscope slide. The slide was exposed to the magnetic field of an ordinary magnet for 5 min, when the MMPs that reached the edge of the drop were directly counted with a phase contrast microscope. Figure 2b shows that the MMPs were located up to 3 cm deep in the microcosm sediment. In the lagoon, we have observed that the MMPs are found mainly at 3–4 cm deep, and the dissolved oxygen penetration into the sediment does not reach deeper than 0.3 cm (J. L. Martins and U. Lins, unpublished observation).

In sample cores collected around the lagoon in a single day, no MMPs were observed at Rio das Moças estuary, where the salinity was 23.7 ppt, the lowest observed in the samples (Fig. 3). Large populations of MMPs are found at Baleia, Iguabinha and Rebolo beaches, with salinities of 56, 60, and 54.6‰, respectively (Fig. 3). This pattern suggests that the MMPs of Araruama Lagoon occur only in hypersaline conditions (J. L. Martins and U. Lins, unpublished observations).

Denaturing gel gradient electrophoresis (DGGE) profiles of a 16S rDNA fragment showed only one band in all samples of MMPs analyzed (Fig. 4), strongly indicating that genetically identical cells form the MMPs from Araruama and Encantada Lagoons. Further, the same melting behavior observed for the 16S rDNA fragment obtained from MMPs collected at Encantada and Araruama Lagoons (Fig. 4) indicates that these lagoons probably contain the same organism (J. L. Martins and U. Lins, unpublished observations).

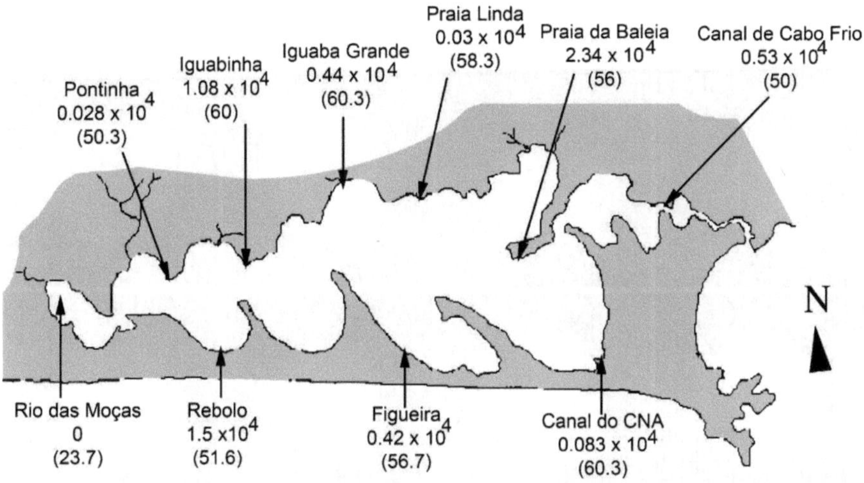

Fig. 3 Distribution of MMPs around Araruama Lagoon in the first 5 cm of sediment, as evaluated from core samples the day after collection. The numbers without parenthesis are the MMP number/mL. The numbers between parenthesis are the salinity (‰) evaluated with a manual refractometer. The city of Rio de Janeiro is located about 50 Km west of this region in the map

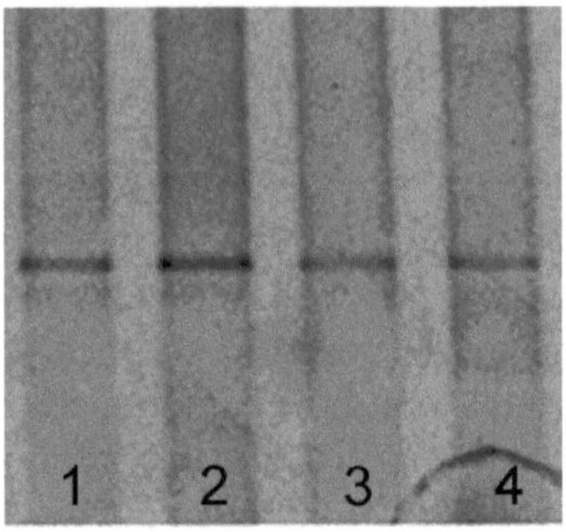

Fig. 4 DGGE profile of a 16S rRNA fragment from magnetically concentrated samples from Encantada (samples 1 and 2) and Araruama (samples 3 and 4) Lagoons. The *single band* in all lanes is indicative of the genetic similarity within and between the samples

4
Morphology of Multicellular Magnetotactic Prokaryotes

MMPs are spherical organisms composed of several cells arranged in a roughly helical disposition (Fig. 1e–g) (Farina et al. 1983, 1990; Lins de Barros et al. 1990a; Lins and Farina 1999; Keim et al. 2004a). The cells are arranged radially around an acellular internal compartment found at the center of the organism (Figs. 5 and 6) (Keim et al. 2004a,b). The radial arrangement of the cells around the internal compartment causes the cells to assume a pyrami-

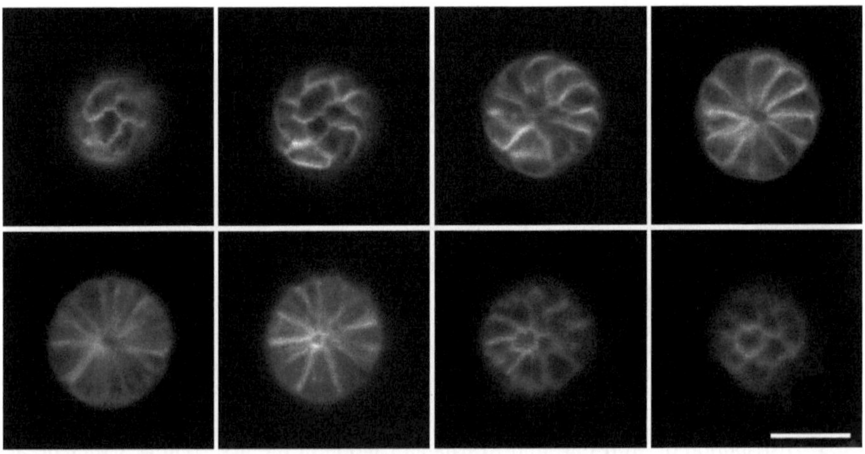

Fig. 5 Confocal laser scanning microscopy (Zeiss 510-META) series of a MMP stained with a lipophylic dye (FM 1-43) that outlines the cell contours. Note that cells are highly elongated radially, with the pointed end in contact with the internal compartment. The *scale bar* represents 5 μm

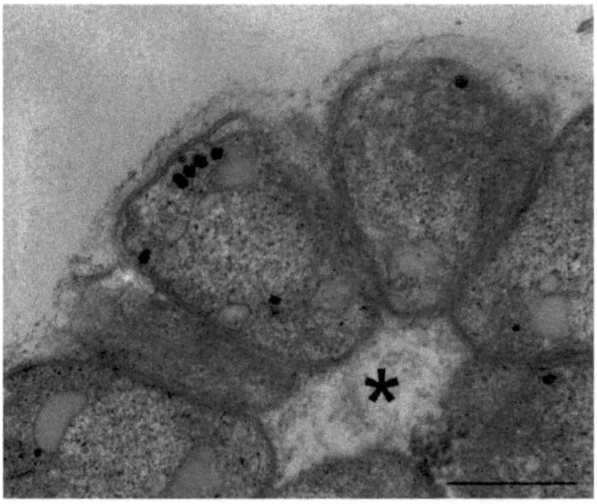

Fig. 6 Ultra-thin section of part of a MMP from Rodrigo de Freitas Lagoon. Note the magnetosomes (*dark regions*), the capsule composed of radial fibers at the organism surface, the lipid droplets (*large inclusions*) and the internal compartment (*asterisk*). The *scale bar* represents 1 μm

dal shape with the bases of the pyramids forming the outer surface of the organism and the apexes lining the internal compartment (Fig. 7a). This compartment contains small membrane vesicles and filaments linking the apex of the pyramids. The pyramid faces are present where the cell is in contact with a neighbor cell, and the edges are present where the cell is in contact with two or more cells (Fig. 7). Accordingly, the pyramid base, which is the part of the cell that faces the environment, is rounded (Fig. 7) (Keim et al. 2004a).

When soap bubbles aggregate to cover space, after a certain number of bubbles, the addition of a new sphere leads to a configuration where one bubble is completely surrounded by others (Williams 1979). Thus we would expect to observe internal cells in the MMPs if only stability of the whole structure, caused by the reduction in surface tension, is considered. In MMPs, specific structures must bind the cells together in specific regions, causing tensions that lead the cells to be elongated in the direction of the center of the organism. Although the internal compartment seems to be a consequence of the organism's architecture, as evidenced by the absence of internal cells and the high level of symmetry of the cells within the microorganisms, it contains fibers that could be involved in binding cells together, and also vesicles that could be involved in cell-to-cell communication. Both functions are supported by the fact that all cells have contact with the internal compartment. Besides, this compartment could be used in cell-to-cell communication through diffusion of soluble molecules. An obvious advantage for an organism having no internal cells is that all cells are in contact with the outside environment, where they take nutrients (Keim et al. 2004a).

MMPs collected in Rodrigo de Freitas Lagoon (Farina et al. 1983, 1990; Lins de Barros et al. 1990a; Lins and Farina, 1999), Itaipu Lagoon (Lins et al., submitted) and Encantada Lagoon (Martins et al., submitted) show the same

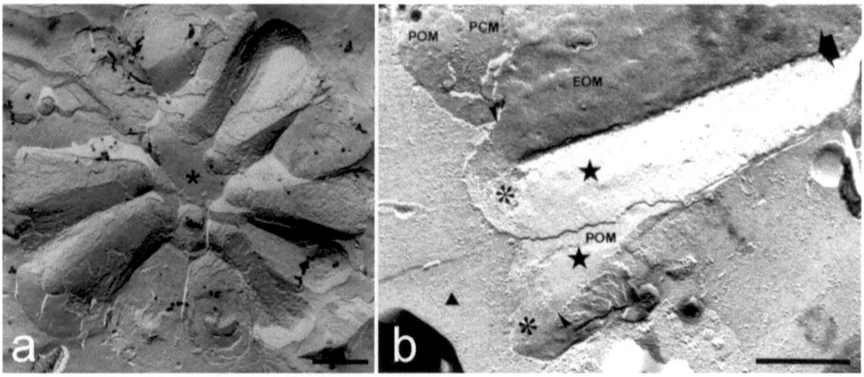

Fig. 7 Freeze-fracture images of MMPs from Araruama Lagoon. **a** The radial arrangement of the cells, the flat membrane surfaces, and the radial edges give the cells a pyramidal shape. The acellular internal compartment fills the central region of the organism (*asterisk*). **b** There are flat membrane surfaces in the region of contact between two cells and a straight groove (*arrow*) in the E-face of the outer membrane (EOM) in the contact region between three cells. Note that the rounded regions, which have direct contact with the environment, present a high concentration of intramembrane particles in both the E-face (*EOM*) and the P-face (*POM*) of the outer membrane (*asterisks*) in comparison with the region of contact between two cells (*stars*). The *triangle* marks the outer environment. The *scale bar* represents 1 μm for (**a**) and 0.5 μm for (**b**). Reproduced from Keim et al., 2004a with permission from Elsevier

general cell organization as those collected in Araruama Lagoon, with the cells arranged around the internal compartment (Figs. 5–7). MMPs from Rodrigo de Freitas Lagoon frequently exhibits a more evident helical cell arrangement (Farina et al. 1983; Lins de Barros et al. 1990; Lins et al. 1992; Lins and Farina, 1999), in comparison with those from Araruama (Keim et al. 2004a,b) and Encantada, (Martins et al., submitted) Lagoons.

Electron micrographs published on the MMPs from the northern hemisphere (Rodgers et al. 1990a,b) do not enable such comparisons. However, based on the genetic similarity (Keim et al. 2004b), the similarity of the light micrographs (Rodgers et al. 1990a,b; DeLong et al. 1993; Greenberg et al. 2005; Simmons et al. 2004, 2006), and whole mount TEMs (Mann et al. 1990; Bazylinski et al. 1993; Penninga et al. 1995; Pósfai et al. 1998a,b; Simmons et al. 2004), we propose that the cell arrangement described by Keim et al. (2004a), summarized in this section, also corresponds to the cell arrangement of the MMPs from the northern hemisphere.

4.1 The Cell Envelope

The cell envelope presents the double membrane characteristic of gram-negative bacteria (Fig. 6) (Farina et al. 1983, 1990; Lins de Barros et al. 1990a;

Rodgers et al. 1990a; Lins and Farina 2001; Keim et al. 2004a). Between the cytoplasmic and outer membranes (in the periplasmic space), a thin peptidoglycan layer is sometimes seen (Rodgers et al. 1990a; Keim et al. 2004a). The four membranes of adjacent cells are tightly apposed, which indicates the existence of structures specialized to bind cells together (Keim et al. 2004a). The observation of a constant distance of about 2 nm between the outer membranes of adjacent cells led to the proposition that cell junctions could be present in these membranes (Rodgers et al. 1990a). Observation of partially disaggregated MMPs by scanning electron microscopy (SEM) (Lins and Farina 1999) and of whole mounts by TEM (Lins et al. 2000) showed regions of focal contact between adjacent cells. No relationships could be found between the constant distance between the outer membranes (Rodgers et al. 1990a, Keim et al. 2004a) and these regions (Lins and Farina 1999, Lins et al. 2000). Furthermore, freeze fracture results did not show any structures that could be definitely responsible for the focal contacts and for the constant distance observed between the outer membranes (e.g., Fig. 7b). On the other hand, the fact that the outline of the underlying cell could be observed in the E-face (inner face of the external leaflet) of the outer membrane shows that some membrane specialization is present, especially between the region of contact with another cell and the region of contact with the environment (Keim et al. 2004a).

The membrane surfaces of MMP cells are different in each cell region as a result of polarization and specialization of the diverse membrane regions. The surface of the pyramid base, which is the part of the cell in contact with the outer environment, presents flagella 15.9 ± 1.4 nm in diameter and 0.9 to 3.8 μm in length (Silva et al., in press), and a capsule formed by radial fibers (Fig. 6) (Farina et al. 1983; Keim et al. 2004a), probably composed of polysaccharides, 94 ± 18 nm thick (Keim et al. 2004a).

Freeze-fracture replicas of MMPs show mainly outer membrane fracture surfaces (Fig. 7). Cell membrane fracture surfaces appear mainly in the central region of the MMPs (Fig. 7a), suggesting a difference in composition of the cell membrane or the outer membrane at these regions. The replicas of the E-face of the outer membrane show a gradient of intramembrane particles growing toward the center of the organism in the part of the cell that has contact with another cell (Fig. 7b). In the P-face (inner face of the protoplasmic leaflet) of the outer membrane, the part of the cell in contact with the environment shows a higher concentration of intramembrane particles than the part of the cell in contact with another cell. The edges or projections (present where the cell is in contact with two or more cells) also present a higher concentration of intramembrane particles, which is continuous with the part of the cell in contact with the environment. In the part of the outer membrane that faces the environment, there is a high concentration of intramembrane particles in both faces (Fig. 7b), possibly related to metabolic exchange. The boundary between the part of the cell in contact with another cell and the part of the cell that faces the environment is evident and continuous in adjacent cells (Fig. 7b) (Keim et al. 2004a). The membrane specializations evidenced by the different distribution of intramembrane particles, capsule fibers, and flagella show that the MMP cells are polarized.

4.2 The Cytoplasm

The cells present a cytoplasm rich in ribosomes (Fig. 6) and sometimes a visible nucleoid (Rodgers et al. 1990a). Crystalline organic structures, as well as a structure similar to polar membranes, are present in the cytoplasm (Farina et al. 1983). The structure similar to polar membranes can also be observed in the organisms collected in another place (Keim et al. 2004a,b). Despite the morphological similarity to polar membranes, the fact that these structures are found in the cytoplasm and not associated with the membranes suggests that they are not polar membranes, but structures unique to MMPs (Silva et al., in press).

Large lipid (Keim et al. 2004a) and/or polyhydroxyalkanoate inclusions (Rodgers et al. 1990b, Keim et al. 2004a) are found in the cytoplasm (Fig. 6). They can be distinguished by their characteristic freeze-fracture patterns and by light microscopy after differential staining. MMPs contain large lipid droplets and, sometimes, small polyhydroxyalkanoate inclusions. Polyphosphate bodies containing mainly P, O, and Mg are sometimes found (Keim, unpublished results).

4.3 Organization of the Magnetosomes

The number of iron sulfide magnetosomes in MMPs was reported to vary from 300 (Lins et al. 1992) to about 1000 (Farina et al. 1983, 1990; Lins de Barros et al. 1990a). Some authors reported that the magnetosomes are uniformly distributed between the cells (Pósfai et al. 1998b), whereas others observed a non-homogeneous magnetosome distribution between the cells (Lins de Barros et al. 1990a). Rodgers et al. (1990a) observed that each individual cell presents 2–65 (average 31) magnetosomes. The size of the magnetosome crystals from MMPs are listed in Table 1. In contrast to the magnetite magnetosomes from unicellular bacteria (Devouard et al. 1998), the size distribution of the iron sulfide crystals from the MMPs show an almost perfect Gaussian curve (Pósfai et al. 2001). Magnetosomes are arranged in planar groups in the cytoplasm (Farina et al. 1983; Lins de Barros et al. 1990a; Lins and Farina 1999; Lins et al., submitted) near the periphery of the microorganism, parallel to the surface (Silva et al., in press; Keim et al. 2004a; Lins et al., submitted). It means that each magnetosome plate is at the largest possible distance from the plates of adjacent cells, considering the spherical symmetry (Silva et al. 2003). Sometimes, the magnetosome chains appear aligned in a whole MMP (Lins et al. 2000; Penninga et al. 1995; Pósfai et al. 1998b). Whether this alignment reflects the in vivo magnetosome arrangement, or is an artifact caused by the magnetic field used to concentrate the bacteria, remains unknown. A structure resembling a magnetosome membrane of unknown nature is sometimes observed coating the iron sulfide crystals (Farina et al. 1983; Lins de Barros et al. 1990a; Mann et al. 1990).

4.4 Iron Minerals

Crystals in the magnetosomes of MMPs usually have irregular equidimensional shapes (Fig. 6) and are composed of iron sulfides (Farina et al. 1990; Mann et al. 1990; Pósfai et al. 1998a,b; Keim et al. 2004a). High resolution TEM images showed that the iron sulfide crystals were crystallographic single domains (Bazylinski et al. 1990; Mann et al. 1990) with irregular surfaces (Bazylinski et al. 1990). Although most crystals were irregular, some of them show roughly cuboidal, parallelepipedal (Bazylinski et al. 1990, 1993), flake (Bazylinski et al. 1990; Lins de Barros et al. 1990a), "barrel" or bullet shapes (Pósfai et al. 1998b). The iron sulfide magnetosomes from MMPs showed substantial amounts of copper, depending on where they were collected (Bazylinski et al. 1993; Pósfai et al. 1998b). Semi-quantitative energy-dispersive X-ray analysis showed that the crystals are composed of 46–61 atom% sulfur, 37–50 atom% iron and up to 10 atom% copper (Pósfai et al. 1998b).

Iron sulfides in MMPs were first described by Farina et al. (1990) and Mann et al. (1990). Pósfai et al. (1998a,b) performed a detailed crystallographic study with high resolution TEM and electron diffraction and found that MMPs contained the ferrimagnetic greigite (Fe_3S_4), previously reported by Mann et al. (1990) as the magnetic phase, together with non-magnetic mackinawite (tetragonal FeS) and cubic FeS. Greigite was reported as the major phase, followed by mackinawite (Pósfai et al. 1998b), in contrast to an earlier report that described greigite as a minor phase (Mann et al. 1990). Mackinawite was found only at the ends of the magnetosome chains (Pósfai et al. 1998b).

Elemental maps obtained by energy filtering electron microscopy showed that crystals from MMPs presented a peripheral region rich in iron and oxygen (Farina et al. 1990; Lins and Farina 2001; Buseck et al. 2003). The non-magnetic phases, mackinawite and cubic FeS, were proposed to be precursor phases of greigite. It was suggested that cubic FeS would be mineralized, converted to mackinawite and then to greigite through solid-state transformations. These transformations would occur by changes in the position of the iron atoms, with the maintenance of the general structure of the sulfur framework. To transform mackinawite (FeS) into greigite (Fe_3S_4), part of the iron must leave the crystalline lattice (due to stoichiometric relations) (Pósfai et al. 1998a). This surplus iron would account for the amorphous or weakly crystalline iron-oxygen mineral observed at the periphery where it can be oxidized (Buseck et al. 2003; Farina et al. 1990; Lins and Farina 2001; Lins de Barros et al. 1990a). The crystalline center has a higher iron concentration, a triangular or cuboidal outline, and is 40–45 nm in diameter, whereas the amorphous halo has a lower iron concentration and is 20–25 nm in width (Rodgers et al. 1990a). The crystallographic analysis performed by Pósfai et al. (1998b, 2001) found regions of non-uniform diffraction contrast in iron sulfide crystals; these regions could result from thickness variations or crystalline defects. Iron-oxygen amorphous minerals, 50–70 nm in diameter, were also found in the cytoplasm, and suggested to be iron storage materials (Lins and Farina 2001).

The fact that diamagnetic mackinawite and paramagnetic cubic FeS were found arranged in chains shows that the crystals are aligned in chains before they acquire a permanent dipole moment. Greigite crystals are preferentially aligned with the [100] direction parallel to the chain (the probable easy axis of magnetization in greigite), and thus the biomineralization of the precursor iron sulfide crystals involves a biochemical machinery to mineralize the crystals in the right position in the cell, independently of the magnetic field of the neighboring crystals (Pósfai et al. 1998b). A similar mechanism was recently proposed for magnetite mineralization in *Magnetospirillum* sp. AMB-1 (Komeili et al. 2004).

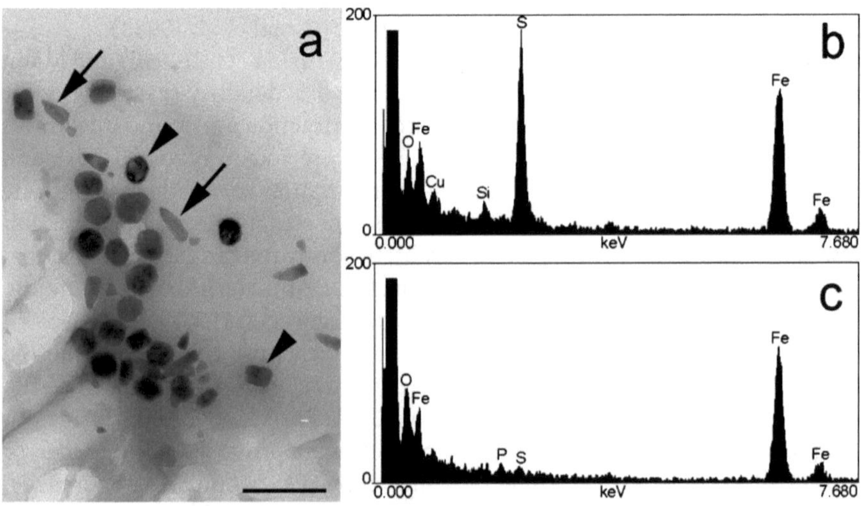

Fig. 8 Observation and elemental analysis of whole mount of a disrupted MMP from Itaipu Lagoon. **a** High magnification of a group of crystals with bullet-shaped (*arrows*) and equidimensional (*arrowheads*) morphologies. The *scale bar* represents 200 nm. **b** Energy-dispersive X-ray (EDX) spectrum of an equidimensional crystal showing iron and sulfur peaks. **c** EDX spectrum of a bullet-shaped crystal showing iron and oxygen peaks. Small P and S peaks are generated by organic constituents of cells

Rarely, magnetite is found in bullet-shaped magnetosomes of MMPs (Keim et al. 2003; Lins et al., submitted), as shown in Fig. 8, and sometimes multiple twins occur (Lins et al. 2000). Bullet-shaped magnetite crystals from MMPs from Itaipu Lagoon are 104 ± 29 nm in length and 42 ± 6 nm in width and are elongated along the [100] direction. Magnetite magnetosomes in MMPs occur alone or together with iron sulfide crystals. MMPs containing only iron sulfides are sometimes found in the same environments as the magnetite-producing MMPs (Lins et al., submitted). Whether magnetite-producing, greigite-producing, and magnetite-greigite-producing MMPs are the same, remains to be discovered. The fact that both types of crystal are found in the same chains suggests that the mechanism of chain assembly is independent of the type of mineral, as suggested for a unicellular magnetotactic bacterium by Bazylinski et al. (1995).

5 Life Cycle

The inheritance of both the magnetic polarity and the movement axis in unicellular MB is epigenetic (Blakemore 1982; Lumsden, 1984; Lins de Barros et al. 1990b; Spring et al. 1998). Considering that the same is occurring in MMPs, which have a more complex body architecture, a complex life cycle is necessary in order to maintain the cell arrangement, the axis of movement, and the magnetic polarity over several generations.

The life cycle of most prokaryotic and eukaryotic multicellular organisms present at least one single-cell stage, i.e., the whole organism can be constructed from a single cell. However, free-swimming individual cells similar to the MMP cells were never observed (Farina et al. 1990; Keim et al. 2004a; Lins and Farina 1999). At least in one environment, MMPs were the only magnetotactic microorganisms found (Keim et al. 2004a,b), excluding the possibility that a small number of magnetotactic unicellular MMP cells could be present in the samples and mistaken for different bacteria. Furthermore, disaggregation in adverse conditions, such as high and low osmolarity, led to a loss of motility (Farina et al. 1990; Lins and Farina, 1999; Lins de Barros et al. 1990a; Mann et al. 1990; Rodgers et al. 1990) and magnetic orientation (Farina et al. 1990). Restoration to the original osmolarity does not restore the motility (Lins de Barros et al. 1990a). Thus, there are no known unicellular stages in the MMP life cycle.

Although most MMPs present a spherical shape, elliptical (Martins et al., submitted; Rodgers et al. 1990) and eight-shaped (Lins and Farina 1999; Martins et al., submitted) organisms are sometimes observed. In fact, these unusual shapes would represent stages of the life cycle of these organisms, as illustrated in Fig. 9. According to this life cycle, MMPs would be spherical for most of their lifetime and would increase in size by increasing the cell

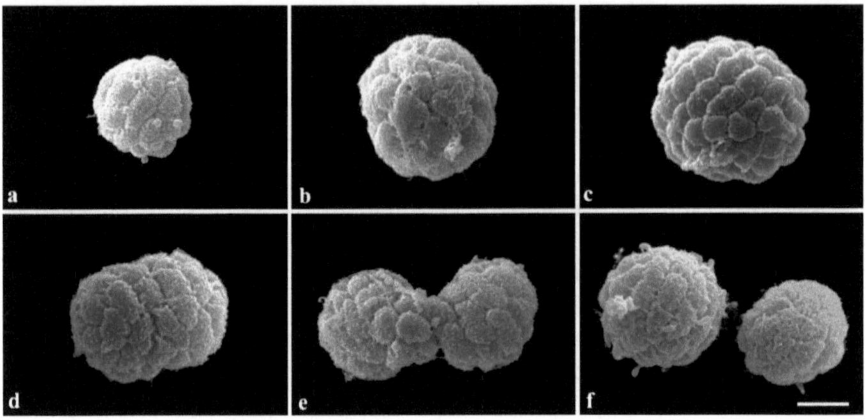

Fig. 9 Sequence of SEM images illustrating the life cycle proposed for MMPs. The MMP is small (**a**) and grows by increasing the cell volume, not the cell number (**b**). Then its cells divide synchronously producing a MMP with twice the cell number (**c**). The organism elongates (**d**) and a constriction appears at the middle (**e**). Then the organism divides into two equivalent organisms (**f**). The *scale bar* represents 4 μm. Reproduced from Keim et al. 2004b with permission from Blackwell

volume, not the cell number (Figs. 9a and b) (Keim et al. 2004b). This is consistent with the fact that small MMPs have smaller cells and large MMPs have larger cells (Lins and Farina 1999). To proceed in the cell cycle, all cells in an organism would divide synchronously but would remain together, resulting in a spherical MMP containing twice the cell number but approximately the same volume (Fig. 9c). The organism would become elliptical (Fig. 9d), and then eight-shaped, as two attached MMPs (Fig. 9e). The constriction between the two daughter MMPs occurs together with a slight torsion between the two halves (Fig. 9e). This torsion solves the topological problem of separating the two halves while maintaining the cell arrangement and the internal compartment. Finally, the eight-shaped MMPs would divide into two equivalent MMPs, returning to the beginning of the cycle (Fig. 9f). The whole cycle seems to last several hours, which makes impossible to follow a single organism over all stages without culturing it. The growth of the constriction (F. Abreu and U. Lins, unpublished observations) and the division of the organism into two equivalent organisms (Keim et al. 2004b; Winklhofer et al., in press) were documented by light microscopy, as well as the presence of septa in several cells of the same MMP by TEM (Keim et al. 2004b). The fact that the magnetic moment is optimized in the whole organism can be explained by the life cycle described above, which accounts for the epigenetic inheritance of the magnetic polarity of the cells and of the whole organism (Winklhofer et al., in press)

Cell division septa seem to occur concomitantly in all cells of a MMP. The septa always begin at the outer surface and occur radially in relation to

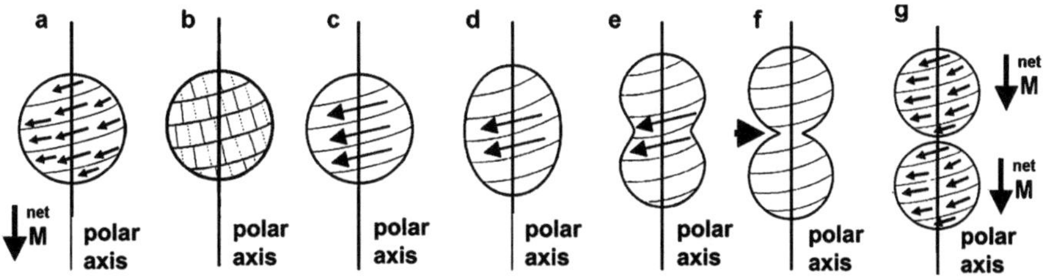

Fig. 10 Schematic drawings of magnetotactic multicelular organisms. **a** Possible distribution of the magnetic moment in a spherical organism (only the surface facing the observer is represented). The cells are arranged in a spiral, and the magnetic moment of each cell is at a fixed angle to the spiral (an example is given by the *small arrows*). The net magnetic moment of the organism is aligned to the polar axis (*large arrow*). **b** The cell division invaginations would be aligned to planes perpendicular to the laps of the spiral (*dashed lines*). **c–d** Cell movements during organism division. **c** The cells in neighboring laps of the spiral slide in relation to each other in the direction illustrated by the *arrows*, making the organism elongate in the direction of the polar axis. **d** After elongation, only the cells at the middle of the organism continue the movements (*arrows*), leading to a constriction. **e** Only a few cells at the middle continue the movement, until two spherical bodies are formed, linked by a single cell. **f** Now, the terminal cells separate from each other, generating two equivalent organisms similar to the parental one. **g** This process is able to generate two equal organisms with the same general cell arrangement and direction of magnetic moment as the parental one

the MMP. This special septa disposition maintains the general cell organization, i.e., all cells arranged around the internal compartment and in contact with the outer environment (Keim et al. 2004b). Furthermore, the magnetosomes, which are found in the cytoplasm mainly near the MMP outer surface (Keim et al. 2004a), and the flagella, only found in the outer surface of MMPs (Rodgers et al. 1990a), can be equally distributed between the daughter cells (Keim et al. 2004b).

The organization of cells in the MMPs in a helical pattern (Lins de Barros et al. 1990a; Keim et al. 2004b), and the torsion observed in the "eight-shaped" stage of the life cycle suggest a series of cell movements to divide one MMP into two new MMPs. The cells would be arranged as a string coiled around the internal compartment. The cells would bind less strongly to cells of other laps of the string, and the different laps would slide in relation to each other during the elongation of the organism, leading to the elliptical and eight-shaped stages, as illustrated in Fig. 10.

6
A Proposed Mathematical Model for Division and Magnetization of MMPs

To obtain a unique interpretation for all magnetic properties and movement characteristics compatible with the life cycle, an appropriate topological model can be proposed. This model is based in an appropriate parametrized family of curves (Cassini's oval). These curves have a very simple analytical expression in polar coordinates:

$$\mathbf{r}^4 + \mathbf{a}^4 - 2\mathbf{a}^2\mathbf{r}^2 \cos(2\boldsymbol{\theta}) = \mathbf{b}^4 \quad (1)$$

where **r** and $\boldsymbol{\theta}$ are the usual polar coordinates and **a** and **b** parameters. The parameter **b** is associated with the size of the curve, and **a** is associated with the shape of the curve. When **a** = 0 the Cassini's oval is reduced to a circumference of radius **b** (i.e., **r** = **b**). When **a** = **b**, the Cassini's oval is the leminiscate of Bernoulli. Using the Cassini's oval as generator of a revolution surface and making the convolution with a cylindrical helix, a three-dimensional spiral is obtained, which changes form as a function of the parameters **a** and **b**.

The life cycle stages of MMPs observed in SEM images (Fig. 9) can be emulated by two steps in the mathematical model:

1. Growing **b**, with **a** = 0, the radius of the sphere grows; e.g., variation of parameter **b** from 0.79 to 1, increases the volume twofold. This corresponds to the observed growing in size of each individual cell during the initial phase of the life cycle.
2. Variation of parameter **a** from 0 to 1.2, maintaining **b** constant (Fig. 11). Changing **a**, changes the revolution surface and gives rise to two spheres. The shape of the surface changes (from a sphere to a melon, when **a** is varied from 0 to 0.999; when **a** = 1 the surface has a singularity; for **a** > 1 two separate surfaces are generated).

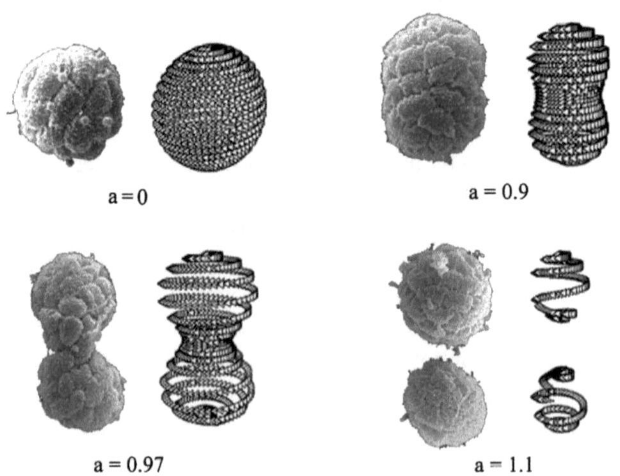

Fig. 11 Mathematical model proposed to interpret the magnetic properties and movement characteristics compatible with the observed life cycle of MMP. The Cassini's oval generated a revolution surface that was convoluted with a cylindrical helix. A three-dimensional spiral is generated, which changes form as a function of parameters **a** and **b**. In the present simulations **b** = 1

This model assumes that the cells are distributed in a spiral array in the sphere (Farina et al. 1983, 1990; Keim et al. 2004b; Lins de Barros et al. 1990; Lins and Farina, 1999), that all the cells contact the internal acellular region (Keim et al. 2004a), and that the magnetic crystals are found near the external surface of the cells (Silva et al., in press). In addition, each cell would have a net magnetic moment nearly tangent to the sphere surface, in a direction that follows the trace of the spiral at each point, and thus inclined with respect to the symmetry axis. This distribution of magnetic crystals is a spatial distribution. When the organism is exposed to an intense magnetic field, it is possible to align individual magnetic moments and increase the total magnetization of the organism (Winklhofer et al., in press). When the organism is exposed to a demagnetizing field, each planar array can change the magnetization polarity, and the total magnetic moment of the whole organism may be zero (Rodgers et al. 1990a,b). This division of one sphere into two equal spheres maintains the center region always isolated from the external environment.

Because of the helical symmetry of the spatial distribution of cell magnetic moments, the total magnetic moment of the organism (the vectorial sum of magnetic moments of all cells) is nearly parallel to the symmetry axis. Flagella would obey the same distribution (see Sect. 7.1 below).

7 Motility of MMPs

The study of movement of MMPs in the presence of a known applied magnetic field gives important clues for understanding magnetic properties, flagellar action, and coordination between cells of the organisms. Samples (sediments and water) from Araruama Lagoon were stored in a rectangular glass aquarium of 14 x 28 x 12 (depth) cm in the laboratory.

A pair of Helmholtz coils coupled to a DC source was adapted to the microscope stage. With this device, it was possible to generate a maximum homogeneous magnetic field of the order of 15 Gauss parallel to the glass slide (for a short period of time it was possible to obtain fields of approximately 50 Gauss). A special coil was used to obtain magnetic fields up to 10 Gauss perpendicular to the glass cover slip. The field generated was constant in the observation region, which means that no net magnetic force acts on the organism.

The movement of microorganisms in water is an example of physics at a very low Reynolds number (Purcell 1977). There is no inertial effect and, in a simple way, it is possible to treat the problem considering force as being proportional to the velocity. Acceleration does not appear in the equations. In this extreme regime, the resultant force that acts in the organism is parallel to its trajectory. MMPs present a complex behavior with respect to the movement. We propose four types of motility for them in the presence of a homogeneous magnetic field: free motion, rotation, escape motility (or ping-pong), and walking.

7.1 Free Motion

When a uniform magnetic field is applied, MMPs swim in either straight or helical trajectories (Keim et al. 2004a). Most MMPs swim in a very elongated helical trajectory when compared to the unicellular MB (H. Lins de Barros, unpublished observation). When no obstacle is present, MMPs from Araruama Lagoon swim at 90 ± 20 μm/s. However, velocity can change from one sample to the other, or after a certain period of observation time under the light microscope. For example, after about 20 min, the velocity of the same population decreased to 64 ± 23 μm/s (Almeida et al. 2005).

Observations made with the perpendicular applied field (parallel to the optical axis of the microscope) shows that MMPs swim in a helical trajectory with the symmetry axis aligned to the magnetic field, with a typical radius of the order of some tens of microns, and pitch varying from one sample to the other. Both the helix and the MMP body rotate in the same sense of the trajectory, i.e., clockwise (in accordance to Simmons et al. 2004). Moreover, it was observed that, in one pitch, the body rotates 2π (Fig. 12a). This rotation cannot be explained by assuming that flagella form a single bundle (in this case, by the application of Newton's third law, the sense of rotation of a body would

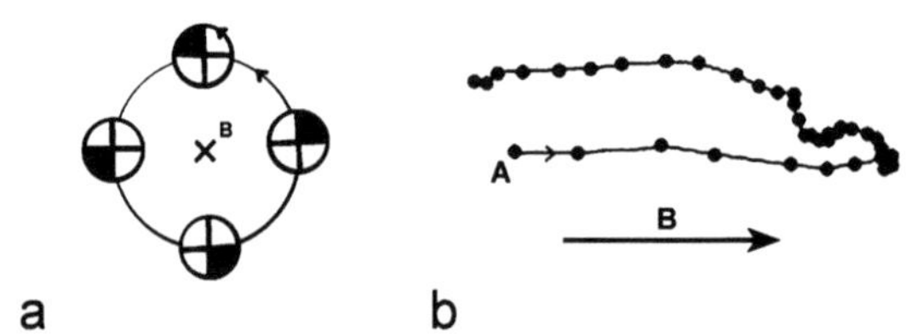

Fig. 12 Trajectories of MMPs. **a** Schedule of the trajectory (*large circle*) and rotation of the MMP (*small circles*) during motion in the direction perpendicular to the image plane (approaching the observer), based on microscopy observation through the vertical axis. The magnetic field *B* points inwards (opposite to the displacement of the south-seeking magnetotactic organism). Note that both the trajectory and the organism's body have the same sense of rotation and the same pitch. **b** Escape motility (or ping-pong) observed near the edge of a drop (*left side* in the figure). The trajectory of one MMP was projected in the xy plane. The image represents an organism swimming approximately in the horizontal plane. Note that the movement in the backward direction (to the *right*, beginning in *A*) is decelerated while the movement in the forward direction is accelerated. The distance between two *dots* in the figure represents the distance swam by the organism in approximately 1/12 s. Applied field $B = 2$ Gauss. The length of the backward movement is approximately 50 μm

be opposite to the sense of rotation of the helix of the trajectory). It is possible that the short flagella observed act at a tangent to or obliquely to the surface of the organism, producing a resultant force that presses around the symmetry axis, making the sense of rotation of the body organism the same as that of the helical trajectory (Fig. 12a). All MMPs observed had the same sense of rotation of the body and the trajectory (H. Lins de Barros, unpublished observations).

7.2 Rotation

MMPs magnetically concentrated at the edge of a drop of water rotate around an axis that passes through the center of the organisms. Observation using a coil to produce a perpendicular magnetic field shows that when the organism is bound by an obstacle (like a small grain of sand or a fragment of sediment), the same rotation pattern is observed. Thus the edge of the drop seems to act as a mechanical stimulus to rotation. In some cases, the rotation of the body is opposite to the rotation performed during free motion (H. Lins de Barros, unpublished observation). This could be associated with a reaction of the organism to the mechanical stimulus.

7.3 Escape Motility

Perhaps the most peculiar behavior of MMPs with respect to motility is the so called "ping-pong", excursion (Greenberg et al. 2005; Keim et al. 2004a; Lins and Farina 1999; Lins de Barros et al. 1990a; Rodgers et al. 1990a; Simmons et al. 2004) or escape motility (a denomination considered in this work to be more appropriate for describing this type of motion). This movement consists of a backward movement (south-seeking in the northern hemisphere and north-seeking in the southern hemisphere) for some tens or hundreds of micrometers, followed by a forward movement (Fig. 12b) (Greenberg et al. 2005; Keim et al. 2004a; Lins de Barros et al. 1990a; Lins and Farina, 1999, 2001; Rodgers et al. 1990a). The backward movement decelerates continuously with time, whereas the forward movement that follows shows uniform acceleration (Fig. 12b) (Greenberg et al. 2005).

Observation in slow motion video shows that MMPs maintain the same orientation with respect to the magnetic field lines when they invert the sense of rotation of the body, which suggests that a change in the rotation of flagella is responsible for the escape motility (H. Lins de Barros, unpublished observation). Thus, the forward movement is clockwise, and the backward movement is counterclockwise, as also observed by Simmons et al. (2004). During the escape motility, the MMPs achieve velocities of $170 \pm 25\ \mu m/s$. The deceleration observed during the backward movement (Greenberg et al.

2005) can be caused by a gradual reduction in the flagellar rotation or in the number of active flagella, producing an uniform decrease in the propelling force. Then, the flagella rotate in the usual direction, and the organism swims as in free motion. Thus, the observed movement is caused by the coordinated action of flagella, meaning that the velocity varies during this movement as a consequence of the change in flagellar action.

Because the probability for an MMP to begin a backward excursion increases with the magnetic field strength, it was proposed that they present a magnetoreception mechanism (in magnetic fields larger than the Earth's) (Greenberg et al. 2005).

7.4 Walking

At magnetic fields of the order of magnitude of the Earth's magnetic field, MMPs swim at constant velocity in a looping motion near the water–air interface (Greenberg et al. 2005). The use of an applied magnetic field perpendicular to the glass slide allowed us to analyze the trajectories during these loops. It was observed that when MMPs reach the air–water interface on the top of the drop, they walk freely in a complex trajectory, maintaining, however, the same sense of body rotation. In this situation, the organism did not present escape motility (H. Lins de Barros, unpublished observation).

8 Magnetic Properties of MMPs

MMPs collected in Araruama Lagoon are south-seeking magnetotactic organisms. The process of magnetic concentration in the laboratory (described in detail by Lins et al. 2003) can artificially introduce a bias in the observations. However, using the same method but changing the polarity of magnets, a very small number of MMPs is observed with a magnetic response that corresponds to north-seeking organisms. Observations made with very rich populations, collected as described above, show that a very small number of MMPs with opposite polarity are present in the sample.

The most striking characteristic of magnetotactic microorganisms is their capacity to align to the lines of a magnetic field. Magnetotaxis can be explained by using the direct application of dipole-magnetic field interaction theory. The simplest method for estimating the total magnetic moment of a magnetotatic microorganism is the "U-turn" method (Lins de Barros et al. 1990b). This method is based on the assumption that, statistically, a population of magnetotactic organisms can be treated as a population of non-interacting dipoles, and the mean orientation of the dipoles is given by the theory of paramagnetism of Langevin (Frankel and Blakemore 1980). The

U-turn method gives the relation between the total magnetic moment of the organism, m, and the diameter of the trajectory (U-turn), L, when the applied field is suddenly reversed. The method also gives the time, T_u, that the organism needed to perform the U-turn. Comparison between T_u and the measured time, T_{exp}, is an important way to guarantee the precision of the estimate of m. The trajectory diameter, L, and the time of the U-turn, T_u, are given by the following equations (assuming a spherical body with radius R translating with migration velocity v, in a medium with viscosity η in a external magnetic field B at an absolute temperature T):

$$m = 8\pi^2 \eta R^3 v / LB \tag{2}$$

$$T_u = (8\pi\eta R^3/\mathrm{m}B) ln(2\mathrm{m}B/\mathrm{k}T) \tag{3}$$

where k is the Boltzmann constant. The total magnetic moment, m, depends on R^3.

Simultaneous measurements of v, R, B, L and T_{exp} can be made using a video recorder and analyzing the images frame by frame. Considering only values of m that give T_u with an error of less than 10% when compared with T_{exp}, it is observed that the total magnetic moment of MMPs has a distribution with two peaks, one centered at $(1.0 \pm 0.3) \times 10^{-10}$ emu, and the other at $(1.8 \pm 0.3) \times 10^{-10}$ emu.

8.1 Magnetization, Demagnetization and Remagnetization

Experiments on the alignment of the MMPs from Araruama Lagoon to a laboratory magnetic field of the order of 10 Gauss and submitting the organism to an applied pulse (of the order of 1 millisecond) of circa 1000 Gauss showed that the net magnetization of the organism grows only about 20% with respect to the natural magnetization. This indicates that the crystal magnetic distribution is optimized in the MMPs (Winklhofer et al., in press).

When MMPs were exposed to a 60 Hz AC magnetic field, they could be remagnetized (many originally north-seeking MMPs could migrate as south-seeking microorganisms), indicating that the intact organism has an axis of motility (Rodgers et al. 1990a,b), similar to unicellular magnetotactic bacteria (Blakemore et al. 1980). Unlike the cells of *Magnetospirillum magnetotacticum*, a cultivated magnetite producer, the MMP hysteresis loops were not square, indicating that they could be demagnetized (Penninga et al. 1995). Accordingly, after exposure to a 60 Hz AC magnetic field some organisms were demagnetized and subsequently failed to respond to changes in the magnetic field, and swam in random directions (Rodgers et al. 1990a).

Demagnetization experiments were made in magnetically concentrated samples from Araruama Lagoon. A commercial tape recorder demagnetizer (60 Hz, 600 Gauss) was used. The samples were exposed for a few seconds and observed in the microscope. After the demagnetization process, the MMPs presented no response to an applied magnetic field. This could be a consequence of the complex spatial distribution of magnetic crystals that produces a multipolar field. The interaction between crystal arrays of different cells of the same organism is insufficient to change the polarity of the arrays. However, after demagnetization, each cell can present a different polarization, giving a net magnetic moment of the whole MMP near zero. When demagnetized samples were exposed to an intense field produced by a magnet (Sm Co industrial magnet), remagnetization of the organisms was observed. This remagnetization produced predominantly organisms of the same original polarity (H. Lins de Barros, unpublished observation).

9
Are MMPs Really Multicellular Organisms?

The answer to the above question depends on the concept used to define a multicellular organism. Traditional concepts based on animal and plant studies include the presence of different cell types and a characteristic cell organization in tissues and/or organs (Gilbert, 2003). "Mid-way" concepts require specific organism shape and cell organization and lack of cell autonomy and competition in order to consider an organism multicellular (Carlile 1980). Current concepts arose as a consequence of new discoveries that challenged the conventional way of thinking. What once was based mainly on morphology now incorporates cell-to-cell signaling and behavior. For example, Shapiro (1998) and Kaiser (2001) proposed that multicellular organisms have many cells in close contact that coordinate growth, movements, and biochemical activities.

MMPs do not seem to have different cell types, and thus do not fulfill the traditional requirements for a multicellular organism. On the other hand, with the present knowledge, MMPs perfectly fit the mid-way concept proposed by Carlile (1980) and the modern concepts of Shapiro (1998) and Kaiser (2001): their cells are closely apposed and organized in a characteristic way, are not autonomous, seem not to compete with each other, and coordinate cell division and movements. Furthermore, the MMPs were never observed as single cells (Farina et al. 1990, Lins de Barros et al. 1990; Lins and Farina 1999; Mann et al. 1990; Rodgers et al. 1990).

10
Conclusions

In this chapter we have discussed several aspects of the biology and biophysics of the multicellular magnetotactic prokaryotes. It was shown that after the first description of MMPs (Farina et al. 1983), it took 7 years for the appearance of papers showing the presence of a new biomineralized magnetic material inside their cytoplasm: the iron sulfide magnetic crystals, different from the iron oxide magnetite known at that time in magnetosomes (Farina et al. 1990; Mann et al. 1990). Eight more years were needed to build a consistent model to explain the reaction sequence related to the production of the magnetic iron sulfides in MMPs (Pósfai et al, 1998a,b). Only recently, has the morphology of MMPs been studied in detail (Keim et al. 2004a) and a cell cycle proposed (Keim et al. 2004b). A detailed study on the motility of the organisms in applied magnetic fields was also performed (Greenberg et al. 2005).

Curiously, only one paper exists on the phylogenetic affiliation of the MMP (DeLong et al. 1993). From this comprehensive review of the literature on MMPs, we now see that it is time for the contributions that can be achieved using a molecular biology approach. The first deep insights into the structure of MMPs were obtained from high resolution electron microscopy studies of the iron mineral inclusions, which meant solving questions directly related to biomineralization. Nowadays, fundamental questions on the biology of the MMPs arise. These are mainly related to multicellularity in prokaryotes, cell division and magnetoreception, and information transduction. There are also new biomineralization questions related to MMPs after the finding that magnetic crystals inside MMPs from the same environment can be composed of iron sulfides, iron sulfides together with iron oxides, or only iron oxides.

We have presented in this chapter a unified hypothesis linking cell organization, proliferation, magnetism, and motility of MMPs, which is based on experiments and direct observation of the microorganism. Further work is necessary to strengthen or correct some points, but all knowledge accumulated up to now is in accordance with our hypothesis. We believe that further work on 16S rDNA phylogeny, ecology, behavior, magnetic properties, and ultrastructure will increase the breadth of a unified view of these interesting microorganisms.

Acknowledgements CK acknowledges support from FAPERS and FUSB. UL acknowledges partial support from CNPq (Pronex). MF acknowledges CNPq and FAPERJ Brazilian Agencies. We are indebted to the staff of "Laboratório de Ultraestrutura Celular Hertha Meyer" at the Biophysical Institute, Federal University of Rio de Janeiro, for the electron microscopy facilities.

References

Almeida FP, Farina M, Keim CN (2005) Angular dispersion of magnetotactic multicellular organisms trajectories in an applied magnetic field. Braz J Morphol Sci (suppl. 2005), p 207

Bazylinski DA, Frankel RB (2004) Magnetosome formation in prokaryotes. Nature Rev Microbiol 2:217–230

Bazylinski DA, Frankel RB, Garratt-Reed AJ, Mann S (1990) Biomineralization of iron sulfides in magnetotactic bacteria from sulfidic environments. In: Frankel RB, Blakemore RP (eds) Iron biominerals. Plenum, New York, pp 239–255

Bazylinski DA, Garratt-Reed AJ, Abedi A, Frankel R (1993) Copper association with iron sulfide magnetosomes in a magnetotactic bacterium. Arch Microbiol 160:35–42

Bazylinski DA, Frankel RB, Heywood BR, Mann S, King JW, Donaghay PL, Hanson AK (1995) Controlled biomineralization of magnetite (Fe_3O_4) and greigite (Fe_3S_4) in a magnetotactic bacterium. Appl Environ Microbiol 61:3232–3239

Blakemore RP (1975) Magnetotactic bacteria. Science 190:377–379

Blakemore RP (1982) Magnetotactic bacteria. Annu Rev Microbiol 36:217–238

Blakemore RP, Frankel RB, Kalmijn AJ (1980) South-seeking magnetotactic bacteria in the Southern Hemisphere. Nature 286:384–385

Buseck PR, Pósfai M, Dunin-Borkowski RE, Weyland M (2003) Iron oxide and sulfide nanocrystals as biomarkers. Microsc Microanal 9(suppl 2):242–243

Carlile MJ (1980) From prokaryote to eukaryote: gains and losses. In: Gooday GW, Lloyd D, Trinci APJ (eds) The eukaryotic microbial cell. Cambridge University Press, pp 1–38

DeLong EF, Frankel RB, Bazylinski DA (1993) Multiple evolutionary origins of magnetotaxis in bacteria. Science 259:803–806

Devouard B, Posfai M, Hua X, Bazylinski DA, Frankel RB, Buseck PR (1998) Magnetite from magnetotactic bacteria: Size distributions and twinning. Amer Mineral 83:1387–1399

Enrich-Prast A, Bozelli RL, Esteves FA, Meirelles FP (2003) Lagoas Costeiras do Parque Nacional da Restinga de Jurubatiba: descrição de suas variáveis limnológicas. In: Rocha CFD, Esteves FA, Scarano FR (eds) Ecologia, história natural e conservação do Parque Nacional da Restinga de Jurubatiba

Farina M, Lins de Barros HGP, Esquivel DMS, Danon J (1983) Ultrastructure of a magnetotactic microorganism. Biol Cell 48:85–88

Farina M, Esquivel DMS, Lins de Barros HGP (1990) Magnetic iron-sulphur crystals from a magnetotactic microorganism. Nature 343:256–258

Flies CB, Peplies J, Schüler D (2005) Combined approach for characterization of uncultivated magnetotactic bacteria from various aquatic environments. Appl Environ Microbiol 71:2723–2731

Frankel RB, Blakemore RP, Wolfe RS (1979) Magnetite in freshwater magnetotactic bacteria. Science 203:1355–1356

Frankel RB, Blakemore RP (1980) Navigational compass in magnetic bacteria. J Magnetism Magnetic Mater 15–18(Part 3):1562–1564

Frankel RB, Bazylinski DA, Johnson MS, Taylor BL (1997) Magneto-aerotaxis in marine coccoid bacteria. Biophys J 73:994–1000

Gilbert SF (2003) Developmental biology, 7th edn. Sinauer, Sunderland, MA, USA

Greenberg M, Canter K, Mahler I, Tornheim A (2005) Observation of magnetoreceptive behavior in a multicellular magnetotactic prokaryote in higher than geomagnetic fields. Biophys J 88:1496–1499

Kaiser D (2001) Building a multicellular organism. Annu Rev Genet 35:103–123

Keim CN, Lins U, Farina M (2003) Iron oxide and iron sulphide crystals in magnetotactic multicellular aggregates. XIX Congress of the Brazilian Society for Microscopy and Microanalysis. Acta Microsc 12(suppl B):3–4

Keim CN, Abreu F, Lins U, Lins de Barros HGP, Farina M (2004a) Cell organization and ultrastructure of a magnetotactic multicellular organism. J Struct Biol 145:254–262

Keim CN, Martins JL, Abreu F, Rosado A, Lins de Barros HGP, Borojevic R, Lins U, Farina M (2004b) Multicellular life cycle of magnetotactic prokaryotes. FEMS Microbiol Lett 240:203–208

Kirschvink JL (1980) South-seeking magnetic bacteria. J Exp Biol 86:345–347

Kjerfve B, Schettini CAF, Knoppers B, Lessa G, Ferreira HO (1996) Hydrology and salt balance in a large, hypersaline coastal lagoon: Lagoa de Araruama, Brazil. Estuar Coast Shelf Sci 42:701–725

Lins de Barros HGP, Esquivel DMS, Farina M (1990a) Biomineralization of a new material by a magnetotactic microorganism. In: Frankel RB, Blakemore RP (eds) Iron biominerals. Plenum, New York, pp 257–268

Lins de Barros HGP, Esquivel DMS, Farina M (1990b) Magnetotaxis. Sci Progress 74:347–359

Lins U, Farina M (1999) Organization of cells in magnetotactic muticellular aggregates. Microbiol Res 154:9–13

Lins U, Farina M (2001) Amorphous mineral phases in magnetotactic multicellular aggregates. Arch Microbiol 176:323–328

Lins U, Farina M, Lins de Barros HGP (1992) Contribution of electron spectroscopic imaging to the observation of magnetic bacteria magnetosomes. Microsc Eletr Biol Cel 16:151–162

Lins U, Freitas F, Keim CN, Farina M (2000) Electron spectroscopic imaging of magnetotactic bacteria: magnetosome morphology and diversity. Microsc Microanal 6:463–470

Lins U, Freitas F, Keim CN, Lins de Barros H, Esquivel DMS, Farina M (2003) Simple homemade apparatus for harvesting uncultured magnetotactic microorganisms. Braz J Microbiol 34:111–116

Lumsden CJ (1984) Dual inheritance in haploid organisms: a model of magnetotactic bacteria. J Theor Biol 111:1–16

Mann S, Sparks NHC, Frankel RB, Bazylinski DA, Jannasch HW (1990) Biomineralization of ferrimagentic greigite (Fe_3S_4) and iron pyrite (FeS) in a magnetotactic bacterium. Nature 343:258–261

Penninga I, Waard H, Moskowitz BM, Bazylinski DA, Frankel RB (1995) Remanence measurements on individual magnetotactic bacteria using a pulsed-field magnetic field. J Mag Mag Mater 149:279–286

Pósfai M, Buseck PR, Bazylinski DA, Frankel RB (1998a) Reaction sequence of iron sulfide minerals in bacteria and their use as biomarkers. Science 280:880–883

Pósfai M, Buseck PR, Bazylinski DA, Frankel RB (1998b) Iron sulfides from magnetotactic bacteria: structure, composition, and phase transitions. Am Mineral 83:1469–1481

Pósfai M, Cziner K, Márton E, Márton P, Buseck PR, Frankel RB, Bazylinski DA (2001) Crystal-size distributions and possible biogenic origin of Fe sulfides. Eur J Mineral 13:691–703

Purcell EM (1977) Life at low Reynolds number. Am J Physics 45:3–11

Rodgers FG, Blakemore RP, Blakemore NA, Frankel RB, Bazylinski DA, Maratea D, Rodgers C (1990a). Intercellular structure in a many-celled magnetotactic prokaryote. Arch Microbiol 154:18–22

Rodgers FG, Blakemore RP, Blakemore NA, Frankel RB, Bazylinski DA, Maratea D, Rodgers C (1990b) Intercellular junctions, motility and magnetosome structure in a multicellular magnetotactic prokaryote. In: Frankel RB, Blakemore RP (eds) Iron biominerals. Plenum, New York, pp 231–237

Shapiro JA (1998) Thinking about bacterial populations as multicellular organisms. Annu Rev Microbiol 52:81–104

Silva KT, Keim CN, Abreu F, Martins JL, Rosado AS, Farina M, Lins U (2003) Spatial relationships of cells within magnetotactic multicellular aggregates. Acta Microsc 12(suppl. B):367–368

Silva KT, Abreu F, Almeda FP, Keim CN, Farina M, Lins V (2006) Flagellar apparatus of south-seeking many-called magnetotactic prokaryotes. Micros Res Techn, in press

Simmons SL, Sievert SM, Frankel RB, Bazylinski DA, Edwards KJ (2004) Spatiotemporal distribution of marine magnetotactic bacteria in a seasonally stratified coastal salt pond. Appl Environ Microbiol 70:6230–6239

Simmons SL, Bazylinski DA, Edwards KJ (2006). South-seeking magnetotactic bacteria in the northern hemisphere. Science 311:371–374

Spring S, Lins U, Amann R, Schleifer K-H, Ferreira LCS, Esquivel DMS, Farina M (1998) Phylogenetic affiliation and ultrastructure of uncultured magnetotactic bacteria with unusually large magnetosomes. Arch Microbiol 169:136–147

Williams R (1979) The geometrical foundation of natural structure. Dover, New York

Winklhofer M, Abraçado LG, Davila AF, Keim CN, Lins de Barros HGP (2005) Magnetic optimization in a multicellular magnetotactic organism. Biophys J, in press

Microbiol Monogr (3)
D. Schüler: Magnetoreception and Magnetosomes in Bacteria
DOI 10.1007/7171_041/Published online: 11 October 2006

Genetic Analysis of Magnetosome Biomineralization

Christian Jogler · Dirk Schüler (✉)

Max Planck Institute for Marine Microbiology, Celciusstrasse 1, 28359 Bremen, Germany
dschuele@mpi-bremen.de

Abstract Magnetite crystals produced by magnetotactic bacteria (MTB) have uniform species-specific morphologies and sizes, which are mostly unknown from inorganic systems. This indicates that biomineralization in magnetosomes is a process with genetic control over the accumulation of iron, the deposition of the magnetic crystal within a specific compartment, as well as their intracellular assembly and alignment into chain-like structures. Our understanding of the molecular and genetic basis of magnetosome formation has substantially improved during the last few years due to the progress in genome analysis and the development of advanced genetic techniques to study MTB.

In this review, we describe the methods and systems, which have become available for genetic analysis of MTB. In addition, we summarize the current knowledge of genes and proteins controlling magnetosome formation, and give an overview over genomic analysis of MTB with emphasis on the structure and organization of genomic "magnetosome islands" identified in several strains of MTB.

1 Introduction

The most intriguing feature of magnetotactic bacteria (MTB) is their capability to synthesize perfectly shaped, sized and arranged magnetosome crystals. Soon after this discovery, there was realization that the formation of a variety of crystals in diverse bacteria must be under strict genetic control. However, despite reasonable progress in the ecological, ultrastructural, and physicochemical characterization of the bacteria and their magnetosome particles, genetic analysis was lagging behind despite considerable efforts by a number of researchers. This was due to various reasons including the lack of appropriate bacterial strains available in pure culture, their fastidiousness, and the absence of methods for their genetic manipulation. Accordingly, the mechanism of magnetosome biomineralization has remained poorly understood at the molecular level, and until very recently, progress has been slow in the past 30 years after the discovery of MTB by Richard Blakemore (Blakemore 1975). However, during the last few years our knowledge of the molecular, biochemical and genetic basis of magnetosome formation has increased tremendously, thanks to the progress in genome analysis and the development of new techniques for the genetic manipulation of various strains in several laboratories. In this review, we will first describe the specific techniques available for the genetic analysis of MTB, and will then summarize current knowledge about the molecular and genetic basis of magnetosome formation.

2 The Molecular Tool Box: Genetic Manipulation of Magnetic Bacteria

In many ways, progress in the elucidation of the genetic basis of chemical and biochemical pathways that are involved in magnetosome synthesis, has been limited by the general lack of a workable genetic system for magnetotactic bacteria. Additionally, several problems specifically associated with the biology of MTB had to be overcome. Methods for the genetic manipulation have been described for *Magnetospirillum magneticum* AMB-1 and *M. gryphiswaldense* MSR-1, which will be briefly described in the following section.

2.1
Methods of DNA Transfer, Genetic Markers and Vector Systems

Conjugational transfer of various mobilizable plasmids to MTB was first reported by Matsunaga and coworkers (Matsunaga et al. 1992), and later also established for MSR-1 (Schultheiss and Schüler 2003). Using biparental mating with different *E. coli* donor strains including S17-1 (Simon et al. 1983) and BW29427 (K.A. Datsenko and B.L. Wanner, unpublished results) conjugation frequencies of 1×10^0 and $3–4.5 \times 10^{-3}$ transconjugants per recipient cell were reported for MSR-1 and AMB-1, respectively (Matsunaga et al. 1992; Schultheiss and Schüler 2003). The relatively high efficiency and convenience of conjugation makes it the method of choice for most genetic experiments. Since electroporation protocols were described for several *Aquaspirillum* strains, it was hypothesized that this application should also work in *Magnetospirillum* strains (Eden and Blakemore 1991). Protocols for electroporation of *M. gryphiswaldense* MSR-1 and *M. magneticum* AMB-1 were reported (Okamura et al. 2003; Schultheiss and Schüler 2003). However, the limited efficiency and reproducibility of this method, which has been attributed to the adverse effects of the magnetosome chain in strong electric fields (Okamura et al. 2003), has limited its practical value so far (Schultheiss and Schüler 2003).

Native plasmids were detected in AMB-1 (Okamura et al. 2003) and MSR-1 (D. Schultheiss and M. Kube, unpublished data). A cloning vector based on the endogenous plasmid pMGT of AMB-1 was constructed, which might provide the advantage of stable maintenance in the absence of selective pressure (Okamura et al. 2003). A number of broad host range vectors including those of the IncQ, IncP, and pBBR1 groups are capable of stable replication within MSR-1 and AMB-1 strains (Matsunaga et al. 1992). The small size and high transfer rates observed with pBBR-related vectors make them useful in *Magnetospirillum* species, and various derivatives carrying different antibiotic markers are available (Kovach et al. 1994, 1995). Commonly used suicide vectors such as pK19mobsacB are based on the pMB1 replicon, restricting their host range to *E. coli* and closely related Enterobacteria (Sutcliffe 1979), which enables their use to force genome integration through homologous recombination (Komeili et al. 2004; Schultheiss et al. 2004). A range of various genetic markers can be expressed in MTB to select or screen for genetic events (Matsunaga et al. 1992; Schultheiss and Schüler 2003). *Magnetospirillum* strains are sensitive to commonly used antibiotics, like ampicillin, kanamycin, tetracycline, chloramphenicol, streptomycin and rifampicin. Spontaneous MSR-1 resistant mutants to streptomycin and rifampicin have been isolated and can be used for counterselection against the conjugative donor S17-1, whereas BW29427 can conveniently counterselected by using its diamino pimelic acid auxotrophy (Schultheiss and Schüler 2003). Expression of the *sacB* suicide marker gene leads to efficient cell killing in AMB-1 and MSR-1, enabling counterselection to force gene replacement by homologous recombination (Schultheiss and Schüler 2003; Komeili et al. 2004). Screening for rare double recombinants was also described with vectors harboring *gusA* (Schultheiss et al. 2004), which encodes a β-Glucuronidase. This enzyme facilitates blue/white screening through its conversion of colorless X-Gal into a blue dye (Horwitz et al. 1964). Additionally, several reporter genes have been demonstrated as useful genetic tools in MTBs. For instance, luciferase gene fusions were used to study the expression of magnetosome genes (Nakamura et al. 1995b; Matsunaga et al. 2000). The *green fluorescent protein* (GFP) and its engineered variants, which have become an extremely powerful tools in prokaryotic cell biology in the last few years (Southward and Surette 2002; Valdivia et al. 2006), has been used to study the subcellular localization of magnetosome-related proteins in *Magnetospirillum* strains (Handrick et al. 2004; Komeili et al. 2004; Schultheiss et al. 2004; Scheffel et al. 2005; see also A. Komeili, 2006, in this volume).

Several experimental approaches require strong gene expression. For this purpose, endogenous promoters from *M. magneticum* AMB-1 were identified by a combination of proteomic and genomic analysis. The promoters of highly expressed genes were cloned and screened in a luciferase reporter gene assay, and the *msp3* (***membrane specific protein** 3*) promoter was shown to support the highest reporter gene expression in the assay (Yoshino and Matsunaga 2005; see also T. Matsunaga et al., 2006, in this volume).

2.2
Isolation and Generation of Mutants

The ability to grow cells as colonies for clonal selection is a seemingly trivial requirement for any genetic experiment, and in particular for the isolation of mutants. However, this is not easily achieved owing to the fastidiousness of MTB to grow from single cells. Colony formation was reported in *M. magneticum* AMB-1 (Matsunaga 1991). A medium containing 0.3% of activated charcoal as a scavenger of toxic oxygen compounds promotes efficient colonial growth of *M. gryphiswaldense* MSR-1 in the presence of micro-oxic atmospheres (Schultheiss and Schüler 2003). As magnetite-forming colonies can be easily distinguished by their dark-brown color from those devoid of magnetosomes, plating on this medium allows straightforward visual screening of colonies for mutants affected in magnetosome formation (Schultheiss and Schüler 2003; Ullrich et al. 2005) (Fig. 1A). While several investigators used the same method for AMB-1 mutants (Fukuda et al. 2006), a different screening method for nonmagnetic AMB-1 clones that cannot be easily recognized by color on plates was described by Komeili and coworkers. This technique relies on the recultivation of single colonies in liquid medium and a subsequent check of magnetic behavior by a specific array of magnets applied to microtiter plates (Komeili et al. 2004). Growth as colonies of the

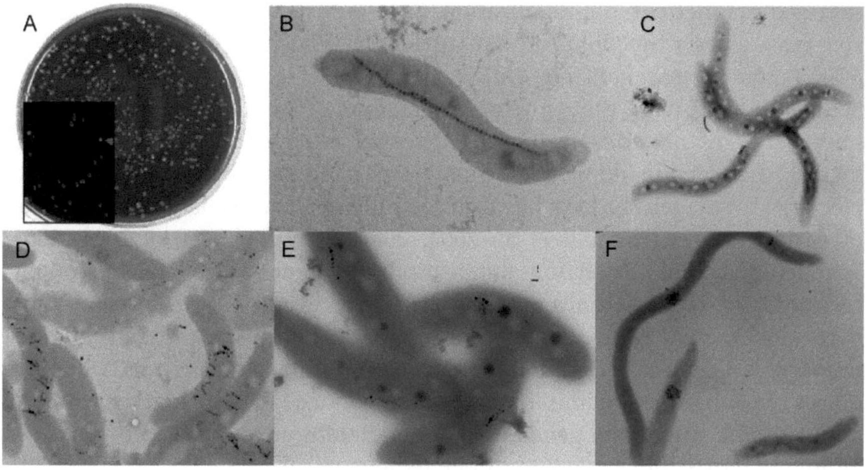

Fig. 1 **A** Appearance of nonmagnetic mutants (*white colonies*) (*arrow*) among magnetic clones of wild-type *Magnetospirillum gryphiswaldense* MSR-1 on charcoal agar plates. The *dark brown* color of wild-type colonies is due to their high intracellular content of the iron oxide magnetite, which is *black* in color. **B–F** Transmission electron micrographs of the wild-type (**B**) and various mutant strains (**C–F**) of *Magnetospirillum gryphiswaldense*. Compared to the wild-type, which forms a straight chain of up to 100 magnetosome crystals, mutants are either devoid of any magnetite crystals (**C**), or affected in the size, shape, number, and intracellular alignment of magnetosomes (**D–F**), indicating that these parameters are under genetic control (micrographs **D** and **E** by Mihály Posfai)

marine magnetotactic vibrio MV-1 on solid media can be accomplished by streaking cells onto the surface of solidified growth medium and incubating under micro-oxic atmospheres. However, deep Petri dishes must be used as the thickness of the agar has to be 2–3 times that of conventional Petri dishes, probably to provide a greater volume for buffering the accumulation of toxic waste products around colonies and to minimize the effects of evaporation (Schüler 2006). By using these techniques, a number of spontaneous mutants of several MTB strains were isolated, which are affected with respect to their capability to form regular magnetosomes (Schübbe et al. 2003; Ullrich et al. 2005). Beside entirely nonmagnetic phenotypes (i.e., the complete absence of magnetosomes), mutants displaying altered sizes, shapes, numbers, organization or alignment of magnetosomes were described, indicating that all these characteristics are under strict genetic control in MTB (Fig. 1B–F). The occurrence of mutants at high frequency indicates that the magnetic phenotype is genetically rather unstable, which can complicate mutagenesis studies and requires careful controls, e.g. by complementation analysis, to confidently correlate the observed phenotype to the introduced mutation.

Random mutagenesis is an essential tool to perform saturating screens in order to identify the total number of genes involved in magnetosome formation and magnetite biomineralization. Several investigators applied transposon mutagenesis by conjugational transfer of Tn5 into *M. gryphiswaldense MSR-1* and *M. magneticum AMB-1* (Wahyudi et al. 2001; Schultheiss and Schüler 2003). With the growing availability of genome information, the importance of methods facilitating directed mutagenesis has dramatically increased. Insertion–duplication mutagenesis and gene deletion by allelic replacement mutagenesis have been successfully applied to analyze the function of various genes (Komeili et al. 2004, 2006; Schultheiss et al. 2004, 2005; Scheffel et al. 2005). Insertion–duplication mutagenesis is relatively straightforward and requires only a single crossover event. However, since most magnetosome genes are arranged in operon structures, the construction of unmarked deletions is essential in order to minimize polar effects, which requires the subsequent elimination of the introduced genetic marker from the chromosome. Successful gene knockout is therefore complicated by the need of double crossover events that can be tedious to select for.

2.3 Heterologous Expression of Magnetobacterial Genes in Different Hosts

The functional expression of *Magnetospirillum* (formerly *Aquaspirillum*) *magnetotacticum* genes in *E. coli* K12 was shown first by Waleh and coworkers, who demonstrated that several *E. coli* mutants defective in amino acid synthesis could be complemented with cosmids from an *A. magnetotacticum* library (Waleh 1988). In another study the functional expression of *M. magnetotacticum* AMB-1 RecA protein in *E. coli* was described (Berson et al. 1989). The complementation of aromatic-metabolite requirements and iron-uptake deficiencies in *E. coli* and *Salmonella typhimurium* by genome fragments of *M. magnetotacticum* was demonstrated by the same group (Berson et al. 1991). Both enterobacterial species lacked a functional *aroD* gene, which is responsible for biosynthetic dehydroquinase activity. Southern hybridization suggested the absence or only minor sequence homology between *aroD* of *E. coli* and *A. magnetotacticum*, indicating that even remotely related genes were able to functionally complement each other (Berson et al. 1991). In summary, these experiments have demonstrated that MTB genes can be functionally expressed in unrelated hosts, which might encourage future strategies for heterologous expression of the entire metabolic pathway from cultivated and uncultivated MTB in surrogate hosts.

2.4 A Wish List for Further Genetic Tools

Despite the recent progress in the development of genetic tools for manipulation of MTB there is still a great need for further improvements. Although the power of deletion mutagenesis in functional analysis has been demonstrated, it has remained cumbersome. Therefore, an improved and extended range of markers, vectors and selection strategy are on the wish list for genetic tools to simplify genetic manipulations. Furthermore, an adjustable expression system based on inducible promoters will be required. The development of a site-specific chromosomal integration system for the introduction of foreign genes in order to study their single-copy expression in a close-to-native manner (e.g. for complementation or localization experiments) will be another challenge for the future. Several strategies could solve this problem, like Tn7-based transposon integration (Koch et al. 2001; Choi et al. 2005), or the design of shuttle vectors enabling homologous recombination into specific genomic regions without disturbing other gene functions (Guiral et al. 2006). Generally, the fact that no single model organism (i.e., a magnetic "*E. coli*") has emerged so far for the study of magnetosome biomineralization, makes it difficult to directly compare contradictory observation, e.g. obtained from the analysis of the closely related, yet physiologically and genetically distinct strains *M. gryphiswaldense* MSR-1 and *M. magneticum* AMB-1. On the other hand it also should be kept in mind that the two genetically tractable *Magnetospirillum* strains by no means represent the vast diversity of cultivated and uncultivated MTB. For instance, almost nothing is known with respect to the genetics of magnetosome formation outside the Alphaproteobacteria, such as in the magnetotactic Deltaproteobacterium *Desulfovibrio* strain RS-1 (DeLong et al. 1993; Sakaguchi et al. 2002), and the successful genetic work recently reported for *Magnetospirillum* species should help stimulate and encourage comparable studies in other cultivated MTB.

3
Molecular Analysis of Magnetosome Formation

The understanding of magnetosome formation at the molecular level has greatly benefited from the combined application of genomic and proteomic approaches. Biochemical and proteomic analysis of the magnetosome membrane led to the cloning and identification of key genes encoding major magnetosome proteins in a reverse genetic approach. This has been complemented lately by the genome-wide analysis of various MTB, as well as mutagenesis studies of individual gene functions, which will be summarized in the following section.

3.1
Biochemical Characterization of the Magnetosome Membrane (MM)

Magnetic crystals synthesized in *Magnetospirillum* species are enveloped by a lipid bilayer described as a "magnetosome membrane" (MM) (Gorby et al. 1988). There is growing evidence that the presence of a MM seems

Table 1 Characteristics of identified MMPs from *M. gryphiswaldense* within the *mms6*-, *mamGFDC*-, *mamAB*-operons, and other areas of the MAI

	Protein	Length	BlastP-homologs present AMB1/MS1/MC1 (e-value)	BlastP-homologs other organisms (e-value)	Characteristics or function
mms6	MamW	138 AS	$5e^{-45}/3e^{-45}/0.92$	mox J motive (1.7)	membrane protein moxJ like
	MgI457	449 AS	$1e^{-136}/1e^{-133}/0.007$	*Rhodospirillum rubrum* ($2e^{-23}$)	TPR motive
	MgI458	347 AS	$2e^{-86}/4e^{-79}/3.0$	*Sinorhizobium meliloti* ($3e^{-05}$)	hypothetical transmembrane protein
	MgI459	125 AS	$2e^{-44}/5e^{-37}/3e^{-17}$	*Thermococcus kodakaraensis* (0.032)	hypothetical membrane protein
	Mms6	136 AS	$7e^{-17}/1e^{-12}/1.5$	–	iron binding
	MgI462	90 AS	–/–/–	–	unknown
mamGFDC	MamG	84 AS	–/–/–	–	unknown
	MamF	111 AS	$1e^{-42}/2e^{-42}/2e^{-17}$	–	unknown
	MamD	314 AS	$2e^{-89}/8e^{-90}/2e^{-30}$	*Alphavirus*; glycoprotein J (1.9)	unknown
	MamC	125 AS	$5e^{-21}/7e^{-21}/2e^{-07}$	–	unknown
mamAB	MamH	428 AS	$0.0/1e^{-171}/1e^{-116}$	*Chlorobium tepidum*; putative transporter ($2e^{-34}$)	Major Facilitator Superfamily
	MamI	77 AS	$4e^{-15}/7e^{-15}/2e^{-05}$	–	unknown
	MamE	772 AS	$0.0/0.0/3e^{-57}$	*Rhodopirellula baltica*; serine protease do-like DEGP ($1e^{-37}$)	PDZ domains; putative Serin protease
	MamJ	426 AS	$3e^{-47}/2e^{-52}/–$	*Streptomyces avermitilis*; hypothetical protein (0.003)	magnetosome chain assembly
	MamK	360 AS	$0.0/3e^{-172}/1e^{-99}$	*Methanopyrus kandleri*; HSP70 class molecular chaperones involved in cell morphogenesis ($3e^{-17}$)	cytoskeleton filament; MreB-like; chain formation

Table 1 (continued)

	Protein	Length	BlastP-homologs present AMB1/MS1/MC1 (e-value)	BlastP-homologs other organisms (e-value)	Characteristics or function
mamAB	MamL	123 AS	$5e^{-32}/8e^{-27}/0.42$	*Takifugu rubripes*; claudin 32a (0.12)	unknown
	MamM	318 AS	$3e^{-173}/2e^{-159}/3e^{-79}$	*Geobacter metallireducens*; predicted Co/Zn/Cd cation transporters ($4e^{-34}$)	iron transporter
	MamN	437 AS	0.0/0.0/0.002	*Rubrobacter xylanophilus*; Na+/H+ antiporter NhaD and related arsenite permeases ($3e^{-31}$)	unknown
	MamO	632 AS	$0.0/0.0/5e^{-79}$	*Caulobacter crescentus*; serine protease ($1e^{-14}$)	PDZ domain; putative Serin protease
	MamP	270 AS	$1e^{-108}/1e^{-108}/1e^{-34}$	*Legionella pneumophila*; Protease ($4e^{-05}$)	unknown
	MamA	217 AS	$5e^{-113}/7e^{-113}/2e^{-37}$	*Methanosarcina mazei*; O-linked N-acetyl-glucosaminetransferase ($6e^{-15}$)	TPR-motifs "activation" of magnetosomes
	MamQ	272 AS	$6e^{-112}/4e^{-106}/1e^{-16}$	*Thermotoga maritima*; lemA protein ($6e^{-16}$)	Putative membrane protein
	MamR	73 AS	$3e^{-31}/3e^{-31}/1.9$	*Thermococcus kodakaraensis*; Predicted site-specific integrase-resolvase (0.015)	hydrophilic protein
	MamB	297 AS	$1e^{-159}/2e^{-159}/1e^{-78}$	*Geobacter sulfurreducens*; cation efflux family protein ($1e^{-39}$)	iron transport
	MamS	180 AS	$5e^{-60}/7e^{-55}/3e^{-13}$	*Irpex lacteus*; Exocellulase (3.6)	putative membrane protein
	MamT	174 AS	$3e^{-87}/4e^{-83}/7e^{-27}$	*Caenorhabditis briggsae;* Hypothetical protein (0.66)	putative CytC binding
	MamU	297 AS	$5e^{-114}/4e^{-116}/2.5$	*Mesorhizobium sp.*; Sphingosine kinase and enzymes related to eukaryotic diacylglycerol kinase ($2e^{-28}$)	unknown
	MgI505	411 AS	$0.002/7e^{-4}/0.01$	*Anaeromyxobacter dehalogenans*; ($3e^{-4}$)	hypothetical protein

to be common to many, if not all magnetite-forming MTB. The membrane can be solubilized by hot SDS treatment or organic solvents after magnetosome isolation, which results in the immediate agglomeration from suspensions of the particles. Biochemical analysis of isolated magnetosomes from *M. gryphiswaldense* MSR-1 revealed that phosphatidylethanolamine and phosphatidylglycerol were the most abundant lipids found in the MM and other cell membranes, while amide-linked fatty acids, which are common in the outer membranes of gram-negative bacteria, were not detected in the MM (Grünberg et al. 2004). This similarity in lipid composition between the MM and the cytoplasmic membrane (CM) and the observed localization adherent to the CM seems to point towards a common cellular origin. Recently, Komeili and co-workers presented direct evidences by cryo-electron-microscopy that magnetosomes vesicles in fact derive from invaginations of the inner membrane (Komeili et al. 2006), which suggests that MM and CM at least transiently form a continuum (Gorby et al. 1988; Bazylinski et al. 2004; see also Komeili, 2006, in this volume).

Isolated magnetosome particles of MS-1 were analyzed with respect to their biochemical composition first by Gorby and coworkers (Gorby et al. 1988), who found that at least two proteins were unique to the magnetosome membrane, but absent from other membrane systems of MS-1. Although the protein patterns of MM preparations from different MTB appear to be dissimilar in 1D- and 2D-analysis, there is growing evidence that homologous MM constituents are present in other related MTB (Grünberg et al. 2001, 2004). Most of the knowledge about the biochemical composition of magnetosome membrane (MM) comes from analysis of the *M. gryphiswaldense* MSR-1 magnetosome subproteom. A number of specific magnetosome-associated proteins (MMPs) were identified using various proteomic techniques in SDS-solubilized extracts of the MM and tryptic digests of isolated magnetosomes (Grünberg et al. 2004) (Fig. 2). Several of the identified polypeptides in MM preparations represent oligomers or post-translational modifications of the same protein, as for instance by proteolytic processing, or covalently bound c-type heme as revealed by peroxidase staining.

Beside a set of major and apparently specific MMPs, a number of minor constituents were occasionally found bound to isolated magnetosomes (Grünberg et al. 2004; Matsunaga et al. 2005). Because they were identified in small amounts and/or represent highly abundant cellular proteins with different, well-defined functions, they most likely result from contaminations by other subcellular compartments. For example, Mms16 which was detected in MM preparations from different *Magnetospirillum* strains was originally suggested to be involved in the formation of magnetosome membrane vesicles in *Magnetospirillum* strain AMB-1 (Okamura et al. 2001). However, recent experimental evidence in *M. gryphiswaldense* MSR-1 demonstrates that Mms16 is not associated with magnetosomes in vivo. Moreover, its function seems entirely unrelated to magnetosome formation, and instead, it is involved in

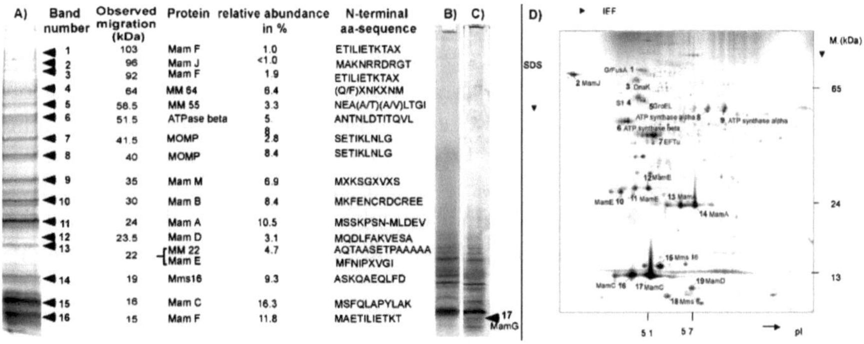

Fig. 2 Proteomic analysis of the magnetosome membrane of *M. gryphiswaldense* (modified after Grünberg et al. 2004). **A–C** Separation of MM-associated proteins by 1D PAGE. Summary of MMPs detected by Coomassie blue staining in 1D SDS–16% PAGE gels. (**A** and **B**) and silver-stained 16% SDS-Tricine gels of MMPs (**C**). **D** Silver-stained 2D-PAGE of MMPs. Proteins from marked spots were excised from the corresponding Coomassie blue-stained gel and identified by electrospray ionization-MS-MS after tryptic digestion; IEF, isoelectric focusing

the formation of PHB granules. Thus, it most likely represents a contamination, which becomes attached to magnetosomes during cell disruption (Handrick et al. 2004; Schultheiss et al. 2005).

In MSR-1, 18 specific bona fide MMPs were identified so far based on their specific and abundant occurrence in preparations of the MM, and the co-localization of their genes. Their characteristics are shown in Table 1. However, it cannot be excluded that further MMPs might have escaped detection due to their low abundance, tight mineral association and/or poor solubilization. On the basis of sequence analysis, MMPs can be assigned to various protein families, which are outlined in the following section.

3.1.1 MamA

MamA described in *M. gryphiswaldense* MSR-1 has also been identified in the MM of several *Magnetospirillum* species (Grünberg et al. 2004). It corresponds to *mms24* which was found in *M. magneticum* AMB-1 (Arakaki et al. 2003) and to Mam22 identified in *M. magnetotacticum* MS-1 (Okuda et al. 1996; Grünberg et al. 2001; Okuda and Fukumori 2001). MamA is an abundant protein of the MM and contains four to five copies of the tetratricopeptide repeat (TPR) motif. These motifs have been identified in a growing number of proteins with diverse functions, and are known to mediate protein–protein interactions (Blatch and Lassle 1999). It therefore has been speculated that MamA is a receptor in the MM that interacts with cytoplasmic proteins or is involved in the assembly of multiprotein complexes within the

MM. Data supporting this notion were obtained via heterologous expression of *mam22* in *E. coli* (Okuda et al. 1996; Okuda and Fukumori 2001). However, MamA-deficient mutants of AMB-1 were not affected in the formation of magnetosome vesicles, but produced magnetite crystals identical in shape and alignment to those in the wild-type, albeit at reduced numbers. These results indicate that MamA is not essential for magnetosome biomineralization, but has a so-far uncharacterized function, which might be in the "activation" of magnetosome vesicles as suggested by Komeili and co-workers (Komeili et al. 2004).

3.1.2 CDF Proteins

Both MamB and MamM were identified as members of the cation diffusion facilitator (CDF) family of metal transporters, which consists of proteins that function as efflux pumps of toxic divalent cations, such as zinc, cadmium, cobalt, and other heavy metal ions. More specifically, MamB and MamM have greatest similarity to the CDF3 subfamily, which was postulated to comprise putative iron transporters (Nies and Silver 1995). It has been speculated that MamB and MamM are involved in the magnetosome-directed uptake of iron (Grünberg et al. 2001), and preliminary evidence obtained from mutant analysis seems to support this assumption (Junge et al., manuscript in preparation). A further CDF gene paralogous to *mamB* and *mamM*, named *mamV* is present in the *mamAB* operon of AMB-1 and MS-1, which, however, is missing in the genomes of other MTB. Intriguingly, FieF, a CDF transporter identified in *E. coli* with remarkable sequence similarity to the MamB and MamM proteins, was recently demonstrated to export iron and zinc over the cytoplasm membrane (Grass et al. 2005), which lends further support to the notion that CDF3 proteins are involved in the transport of iron.

3.1.3 HtrA-Like Serine Proteases

MamE and MamO display sequence similarity to HtrA-like serine proteases, although they share only relatively weak (31%) sequence similarity to each other. The *mamP* gene, encoding a further putative serine protease, is co-located with *mamE* and *mamO* within the same operon, but the gene product of *mamP* has not been identified in the MM. HtrA-like proteins share a conserved trypsin-like protease domain and one or two PDZ domains. They act as molecular chaperones and heat-shock induced proteases, which degrade misfolded proteins in the periplasm (Clausen et al. 2002). It has been suggested that MamE and MamO are involved in magnetosome formation, perhaps by the processing, maturation, and targeting of MMPs during MM assembly (Grünberg et al. 2001).

3.1.4
MamJ

The MamJ protein is particularly rich (18.7%) in repeats of the acidic amino acid residues glutamate and aspartate. Its central domain contains two identical, highly acidic Glu-rich stretches, in which acidic amino acid residues account for 32.4% of the sequence (Scheffel et al. 2005). MamJ was initially assumed to be involved in magnetite crystallization (Schüler 2004), but recent data obtained from a *mamJ* deletion mutant of *M. gryphiswaldense* MSR-1 have revealed that MamJ is involved in the assembly of magnetosome chains by interaction with the magnetosome filament (Scheffel et al. 2005). Specifically, mutant cells of MSR-1 lacking MamJ display a reduced magnetic orientation, although they contain magnetosome crystals identical in number, size and morphology to those in the wild-type. However, in the MamJ-deficient strain magnetosome crystals are no longer aligned in a straight chain as in the wild-type, but instead arrange in compact clusters (Fig. 1F).

3.1.5
MamK

Because of its striking sequence similarity to actin-like MreB proteins (Carballido-Lopez and Errington 2003; Errington 2003), it was speculated that the *mamK* gene product might be involved in the formation of a cytoskeletal structure associated with magnetosome chains (Schüler 2004). Consistent with this notion, the deletion of *mamK* in *M. magneticum* AMB-1 resulted in a perturbation of chain formation and the absence of the "magnetosome filament", which led to the conclusion that MamK itself is forming this filamentous network (Komeili et al. 2006), thereby providing the scaffold to which magnetosomes are bound (for details, see A. Komeili, 2006, in this volume).

3.1.6
Other Proteins with Presumed Functions in Magnetosome Formation

Most of the identified MMPs including, for example, the most abundant MM-associated proteins MamC and MamF have no known homologues in organisms other than MTB and thus represent members of unique, MTB-specific protein families (Grünberg et al. 2004). MamC, for example, was also identified in isolated magnetosomes of *M. magneticum* AMB-1, in which it was termed "Mms13" (Arakaki et al. 2003). Other investigators confirmed the specific localization of MamC within the MM of *M. magnetotacticum* using immunogold-staining techniques (Taoka et al. 2006). Both MamC and MamF are highly hydrophobic proteins with several predicted transmembrane helices.

MamD and MamG are partially identical and share a hydrophobic sequence motive with Mms6 (Grünberg et al. 2004), bearing similarity to LG-rich motives found in self-aggregating proteins of biomineralization systems of animal origin (Sudo et al. 1997; Bochicchio et al. 2001; Zurovec and Sehnal 2002). Mms5, 7 and 8 were also hypothesized to play a role in the magnetite crystallization process (Matsunaga and Okamura 2003). The small Mms6 protein was described recently in *Magnetospirillum* strain AMB-1 as a tightly bound constituent of the MM that exhibited iron-binding activity and had an effect on the morphology of growing magnetite crystals in vitro (Arakaki et al. 2003).

MamN exhibits similarity to several transport proteins, including some H^+-transporting proteins, which makes it tempting to speculate that the MamN protein might mediate H^+ efflux from the magnetosome compartment, as efficient buffering of the protons released during magnetite precipitation is required for the biomineralization process (Cornell 2003; Faivre et al. 2004).

MamT contains two conserved putative cytochrome c heme binding sites (Grünberg et al. 2004) and thus might represent a redox-active protein, which could be responsible for redox mediation of iron within the MM compartment (Fig. 6).

Further MMPs including MamQ, MamR and MamS, display no or only weak similarity with well-characterized proteins in organisms other than MTB (Grünberg et al. 2004), which in the absence of further experimental data precludes any functional prediction.

Different MMPs displayed differences in the strength of association with the MM. While pronase and proteinase K treatment caused total degradation of MM-associated proteins, resulting in agglomeration of particles, MamC, MamF and Mms6 remained associated with the MM after limited tryptic digestion. The incomplete digestion of these proteins might be due to their integral membrane localization and protection against protolytic degradation.

Other proteins that were previously reported to be associated with isolated magnetosomes in the closely related strain AMB-1 are MpsA and MagA (Nakamura et al. 1995a; Matsunaga et al. 2000). Although equivalent genes are present in the genome of MSR-1, they are not encoded within the magnetosome island (MAI), and extensive analysis failed so far to identify them in the MM of MSR-1. So, it seems that neither MpsA nor MagA are universal constituents of the MM and their co-purification with the magnetosomes may have been accidental, resulting from unspecific binding during magnetosome preparation as already shown for Mms16.

3.2
Genome Analysis of MTB

The genome sizes of the marine magnetic vibrios MV-1 and MV-2, and magnetic coccus strain MC-1 by pulsed-field gel electrophoresis were estimated to be approximately 3.7, 3.6 and 4.5 Mb, respectively (Dean and Bazylinski 1999). Genomic DNA of these three strains exhibits a G + C content of 52–57% (Dean and Bazylinski 1999). The genome size of *M. magnetotacticum* MS-1 was estimated to be 4.3 Mb (Bazylinski 1999; Bertani et al. 2001) with a G + C content of 64%. Whereas no evidence for plasmids was found for strains MS-1, MC-1, MV-1, native plasmids were detected in AMB-1 and MSR-1 as discussed before (Sect. 2.1). On the basis of restriction mapping and hybridization analysis, Bertani and coworkers suggested that the putative magnetosome-related gene *magA* is co-localized with the *bfr* gene encoding a subunit of the iron storage protein bacterioferritin in AMB-1 and MS-1 (Bertani et al. 2001).

In 2000 the DOE Joint Genome Institute (USA) started a sequencing effort to decipher the genomes of MC-1 and MS-1. Although as yet only incomplete draft versions of these two genomes are available, this nevertheless already provided important insights, and for the first time allowed the genome-wide identification of magnetosome genes (Grünberg et al. 2001). At the time of writing (May 2006), a draft version of the *M. gryphiswaldense* MSR-1 genome sequence is available with a coverage of more than 90%, which includes a continuous 500 kb contig encompassing the 130 kb magnetosome island (Richter et al., manuscript in preparation). Sequence determination of the first complete genome from a MTB was recently accomplished for *M. magneticum* strain AMB-1, which has a genome size of 4.9 Mb. Its G + C content differs only 1% from *M. magnetotacticum* MS-1 (65.1%), whereas 62.7% G + C was determined for MSR-1 (Sakane 1994).

Besides other noteworthy features described in detail in the work of Matsunaga et al. (in this volume), one striking characteristic is the presence of an unusually large number of chemotaxis transducer sequences (see Sect. 3.8). The complete or draft sequencing of several magnetobacterial genomes has already led not only to an improved understanding of genetic determination of magnetosome formation, but also permitted the reconstruction of other metabolic features and pathways which is described in the work of Bazylinski and Williams, and Frankel in this volume.

3.3
The Magnetosome Genes are Clustered Within a Genomic Magnetosome Island

The first indication that magnetosome genes are not scattered throughout the chromosome, but might be confined to a single genomic region, came from sequence analysis of clones encoding various MMPs (Grünberg et al. 2001), which was further corroborated by the analysis of a spontaneous non-magnetic mutant of *M. gryphiswaldense* termed strain MSR-1B (Schübbe et al. 2003). Southern blot experiments in combination with restriction analysis and pulse field gel electrophoresis indicated a large deletion of 40 kb in MSR-1B encompassing all previously described magnetosome genes. Sequence analysis of a 500 kb genomic fragment from *M. gryphiswaldense* MSR-1 later revealed that the conspicuous region harboring the magnetosome genes may extend over 130 kb (Ullrich et al. 2005). This region displays the following characteristics (i) it harbors all identified magnetosome (*mam*/*mms*) genes, (ii) it contains 42 transposase genes, (iii) most other ORFs are hypothetical genes (Fig. 3). The 130 kb region contains numerous direct and inverted repeats, which mostly correspond to similar copies of transposase genes (Figs. 3 and 5). Its G + C content is slightly distinct from that of the rest of the genome and displays a more heterogeneous distribution. Three *tRNA* genes are present within this region, which is also bounded by an integrase gene fragment. In summary, all these features are strongly reminiscent of those described for genome islands in other bacteria (Dobrindt et al. 2004). The existence of genome islands was discovered first in the case of "pathogenicity islands" (PAIs) during the course of genetic analysis of uropathogenic *E. coli* strains (McDaniel et al. 1995; Falkow 1996). The so-called PAIs were defined as regions of bacterial genomes that include a number of virulence genes and are thought to be acquired through horizontal gene transfer (HGT), since very closely related nonpathogenic strains lacked the corresponding sequences. However, the existence of genomic islands seems to extend beyond pathogenesis and also applies to "environmental" islands, encompassing a variety of functional genes or gene

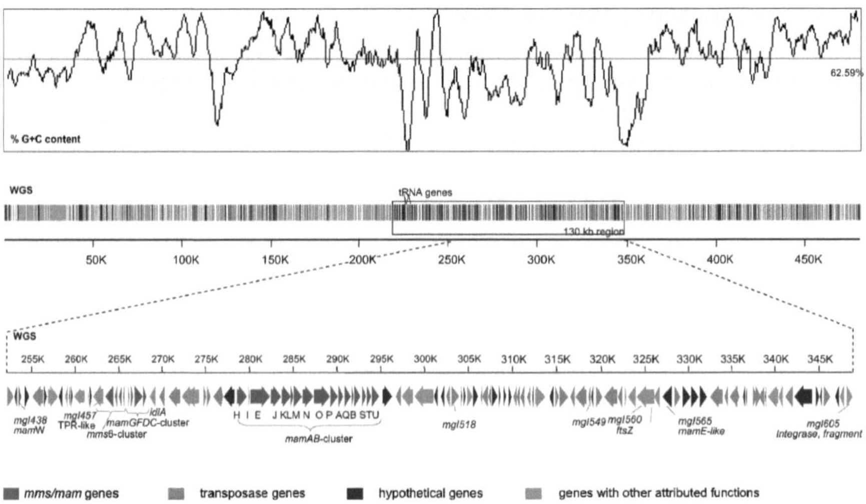

Fig. 3 Schematic representation of the 482-kb WGS genomic region. Magnetosome genes (*mms* and *mam*), hypothetical genes, tRNA genes, genes with assigned functions, and transposase genes are indicated by different colors. The distribution of the G_C content is shown in the *upper panel* (the average value for the 482-kb region is 62.59%). The extent of the conspicuous 130-kb Magnetosome Island (MAI) described in the text is indicated by a box (modified after Ullrich et al. 2005).

clusters that were acquired via HGT in nonpathogenic bacteria. By analogy, it thus seems that the 130-kb region encoding magnetosome genes represents a large genomic "magnetosome island", which may have been acquired by HTG. It is currently not known if the presence of a MAI is common to all MTB. However, there is evidence that the genomic organization of magnetosome genes is conserved in *Magnetospirillum* species and to a lesser degree in the magnetic coccus MC-1 (Schübbe et al. 2003). Most recently, a MAI was identified in *M. magneticum* AMB-1. Although its predicted size of 100 kb was slightly smaller, their molecular structure, gene content and operon organization is very similar to the MAI from MSR-1 (Figs. 3–5). Observed fluctuations in the G + C content, which was 4% below average in AMB-1 MAI, were considered as indicative of HTG events (Matsunaga et al. 2005; Fukuda et al. 2006; see also Matsunaga et al., 2006, in this volume).

3.4 The mam- and mms-Operons Encoding Magnetosome-Associated Proteins

With the exception of MamW, all identified MMPs are encoded within the *mam*- (magnetosome membrane) and *mms*- (magnetic particle membrane-specific) operons, along with several genes of unknown function, which are located within less than 35 kb of the MAI (Grünberg et al. 2001; Schübbe et al. 2003). The colinear organization and close spacing of genes within the three clusters suggested that they each might be transcribed as polycistronic operons from a single promoter. Cotranscription of genes within the *mamAB*, *mamDC* and *mms6* operons was recently experimentally confirmed indicating the presence of long polycistronic transcripts extending over more than 16 kb (Schübbe et al. 2003). In *M. gryphiswaldense* MSR-1, the *mamAB* operon encompasses 17 colinear genes extending 16.4 kb of DNA (Fig. 3). The 2.1 kb *mamGFDC* operon is located 15 kb upstream of the *mamAB* operon and comprises four genes. The 3.6 kb *mms6* operon is located 368 bp upstream of the *mamGFDC* operon and contains five genes. The transcription starting points of the *mamAB*, *mamDC*, and *mms6* operons were mapped closely upstream of the first genes in the operons, respectively. The presence of transcripts was found to be independent from growth phase. However, the expression of the *mam* and *mms* genes was up-regulated under magnetite-forming conditions, i.e., during microaerobiosis and in the presence of iron (Schübbe et al. 2003).

The operon organization of magnetosome genes is conserved among different *Magnetospirillum* strains, and to a lesser extent, also strain MC-1 (Fig. 4). Whereas the overall similarity at protein and DNA level is very high between various magnetospirilla strains, the transcriptional direction of *mms6*- and *mamGFDC*-operons is inversed in *M. gryphiswaldense* MSR-1 compared to *M. magneticum* AMB-1 and *M. magnetotacticum* MS-1. Both, se-

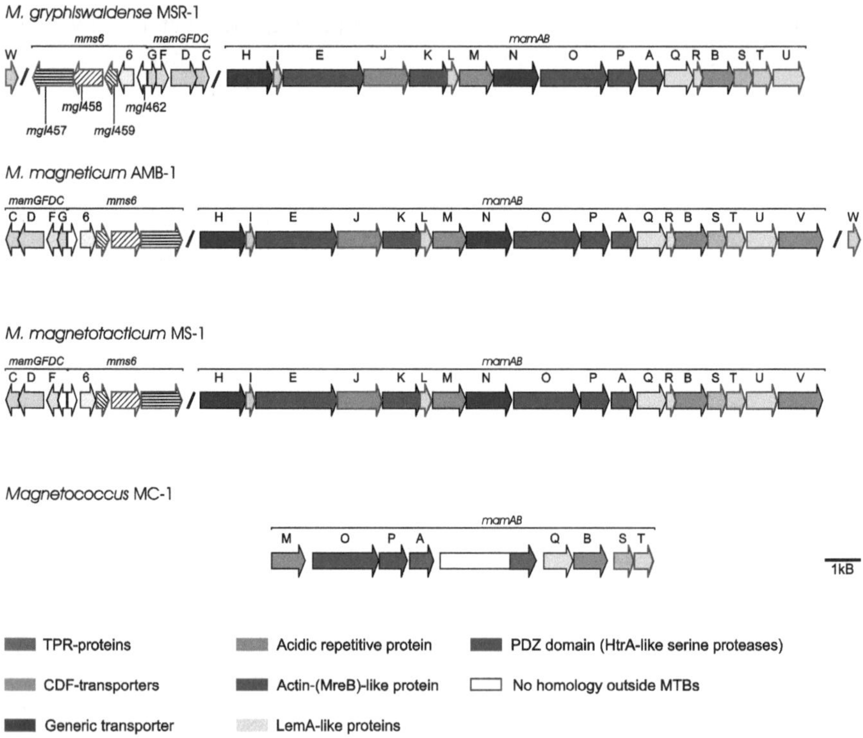

Fig. 4 Molecular organization of MAI genes encoding known magnetosome proteins in *M. gryphiswaldense* and their homologous regions in the genome of *M. magneticum* AMB-1, *M. magnetotacticum* MS-1, and the magnetic coccus strain MC-1. Various gene families shared between diverse MTB strains are indicated by different colors or hatching, equivalent genes are indicated by identical denominations. Although further homologues of *mam* and *mms* genes can be identified in the genome assembly of the magnetic coccus MC-1, only the *mamAB* operon is shown, as the genomic sequence information for other parts is incomplete

quence and organization of magnetosome genes are less conserved in strain MC-1 (Fig. 4).

3.5 The Magnetosome Island is a Highly Unstable Genomic Region and Undergoes Spontaneous Rearrangements

Substantial sequence polymorphism within the MAI was observed between clones from different subcultures of *M. gryphiswaldense* MSR-1 (Ullrich et al. 2005), suggesting that this region undergoes frequent rearrangements during serial subcultivation in the laboratory. Spontaneous mutants affected in magnetosome formation arise at a frequency of up to 10^{-2} after prolonged storage of cells at 4 °C or exposure to oxidative stress (Ullrich et al. 2005). All nonmagnetic mutants exhibited extended and multiple deletions within the MAI and had lost either parts or the entire *mms*- and *mam*-operons encoding magnetosome proteins. Mutations are polymorphic with respect to the sites and extents of deletions, but were all found associated with the loss of various copies of IS elements. These observations indicate that the genomic MAI undergoes frequent transposition events, which lead to subsequent deletions by homologous recombination under conditions of physiological stress (Schübbe et al. 2003; Ullrich et al. 2005). This could be interpreted in terms of adaptation to physiological stress and might contribute to the genetic plasticity and mobilization of the MAI.

The spontaneous loss of the ability to synthesize magnetosomes seems to be a characteristic trait common to diverse MTB and has been reported for different species (Schübbe et al. 2003; Dubbels et al. 2004; Ullrich et al. 2005). For instance, spontaneous nonmagnetic mutants were regularly observed in marine magnetotactic vibrio MV-1. All mutants failed to express a major copper-containing periplasmic protein (ChpA) presumably involved in iron uptake (Dubbels et al. 2004). Although mutants were genetically heterogeneous, they all exhibited two point mutations (transversions) at identical wobble positions within *chpA*, which apparently did prevent the translation of its transcript by an unknown mechanism. This finding suggests that different mechanisms are responsible for the appearance of spontaneous nonmagnetic mutants in different MTB.

Genomic islands are known to be frequently flanked by direct repeats (Dobrindt et al. 2004), which can participate in different RecA-dependent or RecA-independent rearrangements which typically leads to the duplication or deletion of covered sequences (Bzymek and Lovett 2001). Direct repeats flanking gene islands that originate from phage integration have short lengths. Unidirectional recombination between the phage attachment site (*attP*) and the attachment site in the bacterial genome (*attB*) is mediated through phage integrases and typically results in a duplication of *att* sites. Depending on the type of integrases, the resulting direct repeats are only a few base pairs in length (Groth and Calos 2004). Thus, the large direct repeats used for the definition of MAI borders in AMB-1 (Matsunaga et al. 2005; Fukuda et al. 2006) are unlikely to be results of a phage-mediated mechanism as suggested before. However, the direct repeats within the MAI of AMB-1 might be responsible for genetic instability of the MAI in this organism. Fukuda et al. suggested a direct repeat-dependent mechanism of integrase-mediated excision of the MAI followed by transient circularization (Fukuda et al. 2006). Interestingly, the MAI of AMB-1 exhibits a much smaller number of direct repeats than the corresponding region of *M. gryphiswaldense* MSR-1 (Figs. 4 and 5), which might be either due to subsequent duplication events in MSR-1, or to the reduction of complexity in AMB-1 by gradual loss by recombination during repeated subcultivation (Ullrich et al. 2005).

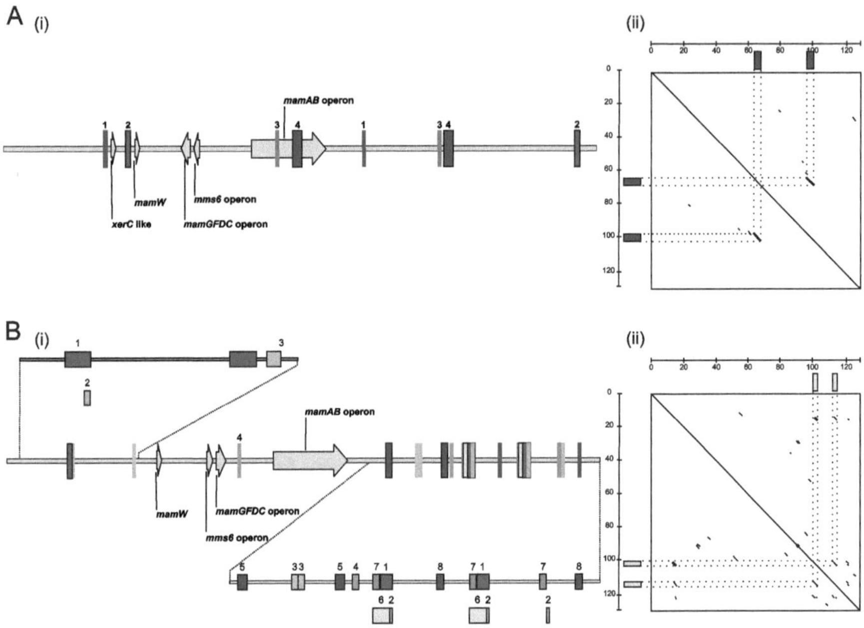

Fig. 5 Selected sequence features of the MAIs of *M. magneticum* AMB-1 (**A**) and *M. gryphiswaldense* MSR-1 (**B**). (i) Positions of the identified magnetosome operons, putative boundaries, and direct repeats are indicated by different boxes. Same colors and numbers indicate identical repeat sequences. (ii) Repeats were identified by dot blot analysis with window size 99 and mismatch limit 5

The discovery of the magnetosome island might also explain the widespread occurrence of magnetic phenotypes among various *Proteobacteria* and the phylum *Nitrospira* (Amann et al. 2004; see also Amann et al., 2006, in this volume). There are two different hypotheses for the phylogenetic origin of magnetosome genes and MTB. The first hypothesis postulated the spontaneous appearance of the magnetic phenotype independently in different lineages (DeLong et al. 1993), whereas the second hypothesis postulates a common ancestor of all magnetic bacteria, despite the distant relationship between different MTB inferred from 16S rDNA-phylogeny. According to the latter suggestion, magnetosome genes evolved only once, and different bacteria dwelling in the same habitat acquired the "magnetic genes" via horizontal gene transfer (Schübbe et al. 2003), which would have been facilitated by clustering of all required genes within a compact "magnetosome island".

3.6
Mutational Analysis of Magnetosome Formation

Random mutagenesis by conjugational transfer of Tn5 transposons into *M. gryphiswaldense MSR-1* and *M. magneticum AMB-1* (Wahyudi et al. 2001; Li et al. 2005) was used to identify genes controlling magnetosome formation (Matsunaga et al. 1992, 2005; Wahyudi et al. 2001, 2003; Calugay et al. 2004). For example, of 5762 identified Tn5 insertants, 69 were found to be defective for magnetosome formation, as reported by Matsunaga and coworkers. Disrupted genes were distributed over the entire genome and could be categorized into functional groups including signal transduction, energy metabolism, cell envelope biogenesis, cell motility, and unknown functions (Matsunaga et al. 2005). For instance, Tn5 mutagenesis led to the destruction of genes encoding an aldehyde ferredoxin oxidoreductase or a periplasmic transport binding protein kinase (Wahyudi et al. 2001; Calugay et al. 2003, 2004; Li et al. 2005; Suzuki et al. 2006). However, none of the mutants were complemented so far, and thus, the roles of these genes require further clarification. Surprisingly, neither of the previously identified *mam* and *mms* genes, which are essential for magnetosome synthesis, nor other genes located within these operons were among the affected genes in the studies mentioned above. In contrast, all transposon mutants obtained in an independent study on the same organism resulted from insertions in the *mamAB* operon (Komeili et al. 2004). Likewise, Li et al. reported a non-magnetic transposon mutant of MSR-1, which mapped within the close neighborhood of *mamW* inside the MAI. The disrupted gene shows similarity to a CheIII-like protein, supporting the notion that putative taxis-related genes are essential for maintenance of the magnetic phenotype (Li et al. 2005). These studies further corroborate the functional link of the MAI to magnetosome formation.

3.7
Reconstruction of Magnetosome Formation from Current Genetic Data

Although magnetosome formation is still not fully understood in detail, a hypothetical model can be postulated based on available physiological, molecular genetic and ultrastructural data (Fig. 6). According to this speculative model magnetosome biomineralization is preceded by vesicle formation, which originates from invaginations of the cytoplasmic membrane (A). The formation of magnetite is coupled to the uptake of large amounts of extracellular ferrous or ferric iron, which is transported into the cell by an energy-dependent process via an unknown uptake system. Supersaturating amounts of ferrous iron are transported by the MamB/MamM-proteins from the cytoplasm into the magnetosome vesicles, which are activated for magnetite formation by the MamA protein. Alternatively, iron transport may proceed via MamB/MamM directly from the periplasmic space into the MM

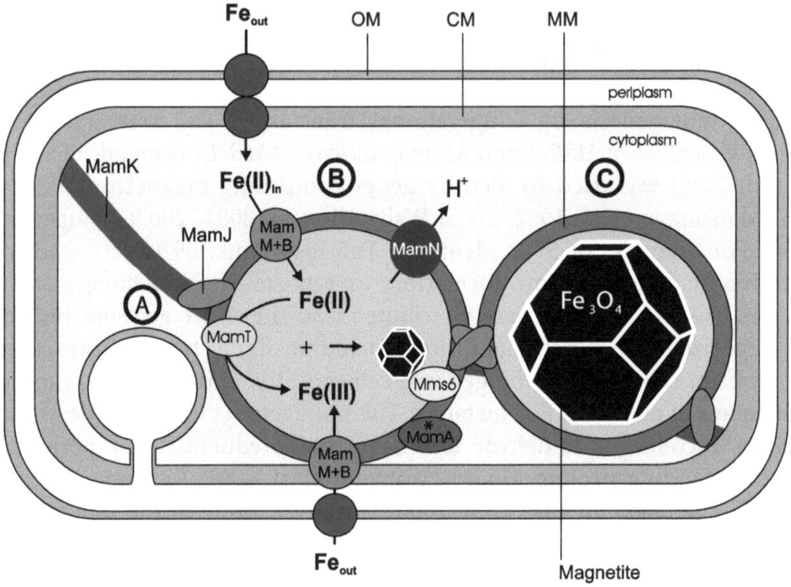

Fig. 6 Hypothetical model of magnetite biomineralization in *Magnetospirillum* strains based on available ultrastructural, genetic, and physiological data. Details are explained in the text

vesicles. Intravesicular iron is then thought to be partially reoxidized, possibly by MamT, to form a highly reactive Fe(III) oxide, which may react with dissolved Fe^{2+} to form magnetite by a via-solution process (Cornell 2003). Within the MM iron is thought to be bound by the Mms6 protein, which promotes nucleation and growth of magnetite crystals. As protons are released in stoichiometric amounts during magnetite precipitation, the solution within the MM compartment has to be sufficiently buffered to ensure that the solubility product of magnetite is always exceeded (Cornell 2003; Faivre et al. 2004). This could be accomplished by the active export of protons, possibly by the MamN protein. (B) Mature MM vesicles, which are attached to the magnetosome filament through the MamJ protein, align themselves along this cytoskeletal structure, probably interconnected and stabilized by the magnetosome membrane (C).

3.8 Genetics of Magnetotaxis: Analysis of Motility and Sensory Transduction

Although magnetotaxis at first glance seems to imply a very simple and merely passive orientation mechanism, at closer inspection MTB display a rather complex behavioral pattern. The passive ability to align cells along Earth's magnetic field lines has to be coordinated with other sensory mechanisms including aerotaxis, redoxtaxis and even phototaxis in order to decide on the "right" swimming direction along the magnetic field (Frankel et al. 1997). In addition, most MTB display an unusually efficient swimming motility based on single or multiple flagella structures (for details, see also Frankel et al., 2006, in this volume).

Although we have started only recently to get an understanding of the genetics behind magnetite crystal biomineralization, even less is known about the control of sensory transduction and motility mechanisms interacting with magnetotaxis. Circumstantial evidence for a possible genetic linkage between magnetosome formation and taxis response comes from the observation that several mutants affected in single *mam* genes or parts of the MAI were also impaired in motility and magnetotaxis (Schübbe et al. 2003; Schüler and coworkers, unpublished observations). Despite uncertainties associated with the annotation and functional prediction of chemotaxis proteins, several genes located within the MAI intriguingly display similarity to, for example, CheY-like receivers, putative signal transduction-related histidine kinases, methylases of chemotaxis methyl-accepting proteins, or contain

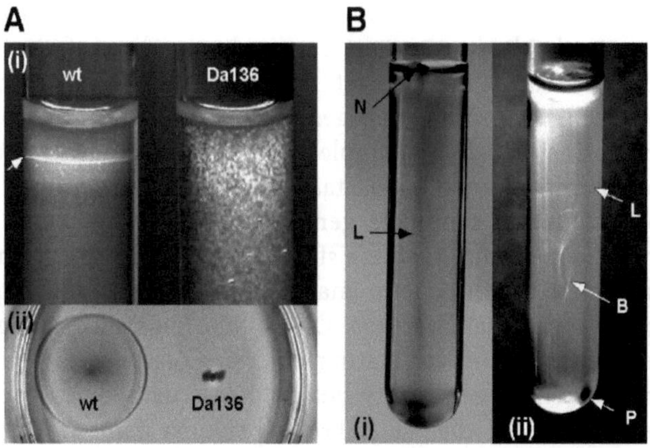

Fig. 7 Phenotype of a ΔflaA-mutant of *M. gryphiswaldense* as demonstrated by different motility assays in semisolid agar (modified after Schultheiss et al. 2004). **A** Growth in the absence of a magnetic field. At the *top* (i), wild-type (wt) *M. gryphiswaldense* grew as sharp microaerophilic bands (*arrow*) in oxygen gradients, while growth of the nonmotile mutant strain Da136 was fuzzy. At the *bottom* (ii), bacteria were stabbed into motility agar in a Petri dish. The wild-type formed a large chemotactic swarming halo (H) with a diameter of approximately 4 cm after 48 h, while no spreading of mutant strain Da136 was visible. **B** Magnetotactic patterns of wild-type *M. gryphiswaldense* in the presence of a horizontal magnetic field. On the *left side* of the *panel*, at the wall facing the magnetic South pole, cells accumulated as a line (L) leading into a nose-like tip (N) close to the agar surface. On the *right side* of the *panel*, at the opposite side (distal to the South Pole of the magnet), a cell pellet (P) was visible close to the bottom of the same tube. In addition, a spherical pattern resembling a bubble (B) was formed in the center of the tube

hemerythrin-like sequence motives, and thus, might play a role in magneto-aerotaxis (see Frankel et al., 2006, in this volume). In a genome-wide analysis Alexandre and coworkers reported a number of 65 genes encoding putative chemotaxis transducers in the genome of *M. magnetotacticum* (Alexandre et al. 2004), and a similar number was found for AMB-1 (Fukuda et al. 2006), compared to only five such genes in the genome of *E. coli.*

Initial experimental approaches to address motility and magnetotaxis were performed in *M. gryphiswaldense* MSR-1. Several unlinked flagellin genes were identified in the genome of this organism. The targeted disruption of the flagellin gene *flaA* resulted in a loss of mobility, flagella formation, and magnetotaxis. Figure 7 illustrates the phenotype of the *flaA* deletion mutant strain DA136 (Schultheiss et al. 2004). However, many details remain to be explored, and therefore the analysis of motility and signal transductions involved in magnetotaxis might represent an interesting future field of research on its own.

4
Conclusions and Future Perspectives

Despite the remarkable progress in our understanding of the genetics of magnetosome formation in the past, many open questions remain to be answered. With the realization that genetic determination of magnetosome biomineralization is amazingly complex and involves many unique and unknown gene functions, an obvious task will be the functional analysis of the numerous identified candidate genes by using a systematic mutational and biochemical approach. Given the still sluggish progress and the technical difficulties associated with genetic analysis in MTB, this will require a collective effort of several laboratories. One of the most intriguing questions to be answered is how the fascinating diversity of species–specific shapes, size and arrangement is genetically determined. Further sequence information will be required of diverse MTB to perform comparative genomics. Provided that the presence of magnetosome islands is a universal feature common to all MTB, their analysis will provide further insights into the questions of genetic diversity and phylogeny of MTB. However, comparative genomic approaches are limited by the fact that only a small minority of the impressive naturally occurring diversity of MTB can be cultivated in the lab. However, MTB can be enriched and collected selectively from environmental samples by magnetic separation without the need of prior axenic cultivation. It has already been demonstrated, that uncultivated MTB can be enriched and collected from environmental samples to virtual homogeneity and in sufficient amounts to permit the isolation of high molecular weight genomic DNA. As biomineralization genes are clustered and conserved between different species, with the advent of advanced metagenomic techniques it now seems feasible to retrieve and identify large fragments, and even entire magnetosome clusters or islands from large-insert clone libraries of uncultivated MTB. Thus, cloning of the "magnetotactic metagenome" eventually may become possible.

The heterologous expression of single genes, entire operons, or ultimately, large parts of the complete MAI will be a further promising option for the future. Cultivated magnetobacterial strains might serve as surrogate hosts for functional expression of alien magnetosome genes. Alternatively, it has been already demonstrated that magnetobacterial genes can be functionally expressed in *E. coli* (Berson et al. 1989, 1991), and other well-developed alphaproteobacterial hosts could be useful to overcome potential limitations associated with heterologous expression of biomineralization pathways in *E. coli*. For example, photosynthetic bacteria of the *Rhodospirillacaea*, which are strikingly similar to magnetotactic Alphaproteobacteria with respect to their phylogenetic position, genetic structure and the presence of intracellular membranous structures, might alternatively serve as surrogate hosts for the expression of magnetosome genes and ultimately, the entire biomineralization pathway.

A complete understanding of magnetosome formation at the molecular level also will be useful for genetic engineering of magnetosomes as biogenic nanoparticles with unique and superior properties that could be applied in biotechnological applications. For instance, genetic knowledge might facilitate in vivo functionalization of the magnetosome membrane by construction of genetic fusions with magnetosome proteins. Successful examples for this approach are documented elsewhere in this volume (see Matsunaga et al., 2006, in this volume). Furthermore, magnetic and crystalline properties of magnetic nanoparticles can be tailored by genetic engineering of crystal size and habits, which has the potential to generate innovative biomaterials with advanced properties.

Acknowledgements We would like to acknowledge our students, colleagues, and numerous collaborators. Research in the author's lab is supported by the Max Planck Society, the Deutsche Forschungsgemeinschaft and the German BMBF.

References

Alexandre G, Greer-Phillips S, Zhulin IB (2004) Ecological role of energy taxis in microorganisms. FEMS Microbiol Rev 28:113–126

Amann R, Rossello-Mora R, Flies C, Schüler D (2004) Phylogeny and in situ identification of magnetotactic bacteria. Wiley, Weinheim

Arakaki A, Webb J, Matsunaga T (2003) A novel protein tightly bound to bacterial magnetic particles in *Magnetospirillum magneticum* strain AMB-1. J Biol Chem 278:8745–8750

Bazylinski DA (1999) Synthesis of the bacterial magnetosome: the making of a magnetic personality. Int Microbiol 2:71–80

Bazylinski DA, Dean AJ, Williams TJ, Long LK, Middleton SL, Dubbels BL (2004) Chemolithoautotrophy in the marine, magnetotactic bacterial strains MV-1 and MV-2. Arch Microbiol 182:373–387

Berson AE, Hudson DV, Waleh NS (1989) Cloning and characterization of the *recA* gene of *Aquaspirillum magnetotacticum*. Arch Microbiol 152:567–571

Berson AE, Hudson DV, Waleh NS (1991) Cloning of a sequence of *Aquaspirillum magnetotacticum* that complements the *aroD* gene of *Escherichia coli*. Mol Microbiol 5:2261–2264

Bertani LE, Weko J, Phillips KV, Gray RF, Kirschvink JL (2001) Physical and genetic characterization of the genome of *Magnetospirillum magnetotacticum*, strain MS-1. Gene 264:257–263

Blakemore R (1975) Magnetotactic bacteria. Science 190:377–379

Blatch GL, Lassle M (1999) The tetratricopeptide repeat: a structural motif mediating protein–protein interactions. Bioessays 21:932–939

Bochicchio B, Pepe A, Tamburro AM (2001) On (GGLGY) synthetic repeating sequences of lamprin and analogous sequences. Matrix Biol 20:243–250

Bzymek M, Lovett ST (2001) Instability of repetitive DNA sequences: the role of replication in multiple mechanisms. Proc Natl Acad Sci USA 98:8319–8325

Calugay RJ, Miyashita H, Okamura Y, Matsunaga T (2003) Siderophore production by the magnetic bacterium *Magnetospirillum magneticum* AMB-1. FEMS Microbiol Lett 218:371–375

Calugay RJ, Okamura Y, Wahyudi AT, Takeyama H, Matsunaga T (2004) Siderophore production of a periplasmic transport binding protein kinase gene defective mutant of *Magnetospirillum magneticum* AMB-1. Biochem Biophys Res Commun 323:852–857

Carballido-Lopez R, Errington J (2003) A dynamic bacterial cytoskeleton. Trends Cell Biol 13:577–583

Choi KH, Gaynor JB, White KG, Lopez C, Bosio CM, Karkhoff-Schweizer RR, Schweizer HP (2005) A Tn7-based broad-range bacterial cloning and expression system. Nat Methods 2:443–448

Clausen T, Southan C, Ehrmann M (2002) The HtrA family of proteases: implications for protein composition and cell fate. Mol Cell 10:443–455

Cornell RMaUS (2003) The iron oxides (Structure, properties, reactions, occurrences and uses). Wiley, Weinheim

Dean AJ, Bazylinski DA (1999) Genome analysis of several marine, magnetotactic bacterial strains by pulsed-field gel electrophoresis. Curr Microbiol 39:219–225

DeLong EF, Frankel RB, Bazylinski DA (1993) Multiple Evolutionary Origins of Magnetotaxis in Bacteria. Science 259:803–806

Dobrindt U, Hochhut B, Hentschel U, Hacker J (2004) Genomic islands in pathogenic and environmental microorganisms. Nat Rev Microbiol 2:414–424

Dubbels BL, DiSpirito AA, Morton JD, Semrau JD, Neto JN, Bazylinski DA (2004) Evidence for a copper-dependent iron transport system in the marine, magnetotactic bacterium strain MV-1. Microbiology 150:2931–2945

Eden PA, Blakemore RP (1991) Electroporation and conjugal plasmid transfer to members of the genus *Aquaspirillum*. Arch Microbiol 155:449–452

Errington J (2003) Dynamic proteins and a cytoskeleton in bacteria. Nat Cell Biol 5:175–178

Faivre D, Agrinier P, Menguy N, Zuddas P, Pachana K, Gloter A, Laval J-Y, Guyot F (2004) Mineralogical and isotopic properties of inorganic nanocrystalline magnetites. Geochim Cosmochim Acta 68:4395–4403

Falkow S (1996) The Evolution of Pathogenicity in *Escherichia*, *Shigella*, and *Salmonella*. ASM Press, Washington, DC

Frankel RB, Bazylinski DA, Johnson MS, Taylor BL (1997) Magneto-aerotaxis in marine coccoid bacteria. Biophys J 73:994–1000

Fukuda Y, Okamura Y, Takeyama H, Matsunaga T (2006) Dynamic analysis of a genomic island in *Magnetospirillum* sp. strain AMB-1 reveals how magnetosome synthesis developed. FEBS Lett 580:801–812

Gorby YA, Beveridge TJ, Blakemore RP (1988) Characterization of the bacterial magnetosome membrane. J Bacteriol 170:834–841

Grass G, Otto M, Fricke B, Haney CJ, Rensing C, Nies DH, Munkelt D (2005) FieF (YiiP) from *Escherichia coli* mediates decreased cellular accumulation of iron and relieves iron stress. Arch Microbiol 183:9–18

Groth AC, Calos MP (2004) Phage integrases: biology and applications. J Mol Biol 335:667–678

Grünberg K, Muller EC, Otto A, Reszka R, Linder D, Kube M, Reinhardt R, Schüler D (2004) Biochemical and proteomic analysis of the magnetosome membrane in *Magnetospirillum gryphiswaldense*. Appl Environ Microb 70:1040–1050

Grünberg K, Wawer C, Tebo BM, Schüler D (2001) A large gene cluster encoding several magnetosome proteins is conserved in different species of magnetotactic bacteria. Appl Environ Microbiol 67:4573–4582

Guiral S, Henard V, Laaberki MH, Granadel C, Prudhomme M, Martin B, Claverys JP (2006) Construction and evaluation of a chromosomal expression platform (CEP) for ectopic, maltose-driven gene expression in *Streptococcus pneumoniae*. Microbiology 152:343–349

Handrick R, Reinhardt S, Schultheiss D, Reichart T, Schüler D, Jendrossek V, Jendrossek D (2004) Unraveling the function of the Rhodospirillum rubrum activator of polyhydroxybutyrate (PHB) degradation: the activator is a PHB-granule-bound protein (phasin). J Bacteriol 186:2466–2475

Horwitz JP, Chua J, Curby RJ, Tomson AJ, Darooge MA, Fisher BE, Mauricio J, Klundt I (1964) Substrates for Cytochemical Demonstration of Enzyme Activity. I. Some Substituted 3-Indolyl-Beta-D-Glycopyranosides. J Med Chem 53:574–575

Koch B, Jensen LE, Nybroe O (2001) A panel of Tn7-based vectors for insertion of the gfp marker gene or for delivery of cloned DNA into Gram-negative bacteria at a neutral chromosomal site. J Microbiol Methods 45:187–195

Komeili A, Li Z, Newman DK, Jensen GJ (2006) Magnetosomes are cell membrane invaginations organized by the actin-like protein MamK. Science 311:242–245

Komeili A, Vali H, Beveridge TJ, Newman DK (2004) Magnetosome vesicles are present before magnetite formation, and MamA is required for their activation. Proc Natl Acad Sci USA 101:3839–3844

Kovach ME, Elzer PH, Hill DS, Robertson GT, Farris MA, Roop RM, Peterson KM (1995) 4 New Derivatives of the Broad-Host-Range Cloning Vector Pbbr1mcs, Carrying Different Antibiotic-Resistance Cassettes. Gene 166:175–176

Kovach ME, Phillips RW, Elzer PH, Roop RM 2nd, Peterson KM (1994) pBBR1MCS: a broad-host-range cloning vector. Biotechniques 16:800–802

Li F, Li Y, Jiang W, Wang Z, Li J (2005) Cloning and functional analysis of the sequences flanking mini-Tn5 in the magnetosomes deleted mutant NM4 of *Magnetospirillum gryphiswaldense* MSR-1. Sci China C Life Sci 48:574–584

Matsunaga S, Tadokoro (1991) Magnetite formation by a magnetic bacterium capable of growing aerobically. Appl Microbiol Biot 35:651–655

Matsunaga T, Nakamura C, Burgess JG, Sode K (1992) Gene transfer in magnetic bacteria: transposon mutagenesis and cloning of genomic DNA fragments required for magnetosome synthesis. J Bacteriol 174:2748–2753

Matsunaga T, Okamura Y (2003) Genes and proteins involved in bacterial magnetic particle formation. Trends Microbiol 11:536–541

Matsunaga T, Okamura Y, Fukuda Y, Wahyudi AT, Murase Y, Takeyama H (2005) Complete genome sequence of the facultative anaerobic magnetotactic bacterium *magnetospirillum* sp. strain AMB-1. DNA Res 12:157–166

Matsunaga T, Togo H, Kikuchi T, Tanaka T (2000) Production of luciferase-magnetic particle complex by recombinant *Magnetospirillum* sp. AMB-1. Biotechnol Bioeng 70:704–709

McDaniel TK, Jarvis KG, Donnenberg MS, Kaper JB (1995) A genetic locus of enterocyte effacement conserved among diverse enterobacterial pathogens. Proc Natl Acad Sci USA 92:1664–1668

Nakamura C, Burgess JG, Sode K, Matsunaga T (1995a) An iron-regulated gene, *magA*, encoding an iron transport protein of *Magnetospirillum* sp. strain AMB-1. J Biol Chem 270:28392–28396

Nakamura C, Kikuchi T, Burgess JG, Matsunaga T (1995b) Iron-regulated expression and membrane localization of the *magA* protein in *Magnetospirillum* sp. strain AMB-1. J Biochem (Tokyo) 118:23–27

Nies DH, Silver S (1995) Ion efflux systems involved in bacterial metal resistances. J Ind Microbiol 14:186–199

Okamura Y, Takeyama H, Matsunaga T (2001) A magnetosome-specific GTPase from the magnetic bacterium *Magnetospirillum magneticum* AMB-1. J Biol Chem 276:48183–48188

Okamura Y, Takeyama H, Sekine T, Sakaguchi T, Wahyudi AT, Sato R, Kamiya S, Matsunaga T (2003) Design and application of a new cryptic-plasmid-based shuttle vector for *Magnetospirillum magneticum*. Appl Environ Microbiol 69:4274–4277

Okuda Y, Denda K, Fukumori Y (1996) Cloning and sequencing of a gene encoding a new member of the tetratricopeptide protein family from magnetosomes of *Magnetospirillum magnetotacticum*. Gene 171:99–102

Okuda Y, Fukumori Y (2001) Expression and characterization of a magnetosome-associated protein, TPR-containing MAM22, in *Escherichia coli*. FEBS Lett 491:169–173

Sakaguchi T, Arakaki A, Matsunaga T (2002) *Desulfovibrio magneticus* sp. nov., a novel sulfate-reducing bacterium that produces intracellular single-domain-sized magnetite particles. Int J Syst Evol Microbiol 52:215–221

Sakane TaAY (1994) Chemotaxonomic investigation of heterotrophic, aerobic and microaerophilic spirilla, the genera *Aquaspirillum, Magnetospirillum* and *Oceanospirillum*. Syst Appl Microbiol 17:128–134

Scheffel A, Gruska M, Faivre D, Linaroudis A, Graumann PL, Plitzko JM, Schüler D (2006) An acidic protein aligns magnetosomes along a filamentous structure in magnetotactic bacteria. Nature 440:110

Schübbe S, Kube M, Scheffel A, Wawer C, Heyen U, Meyerdierks A, Madkour MH, Mayer F, Reinhardt R, Schüler D (2003) Characterization of a spontaneous nonmagnetic mutant of *Magnetospirillum gryphiswaldense* reveals a large deletion comprising a putative magnetosome island. J Bacteriol 185:5779–5790

Schüler D (2004) Molecular analysis of a subcellular compartment: the magnetosome membrane in *Magnetospirillum gryphiswaldense*. Arch Microbiol 181:1–7

Schüler DaDAB (2006) Techniques for studying uncultured and cultured magnetotactic bacteria. ASM, Washington, DC

Schultheiss D, Handrick R, Jendrossek D, Hanzlik M, Schüler D (2005) The presumptive magnetosome protein Mms16 is a poly(3-hydroxybutyrate) granule-bound protein (phasin) in *Magnetospirillum gryphiswaldense*. J Bacteriol 187:2416–2425

Schultheiss D, Kube M, Schüler D (2004) Inactivation of the flagellin gene *flaA* in *Magnetospirillum gryphiswaldense* results in nonmagnetotactic mutants lacking flagellar filaments. Appl Environ Microb 70:3624–3631

Schultheiss D, Schüler D (2003) Development of a genetic system for *Magnetospirillum gryphiswaldense*. Arch Microbiol 179:89–94

Simon R, Priefer U, Puhler A (1983) A Broad Host Range Mobilization System for Invivo Genetic-Engineering - Transposon Mutagenesis in Gram-Negative Bacteria. Bio-Technology 1:784–791

Southward CM, Surette MG (2002) The dynamic microbe: green fluorescent protein brings bacteria to light. Mol Microbiol 45:1191–1196

Sudo S, Fujikawa T, Nagakura T, Ohkubo T, Sakaguchi K, Tanaka M, Nakashima K, Takahashi T (1997) Structures of mollusc shell framework proteins. Nature 387:563–564

Sutcliffe JG (1979) Complete nucleotide sequence of the *Escherichia coli* plasmid pBR322. Cold Spring Harb Symp Quant Biol 43(Pt 1):77–90

Suzuki T, Okamura Y, Calugay RJ, Takeyama H, Matsunaga T (2006) Global gene expression analysis of iron-inducible genes in *Magnetospirillum magneticum* AMB-1. J Bacteriol 188:2275–2279

Taoka A, Asada R, Sasaki H, Anzawa K, Wu LF, Fukumori Y (2006) Spatial Localizations of Mam22 and Mam12 in the Magnetosomes of *Magnetospirillum magnetotacticum*. J Bacteriol 188:3805–3812

Ullrich S, Kube M, Schübbe S, Reinhardt R, Schüler D (2005) A hypervariable 130-kilobase genomic region of *Magnetospirillum gryphiswaldense* comprises a magnetosome island which undergoes frequent rearrangements during stationary growth. J Bacteriol 187:7176–7184

Valdivia RH, Cormack BP, Falkow S (2006) The uses of green fluorescent protein in prokaryotes. Methods Biochem Anal 47:163–178

Wahyudi AT, Takeyama H, Matsunaga T (2001) Isolation of *Magnetospirillum magneticum* AMB-1 mutants defective in bacterial magnetic particle synthesis by transposon mutagenesis. Appl Biochem Biotechnol 91–93:147–154

Wahyudi AT, Takeyama H, Okamura Y, Fukuda Y, Matsunaga T (2003) Characterization of aldehyde ferredoxin oxidoreductase gene defective mutant in *Magnetospirillum magneticum* AMB-1. Biochem Biophys Res Commun 303:223–229

Waleh NS (1988) Functional expression of *Aquaspirillum magnetotacticum* genes in *Escherichia coli* K12. Mol Gen Genet 214:592–594

Yoshino T, Matsunaga T (2005) Development of efficient expression system for protein display on bacterial magnetic particles. Biochem Biophys Res Commun 338:1678–1681

Zurovec M, Sehnal F (2002) Unique molecular architecture of silk fibroin in the waxmoth, *Galleria mellonella*. J Biol Chem 277:22639–22647

Microbiol Monogr (3)
D. Schüler: Magnetoreception and Magnetosomes in Bacteria
DOI 10.1007/7171_042/Published online: 12 September 2006

Cell Biology of Magnetosome Formation

Arash Komeili

Department of Plant and Microbial Biology, Berkeley University of California,
111 Koshland Hall, Berkeley, CA 94720-3102, USA
komeili@nature.berkeley.edu

Abstract Magnetosome chains are the intracellular structures that allow magnetotactic bacteria to align in and navigate along geomagnetic fields (Bazylinski and Frankel 2004). These organelles are typically defined as a unit consisting of a magnetite or greigite crystal surrounded by a lipid bilayer membrane (Balkwill et al. 1980). Although these magnetic minerals are the usual targets of most studies of magnetotactic bacteria it is the magnetosome membrane that fascinates cell biologists. One of the cornerstones of cell biology has been that membrane-bound organelles are unique to eukaryotes. However, it is now known that membranous organelles exist in many prokaryotes raising the possibility that the endo-membrane system of eukaryotic cells might have originated in prokaryotes (Jetten et al. 2003; Seufferheld et al. 2003; Fuerst 2005). Magnetosomes are one of the best-studied examples of these prokaryotic organelles with the potential to be an ideal system for the study of organelle development in prokaryotes. This work provides a review of the current knowledge of magnetosomes from a cell biological perspective focusing on the composition and formation of the magnetosome membrane and the cytoskeletal framework organizing individual magnetosomes into chains.

1 The Magnetosome Membrane

From the very first description of magnetotactic bacteria it was evident that an envelope surrounds their magnetic inclusions (Blakemore 1975). A series of studies have shown that this envelope is a lipid bilayer membrane that originates from the inner membrane and contains a unique complement of proteins. These results and their implications are discussed below.

1.1 Composition and Localization of the Magnetosome Membrane

Early electron microscopic observations of uncultured magnetotactic bacteria suggested that the envelope surrounding the magnetic minerals resembled lipid bilayer membranes in their trilaminate organization and approximate thickness (Blakemore 1975). More detailed examination of cells of *Magnetospirillum magnetotacticum* MS-1 grown under iron-limited conditions revealed that magnetosomes devoid of magnetite as well as partially-filled magnetosomes within chains of fully-filled magnetosomes had the classic trilaminate structure of other lipid bilayer membranes (Gorby et al. 1988). Lipid analysis of purified magnetosomes of MS-1 as well as *Magnetospirillum gryphiswaldense* MSR-1 showed that the magnetosome membrane has a lipid composition similar to the inner membrane and distinct from the outer membrane (Gorby et al. 1988; Grunberg et al. 2004). Furthermore, several imaging techniques including freeze-fracture transmission electron microscopy (TEM) of magnetosomes also revealed that magnetosomes were often close to the cell membrane although a clear connection between the two could not be visualized (Gorby et al. 1988). These results led to a model suggesting that magnetosomes were vesicles derived from the cell membrane.

However, recent imaging of *Magnetospirillum magneticum* AMB-1 using electron cryo-tomography (ECT) has revealed that magnetosomes are not vesicles and are in fact invaginations of the inner membrane (Komeili et al. 2006). In ECT, biological samples are rapidly frozen and imaged in the absence of any fixation, embedding and physical sectioning procedures that can introduce artifacts and obscure important cellular features. The specimen is imaged at various tilt angles and the resulting two-dimensional images are algorithmically assembled into a three-dimensional reconstruction. In ECT reconstructions of AMB-1, all of the clearly imaged magnetosomes (35% of magnetosomes) were invaginations of the cell membrane. These invaginations were apparent for magnetosomes in various parts of the chain containing either no magnetite, small magnetite crystals or fully formed magnetite crystals indicating that invaginations are unlikely to be a developmental step in the formation of a mature magnetosome (Fig. 1). The remaining 65% of magnetosomes are likely to be invaginations as well. However, the imaging of their connection to the cell membrane was hampered due to the "missing wedge" of data resulting from technical limitations that prevent the collection of a full tilt series of images. Together these results were taken to mean that, at least for this organism, magnetosomes are permanent invaginations of the cell membrane and thus should not be classified as vesicles.

Previous studies had dismissed the possibility that magnetosomes could be invaginations for several reasons. A freeze-fracture analysis of MS-1 did not reveal any connections between the magnetosome membrane and the cell membrane (Gorby et al. 1988). However, the invaginations observed in ECT

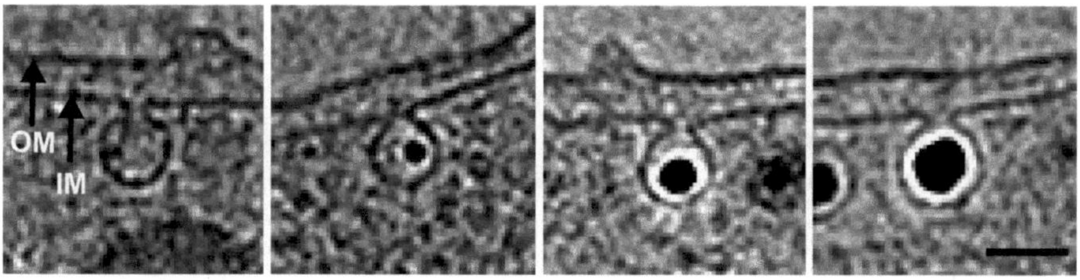

Fig. 1 ECT reconstructions of the AMB-1 reveal that magnetosomes are cell membrane invaginations. Individual magnetosomes with no magnetite, small, medium, and full-sized magnetite crystals are inner membrane (IM) invaginations. The outer membrane (OM) is also marked. Image courtesy of Science magazine

are so small that they would be missed in most fracture angles. Also, since magnetite crystals were not ejected from magnetosomes upon spheroplasting of cells it was assumed that there was no opening between the magnetosome and the periplasm (Gorby et al. 1988). Again the small size of the opening resulting from invaginations would probably prevent the expulsion of magnetite crystals. Furthermore, it is known that magnetosome proteins tightly associated with the magnetite crystal are likely responsible for its biomineralization (Arakaki et al. 2003). It is possible that this organic matrix could hold the magnetite crystal tightly within the magnetosome membrane preventing its expulsion. An additional objection could be that the mixing of the contents between the periplasm and the magnetosome could disrupt the chemical environment required for magnetite production. However, it is possible that there are "purifying" enzymes and transporters within the magnetosome membrane responsible for maintaining the proper chemical environment required for magnetite biomineralization. Also, specific protein complexes not imaged by ECT, could possibly plug the opening between the magnetosome and the periplasm. Finally, it is possible that at a low rate magnetosome membrane invaginations can develop into vesicles and that these rare occurrences were not imaged in this study. Despite these objections, the ECT imaging of AMB-1 shows that the magnetosome membrane originates from the inner membrane and provides the first positive evidence for the existence of magnetosome membrane invaginations. It remains to be seen if this is a common characteristic of magnetosomes in all magnetotactic bacteria. In a separate ECT study the magnetosome chains of MSR-1 were also found to be close to the membrane although it was not reported if they were indeed cell membrane invaginations (Scheffel et al. 2006).

In addition to challenging a long-held view, this finding also provides possible answers to some outstanding questions regarding magnetosomes. It has been long hypothesized that a ferrihydrite precursor of magnetite is formed in the periplasm, transported to the magnetosome and then partially reduced to form magnetite (Frankel et al. 1983). Such a model is problematic since it requires a mineral to be transported across two biological membranes (the cell membrane and the magnetosome membrane). But since magnetosomes appear to be membrane invaginations, any mineral precursors to magnetite formed in the periplasm can be directly shuttled into the magnetosome. Magnetosome invaginations also present a simple method for connecting the magnetosome chain to the cell membrane thus allowing for the passive orientation of cells in magnetic fields.

Finally, the conclusion that magnetosomes are not vesicles complicates the superficial comparisons that have been drawn between magnetosomes and the membrane vesicles and compartments of eukaryotic cells. As a result, a mechanistic understanding of magnetosome membrane formation is required in order to properly evaluate the relationship between magnetosomes and eukaryotic organelles.

1.2 Molecular Mechanisms of Magnetosome Membrane Formation

Two general models can be envisioned for the formation of magnetosome membrane invaginations from the cell membrane (Fig. 2). In the first model, specific proteins assemble on the cytoplasmic face of the inner membrane and bend the membrane to form invaginations. This model is similar to the mechanism of membrane budding during vesicle biogenesis by coat proteins in eukaryotic cells (Bonifacino and Glick 2004). In the second model, periplasmic domains of inner membrane proteins interact like a zipper to invaginate the cell membrane. Examples of this mechanism have been observed in *E. coli* cells overexpressing methyl-accepting chemotaxis proteins and are thought to be responsible for the formation of photosynthetic membranes in various bacteria (Lefman et al. 2004).

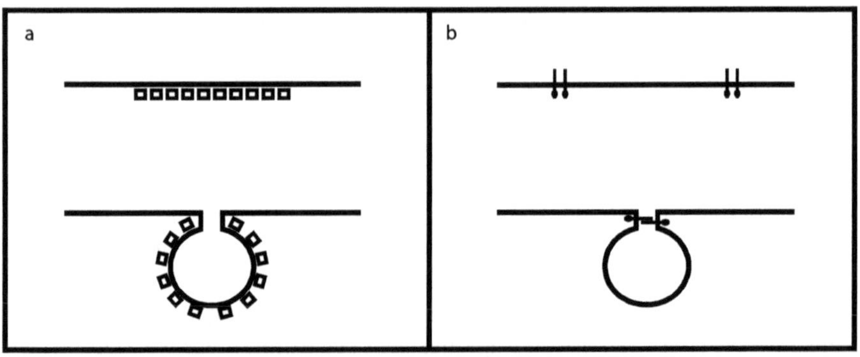

Fig. 2 Two possible models for the formation of magnetosome membrane invaginations. Coat proteins assemble on the inner membrane and bend the membrane to form an invagination (**a**). Interactions between the periplasmic domains of an inner membrane protein provide the force for the formation of the membrane invaginations (**b**)

Clearly, the first step in distinguishing between these two models is to isolate the proteins required for the formation of the magnetosome membrane. Through the convergence of genomic, proteomic and genetic techniques a great number of genes, many of which are grouped within a genomic island, have been predicted to participate in the formation and functioning of the magnetosome (Schubbe et al. 2003; Grunberg et al. 2004; Komeili et al. 2004; Ullrich et al. 2005; Fukuda et al. 2006). The current challenge is to place these genes within the various steps that lead to the formation of a functioning magnetosome. One of the most powerful and definitive methods for assigning more specific functions to these genes is through the disruption of individual genes and subsequent analysis of the mutant phenotype. Although the methodology for making such mutants is well established for both AMB-1 and MSR1, few secondary screens exist to identify their specific dysfunction (Matsunaga et al. 1992; Komeili et al. 2004; Schultheiss et al. 2004). This is especially important in identifying genes that function in the formation of the magnetosome membrane since mutants in these genes will have the same general phenotype (i.e. nonmagnetic cells) as mutants deficient in the production of magnetite crystals. Empty magnetosome membranes in the absence of magnetite are not easily imaged by most electron microscopic techniques making it difficult to distinguish between these two types of mutants. Two different imaging techniques, cryo-ultramicrotomy and ECT, have been used recently to successfully image chains of empty magnetosomes in cells grown under iron-poor conditions (Komeili et al. 2004, 2006). Interestingly, one spontaneous mutant of AMB-1, missing many of the genes within the magnetosome island, was analyzed by ECT and shown to lack the magnetosome membrane suggesting that some of the genes responsible for the formation of the magnetosome membrane lie within this genomic region (Komeili et al. 2006).

It has also been suggested that the Mms16 protein of AMB-1 might play a role in the formation of the magnetosome membrane (Okamura et al. 2001). Mms16 displays GTPase activity in vitro and addition of aluminum fluoride—a compound that can inhibit GTPases in general—to the cells impairs the ability to form magnetite such that gaps are visible in the magnetosome chain in TEMs of whole cells. These observations were taken to mean that Mms16 is required for the formation of the magnetosome membrane (Okamura et al. 2001). However, in these studies no detailed imaging analysis of aluminum fluoride-treated cells was conducted to specifically investigate the presence of the magnetosome membrane and it is unclear if the aluminum fluoride treatment directly targeted Mms16. In addition, deletions of *mms16* in MSR-1 have no effect on magnetosome formation and convincing evidence exists suggesting that it plays a direct role in depolymerization of polyhydroxybutarate granules making it unlikely that Mms16 is necessary for the formation of the magnetosome membrane (Schultheiss et al. 2005).

1.3 Protein Sorting to the Magnetosome Membrane

In addition to its distinctive morphological appearance as a membrane invagination, the magnetosome membrane also has a unique biochemical signature distinguishing it from other cellular compartments. Various studies have shown that approximately 20–40 proteins, including both soluble and predicted transmembrane proteins, are enriched in the magnetosome membrane (see Jogler and Schüler 2006, in this volume, for details) (Grünberg et al. 2004; Matsunaga et al. 2005). These studies have relied on the biochemical fractionation of cell lysates of magnetotactic bacteria followed by N-terminal sequencing or mass spectrometric identification of magnetosome membrane proteins. Furthermore, hybrid proteins created by GFP fusions to MamA and MamK in AMB-1 as well as MamJ in MSR-1 localize as a thin straight line within the cell extending from pole to pole (Komeili et al. 2004, 2006; Scheffel et al. 2006). In spontaneous nonmagnetic mutants lacking the magnetosome island genes (as well as magnetosomes) MamA-GFP and MamJ-GFP fusions fail to localize into a line and are dispersed throughout the cell suggesting that the localization observed in wild-type cells is a magnetosome-specific pattern (Komeili, unpublished results; Scheffel et al. 2006). Additionally, the localization of MamA-GFP in AMB-1 appears to be dynamic switching from a straight line in exponential phase to a punctate pattern in stationary phase. The significance of this dynamic localization behavior with respect to magnetosome function is unknown at the moment. With a list of magnetosome-specific proteins in hand, the next significant task is to identify magnetosome localization sequences on these proteins and isolate the cellular machinery responsible for their sorting to the magnetosome membrane.

2 Assembly of the Magnetosome Chain

Once individual magnetosome membrane invaginations have been formed, they must be assembled and stabilized as chains within the cell. Below, the current knowledge of the magnetosome cytoskeleton, the process of forming a magnetic chain and the possible coordination of these events within the framework of the cell cycle are examined.

2.1 The Magnetosome Cytoskeleton

The observation that magnetosomes are invaginations of the cell membrane implies that the magnetosome chain can be attached to the cell independent of a cytoskeletal framework. However, the assembly of individual magnetosomes into chains and the maintenance of this structure within the cell might rely on a cytoskeleton. Two recent reports combining ECT and genetic analysis have shed light on the possible mechanisms organizing magnetosomes into chains within the cell.

In these reports, ECT imaging of both AMB-1 and MSR-1 revealed the presence of cytoskeletal filaments in close association with magnetosome chains (Komeili et al. 2006; Scheffel et al. 2006). These filaments run alongside multiple magnetosomes and had dimensions stereotypical of the bacterial actin-like protein MreB. MreB has been implicated in the establishment of cell polarity, cell shape determination and chromosome segregation in a variety of bacteria (Gitai 2005). In vivo MreB homologues seem to form filaments within cells and purified MreB can form filaments in vitro. The crystal structures of MreB and its homologue ParM look remarkably similar to actin implying that these bacterial proteins are the evolutionary ancestors of eukaryotic actin (van den Ent et al. 2002). True homologues of *mreB* can be found in the sequenced genomes of *Magnetospirillum magnetotacticum* MS-1, *Magnetococcus* sp. MC-1 and AMB-1 in predicted operons with other *mre* genes. A homologue of *mreB* called *mamK* is also found within the magnetosome islands of MC-1, MS-1, MSR-1 and AMB-1 making it a logical choice for the magnetosome-specific cytoskeleton. Strains harboring deletions of *mamK* produced magnetosome membrane invaginations capable of forming magnetite but were defective in organizing these magnetosomes into chains (Komeili et al. 2006). Instead, individual magnetosomes were dispersed throughout the cell and no filaments were associated with these magnetosomes (Komeili et al. 2006). These results suggest that MamK is either the structural component of the magnetosome cytoskeleton or important in its formation.

An independent set of experiments, focused on MamJ, a protein encoded by the gene immediately preceding *mamK* (Scheffel et al. 2006). Although MamJ does not share significant overall homology with any other protein it is similar to biomineralization proteins in that it contains repeats of acidic amino acids. Surprisingly, MSR-1 cells harboring deletions in *mamJ* were still able to make magnetite crystals but had an unusual morphology to their magnetosome chains. In these mutants the magnetosome chain was collapsed into an aggregate of magnetite-containing magnetosomes within the cell. ECT reconstructions of MSR-1 revealed that long cytoskeletal filaments were present alongside magnetosomes similar to those observed in AMB-1. However, in the *mamJ* deletion strain magnetosomes were not in contact with these filaments. Previous studies have shown that MamJ is localized to the magnetosome membrane leading to the simple and attractive model that MamJ links magnetosomes to the cytoskeletal filaments observed in ECT (Scheffel et al. 2006).

These two studies represent a significant step forward in defining the cell biological parameters that control the formation of a magnetosome chain. They also raise several intriguing questions. First, although MamJ localizes to the magnetosome membrane it is predicted to be a soluble protein. Therefore, it is likely that other magnetosome membrane proteins might be involved in anchoring MamJ to the magnetosome membrane. Second, if MamK and MamJ were interacting proteins working together to align the magnetosomes into chains then their deletions should produce the same phenotype. Instead, in the Δ*mamK* mutant of AMB-1 magnetosomes are dispersed throughout the cell whereas in the Δ*mamJ* mutant of MSR-1 magnetosomes are clustered as subcellular aggregates (Komeili et al. 2006; Scheffel et al. 2006). This might indicate that the two proteins have distinct functions with respect to magnetosome chain formation. On the other hand, as discussed in the next section, this phenotypic difference might be due to inherent differences in the process of chain formation in *Magnetospirillum magneticum* AMB-1 and *Magnetospirillum gryphiswaldense* MSR-1.

2.2
Formation of a Magnetic Chain

The ECT studies on AMB-1 and MSR-1 revealed many of the common ultrastructural similarities between these two closely related organisms such as the presence of empty magnetosomes and a magnetosome-associated cytoskeleton. However, these two studies also revealed fundamental differences in the process of magnetosome chain formation. AMB-1 cells grown in iron-poor conditions are nonmagnetic yet they contain long chains of empty magnetosomes. When iron is added to these cultures magnetite formation is rapidly initiated within multiple adjacent magnetosomes and proceeds at the same rate until full-sized magnetite crystals are formed (Komeili et al. 2004). The process of chain formation appears to be very different in MSR-1. First, in wild-type cells grown in intermediate iron-levels magnetosomes are seen throughout the cytoplasm with a magnetite-containing chain in the middle of the cell (Scheffel et al. 2006). In AMB-1, however, magnetosomes are rarely observed outside of the magnetosome chain. Second, upon the addition of iron to iron-starved cultures of MSR-1, magnetite is initially formed in magnetosomes dispersed throughout the cell and as magnetite crystals grow in size magnetic interactions bring the magnetite-containing magnetosomes to the center of the cell forming a magnetosome chain (Scheffel et al. 2006). Thus, in AMB-1 the magnetosome chain can form independent of magnetite production whereas in MSR-1 magnetic interactions are required for proper chain formation (Fig. 3).

Variability in magnetosome chain morphology is one of the common features of magnetotactic bacteria in nature. There are many examples of magnetotactic bacteria that contain more than one chain per cell localized either

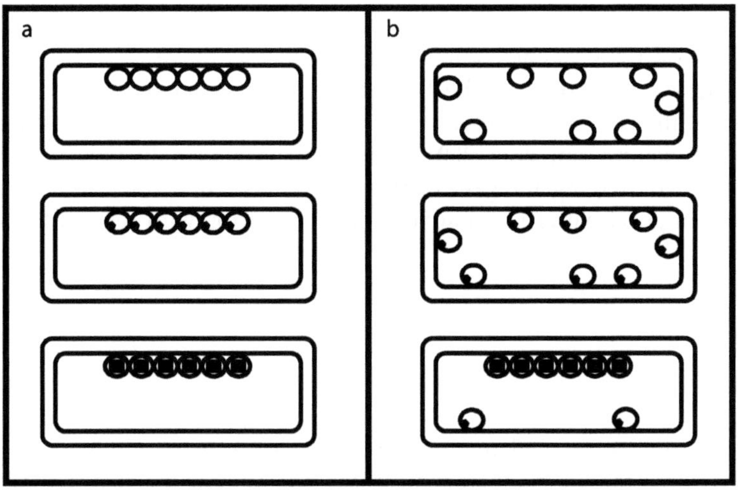

Fig. 3 Schematic of chain formation in AMB-1 and MSR-1. In AMB-1, empty magnetosomes are arranged as a chain prior to magnetite formation (**a**). In contrast, magnetosomes of MSR-1 are more dispersed and require magnetic interactions to form a chain (**b**)

on the same side or on opposite sides of the cell (Blakemore 1982). The differences observed between MSR-1 and AMB-1 can be due to variations in growth conditions between the two organisms or they might be reflective of more fundamental differences in the process of chain formation. Identification of the source of these differences might provide a valuable clue into the environmental and molecular mechanisms of the widespread variation in natural samples of magnetotactic bacteria.

2.3 Cell Cycle Control of Magnetosome Formation

In the above experiments, shifts from iron-poor to iron-rich conditions provided a reliable method of examining the dynamics of magnetite formation within magnetosomes. These experiments, however, are not appropriate for studying the formation of the magnetosome membrane invaginations since the magnetosome membrane is already present in iron-poor conditions. In fact, budding intermediates of a mature magnetosome membrane have never been observed microscopically. It is known that the number of magnetosomes is relatively constant per cell and that the magnetosome chain is divided down the middle during the cell division. Thus, newly divided cells must double their magnetosomes during each cell cycle. The dynamics of magnetite formation were examined across the cell cycle in cultures of AMB-1 synchronized by repeated cycles of alternating incubations in cold and room temperature conditions (Yang et al. 2001). The synchronized cells were examined for the presence of magnetosomes as well as the expression of *magA*, a gene encoding a magnetosome protein with a putative role in iron transport. These cells contained the highest number of magnetite crystals near the end of the cell cycle whereas the expression of *magA* was lowest just before cell division (Yang et al. 2001). These intriguing results pave the way for more detailed imaging and gene expression analyses to define the precise spatial and temporal pattern of magnetosome formation with respect to the cell cycle.

3
Discussion

The experiments detailed in this review mark the beginnings of an exciting new direction in magnetosome research. They show that a functional magnetosome membrane is present prior to magnetite formation and that individual magnetosomes are assembled into chains along cytoskeletal filaments through the combined action of the acidic protein MamJ and the actin-like protein MamK. From this perspective, it is clear that the magnetosome membrane is an independent organelle that provides the proper biochemical environment for magnetite biomineralization. Despite these promising advances, there are many important questions surrounding the cell biology of magnetosomes that remain unanswered. The molecular mechanisms of magnetosome membrane formation and protein sorting, the specific function and dynamics of the magnetosome cytoskeleton and the precise timing of magnetosome formation across the cell cycle are unknown at the moment.

Although answering these questions requires an almost exclusive focus on the "organic" phase of magnetosomes, it is a perfect complement to ongoing efforts aimed at understanding and exploiting its "inorganic" phase. For instance, the discovery that the magnetosome membrane is a cell membrane invagination provides clues into the mechanisms of iron transport into and mineral formation within magnetosomes.

The cell biological approach to understanding magnetosomes has also shown promise in the study of general prokaryotic and eukaryotic cell biology. MamK is a member of the MreB family of bacterial actin-like proteins that have been implicated in a variety of cellular functions including cell shape determination and chromosome segregation. Unlike most MreB proteins, MamK is not essential for survival meaning that it can be used as a model system to investigate the in vivo dynamics of bacterial actin proteins. The peculiar lateral organization of the magnetosome chain also provides an opportunity to study the mechanisms of symmetry breaking in prokaryotes. Finally, a more complete picture of magnetosome cell biology will allow for meaningful comparisons to be drawn between magnetosomes and other prokaryotic and eukaryotic organelles.

References

Arakaki A, Webb J, Matsunaga T (2003) A novel protein tightly bound to bacterial magnetic particles in Magnetospirillum magneticum strain AMB-1. J Biol Chem 278:8745–8750

Balkwill DL, Maratea D, Blakemore RP (1980) Ultrastructure of a magnetotactic spirillum. J Bacteriol 141:1399–1408

Bazylinski DA, Frankel RB (2004) Magnetosome formation in prokaryotes. Nat Rev Microbiol 2:217–230

Blakemore R (1975) Magnetotactic bacteria. Science 190:377–379

Blakemore RP (1982) Magnetotactic bacteria. Annu Rev Microbiol 36:217–238

Bonifacino JS, Glick BS (2004) The mechanisms of vesicle budding and fusion. Cell 116:153–166

Frankel RB, Papaefthymiou GC, Blakemore RP, Obrien W (1983) Fe_3O_4 Precipitation in Magnetotactic Bacteria. Biochimica Et Biophysica Acta 763:147–159

Fuerst JA (2005) Intracellular Compartmentation in Planctomycetes. Annu Rev Microbiol 59:299–328

Fukuda Y, Okamura Y, Takeyama H, Matsunaga T (2006) Dynamic analysis of a genomic island in Magnetospirillum sp. strain AMB-1 reveals how magnetosome synthesis developed. FEBS Lett 580:801–812

Gitai Z (2005) The new bacterial cell biology: moving parts and subcellular architecture. Cell 120:577–586

Gorby YA, Beveridge TJ, Blakemore RP (1988) Characterization of the bacterial magnetosome membrane. J Bacteriol 170:834–841

Grünberg K, Müller EC, Otto A, Reszka R, Linder D, Kube M, Reinhardt R, Schüler D (2004) Biochemical and proteomic analysis of the magnetosome membrane in Magnetospirillum gryphiswaldense. Appl Environ Microbiol 70:1040–1050

Jetten MS, Sliekers O, Kuypers M, Dalsgaard T, van Niftrik L, Cirpus I, van de Pas-Schoonen K, Lavik G, Thamdrup B, Le Paslier D, Op den Camp HJ, Hulth S, Nielsen LP, Abma W, Third K, Engstrom P, Kuenen JG, Jorgensen BB, Canfield DE, Sinninghe Damste JS, Revsbech NP, Fuerst J, Weissenbach J, Wagner M, Schmidt I, Schmid M, Strous M (2003) Anaerobic ammonium oxidation by marine and freshwater planctomycete-like bacteria. Appl Microbiol Biotechnol 63:107–114

Komeili A, Vali H, Beveridge TJ, Newman DK (2004) Magnetosome vesicles are present before magnetite formation, and MamA is required for their activation. Proc Natl Acad Sci USA 101:3839–3844

Komeili A, Li Z, Newman DK, Jensen GJ (2006) Magnetosomes are cell membrane invaginations organized by the actin-like protein MamK. Science 311:242–245

Lefman J, Zhang P, Hirai T, Weis RM, Juliani J, Bliss D, Kessel M, Bos E, Peters PJ, Subramaniam S (2004) Three-dimensional electron microscopic imaging of membrane invaginations in Escherichia coli overproducing the chemotaxis receptor Tsr. J Bacteriol 186:5052–5061

Matsunaga T, Nakamura C, Burgess JG, Sode K (1992) Gene transfer in magnetic bacteria: transposon mutagenesis and cloning of genomic DNA fragments required for magnetosome synthesis. J Bacteriol 174:2748–2753

Matsunaga T, Okamura Y, Fukuda Y, Wahyudi AT, Murase Y, Takeyama H (2005) Complete Genome Sequence of the Facultative Anaerobic Magnetotactic Bacterium Magnetospirillum sp. strain AMB-1. DNA Res 12:157–166

Okamura Y, Takeyama H, Matsunaga T (2001) A magnetosome-specific GTPase from the magnetic bacterium Magnetospirillum magneticum AMB-1. J Biol Chem 276:48183–48188

Scheffel A, Gruska M, Faivre D, Linaroudis A, Plitzko JM, Schuler D (2006) An acidic protein aligns magnetosomes along a filamentous structure in magnetotactic bacteria. Nature 440:110–114

Schübbe S, Kube M, Scheffel A, Wawer C, Heyen U, Meyerdierks A, Madkour MH, Mayer F, Reinhardt R, Schüler D (2003) Characterization of a spontaneous nonmagnetic mutant of Magnetospirillum gryphiswaldense reveals a large deletion comprising a putative magnetosome island. J Bacteriol 185:5779–5790

Schultheiss D, Kube M, Schüler D (2004) Inactivation of the flagellin gene flaA in Magnetospirillum gryphiswaldense results in nonmagnetotactic mutants lacking flagellar filaments. Appl Environ Microbiol 70:3624–3631

Schultheiss D, Handrick R, Jendrossek D, Hanzlik M, Schüler D (2005) The presumptive magnetosome protein Mms16 is a poly(3-hydroxybutyrate) granule-bound protein (phasin) in Magnetospirillum gryphiswaldense. J Bacteriol 187:2416–2425

Seufferheld M, Vieira MC, Ruiz FA, Rodrigues CO, Moreno SN, Docampo R (2003) Identification of organelles in bacteria similar to acidocalcisomes of unicellular eukaryotes. J Biol Chem 278:29971–29978

Ullrich S, Kube M, Schübbe S, Reinhardt R, Schüler D (2005) A hypervariable 130-kilobase genomic region of Magnetospirillum gryphiswaldense comprises a magnetosome island which undergoes frequent rearrangements during stationary growth. J Bacteriol 187:7176–7184

Van den Ent F, Moller-Jensen J, Amos LA, Gerdes K, Lowe J (2002) F-actin-like filaments formed by plasmid segregation protein ParM. Embo J 21:6935–6943

Yang CD, Takeyama H, Tanaka T, Hasegawa A, Matsunaga T (2001) Synthesis of bacterial magnetic particles during cell cycle of Magnetospirillum magneticum AMB-1. Appl Biochem Biotechnol 91–93:155–160

Microbiol Monogr (3)
D. Schüler: Magnetoreception and Magnetosomes in Bacteria
DOI 10.1007/7171_043/Published online: 8 September 2006

Mineralogical and Isotopic Properties of Biogenic Nanocrystalline Magnetites

Damien Faivre[1] (✉) · Pierpaolo Zuddas[2]

[1]Max Planck Institute for Marine Microbiology, Celsiusstrasse 1, 28359 Bremen, Germany
dfaivre@mpi-bremen.de

[2]Institut de Physique du Globe de Paris, 4 place Jussieu, 75252 Paris Cedex 5, France

Abstract Determination of the origin of magnetite nanocrystals is of primary importance because of their significance as biomarkers for extraterrestrial life and as environmental indicators. A critical analysis of the literature indicates that morphology and magnetic properties of the crystals do not necessarily quantitatively allow differentiation of biogenic from abiotic nanomagnetite crystals. Mineralogical properties of magnetosomes and of inorganic crystals such as size and shape factors and their distributions, morphology and defects and twinning are presented and compared in this chapter. Isotopic properties and the fractionation of oxygen and iron isotopes of the nanosized particles are reviewed. These properties are then examined as potential tools if the process and conditions formation responsible for their genesis are known. Exploration of properties such as crystal size distributions and oxygen isotope fractionation at given temperature seems to allow the discrimination of biogenic from abiotic nanocrystals of magnetite.

1 Life on Mars or the "Boosting Story" for Studying Magnetosomes from Magnetotactic Bacteria

The possibility of extraterrestrial life has fascinated human beings for centuries. In 1908, the astronomer Percival Lowell erroneously thought he had discovered hundreds of canals on the surface of Mars, which was the beginning of an exciting century of search for life on Mars. So it came as a great disappointment when the Viking spacecraft failed to detect any traces of life during the first visit to Mars in 1976. In 1996, however, McKay et al. (1996) proposed that nanoparticles of magnetite found in the Martian meteorite ALH84001 could have a biogenic origin. This news received much public attention, but also strongly increased research interest in magnetotactic bacteria. These, which were supposedly the living organisms that had made those tiny magnetic crystals, ironically, had not gained comparable publicity even after the report of their discovery in 1975 by Blakemore (1975).

In fact, although the ancient life on Mars hypothesis has been very extensively challenged, it has inspired numerous studies in the fields of geology, crystallography and mineralogy on these nanomagnetite crystals formed by magnetotactic bacteria (see among others the following contradictory papers: Golden et al. 2004; Thomas-Keprta et al. 2000). Nowadays, the interests are much broader, reaching biology and medicine because of possible application in biotechnology and nanotechnology (Bahaj et al. 1998a, b; Lowe 2000; Safarik and Safarikova 2002; Sarikaya et al. 2003; Tartaj et al. 2003) and because magnetosomes may serve as a model in understanding magnetic nanoparticles as magnetic field receptors in higher organisms (Courtillot et al. 1997; Deutschlander et al. 1999; Lohmann and Johnsen 2000; Maher 1998; Wiltschko et al. 2002; Winklhofer et al. 2001) or even humans (Dobson and Grassi 1996; Dunn et al. 1995).

It needs to be stressed that magnetotactic bacteria are not the only microorganisms able to form nanocrystalline magnetite. Extracellular or even intracellular bioinduced formation of magnetite nanoparticles has also been reported (for reviews, see Frankel and Bazylinski 2003 and and Coker et al., this volume). Here, we will focus on magnetosomes (i.e. the nanomagnetite crystals and their membrane envelope) and especially those which have magnetite as the crystalline form. Since it was difficult for a long time to work with the biological crystals formed under controlled conditions, we will first present what can be learnt from inorganic precipitation of magnetite. We will subsequently focus on the mineralogical properties of the magnetosomes, and particularly their size, size distribution and shape factor, on the one hand, and their morphology, on the other hand. Then, we will present some reports on the isotopic fractionation of the oxygen and the iron isotopes of biogenic magnetite particles. Finally, we will briefly discuss whether these properties can be used as a criterion to determine the origin of a given crystal. We conclude that through the association or the combination of these approaches, it will be possible to better interpret magnetite fossil traces in sediments and also in the Martian meteorite ALH 84001.

2 "Beyond Biomimetism" or What Can We Learn from Inorganic Synthesis?

Human beings were always amazed by the high degree of sophistication and miniaturization found in natural materials. Nature is a school for materials science and its associated disciplines such as chemistry, biology, physics and engineering. Trying to find synthetic pathways to obtain materials that have properties similar to those produced naturally is therefore called "biomimetism". We called our approach "beyond biomimetism", because instead of only trying to obtain inorganic crystals that are similar to the biogenic ones, we also aimed to deduce the conditions in which biogenic crystals are formed. Therefore, we assumed that if inorganic crystals present similarities to biogenic ones, this implied that they were obtained in analogous environments (Faivre et al. 2004).

Abundant synthesis methods allowing inorganic magnetite formation in aqueous media without any control have been reported (Cornell and Schwertmann 1996, 2003; Devouard et al. 1998). However, specific crystal properties cannot, in this case, be correlated to the formation conditions as the chemical affinity changes during the crystal growth. The chemical affinity is in fact a key parameter to be controlled as it represents the capacity of a reaction to happen. A slight change in chemical affinity might therefore have a drastic effect on the reaction products and their properties. Particles obtained by those synthetic pathways have, for example, various morphologies and dimensions that lead to broad crystal size distributions (CSDs) (Devouard et al. 1998). The chemical affinity is, therefore, the key parameter for the synthesis control to better understand in which conditions the magnetotactic bacteria form their magnetosomes.

Hereafter we give a brief theoretical explanation for this beyond-biomimetism approach. The rate and the mechanism of heterogeneous reactions are a function of the reaction chemical affinity A. Classically, mineral dissolution and precipitation reaction rates have most often been expressed in terms of a functional dependence. Using the formalism (Lasaga 1998), we can link the rate of reaction R_{net} to the temperature by

$$R_{\text{net}} = R_{+}\left(1 - e^{-A/RT}\right) , \tag{1}$$

where R is the perfect gas constant, R_{+} the rate for the forward reaction in Eq. 2 and T is the temperature (in Kelvin).

A variation of the chemical affinity might have an important effect on the global rate of the reaction affecting the reaction pathway and the mechanism of mineral formation. These changes in reaction mechanism and rates will in turn affect the geochemical properties of the product of the reaction, in our case the mineral formed.

A series of controlled experimental magnetite precipitations can be carried out by coprecipitation of ferrous and ferric ions in aqueous solution under a constant-pH condition (Faivre et al. 2004). This methodology allows us to form magnetite nanocrystals under constant and controlled chemical affinity conditions. At a given constant pH, the overall reaction of magnetite $[Fe^{3+}{}_{T_d}(Fe^{2+}Fe^{3+})_{O_h}O_4]$ (where T_d and O_h are tetrahedral and octahedral lattice position) precipitation can be schematically represented by a simplified mass balance equation:

$$2Fe^{3+} + Fe^{2+} + 4H_2O \underset{R^-}{\overset{R^+}{\rightleftharpoons}} Fe_3O_4 + 8H^+ , \tag{2}$$

where R_+ is the rate of the forward reaction and R_- the rate of the reverse reaction. Equation 2 indicates that the reaction produces protons. For controlled experimental conditions, the pH can be held constant by automated addition of sodium hydroxide solution. These experimental conditions are suitable for forming magnetite at low temperature under constant chemical affinity. The possibility to form magnetite crystals depends on the saturation state of the parent solution, Ω, classically defined by

$$\Omega = \frac{a_{Fe^{2+}} a^2_{Fe^{3+}}}{a^8_{H^+} K_s} , \tag{3}$$

where a_X is the activity of species X and K_s is the thermodynamic magnetite solubility. Typical values for K_s are in the range between $10^{9.61}$ and $10^{12.02}$ (Cornell and Schwertmann 1996; Sweeton and Baes 1970). Operationally it is, however, difficult to estimate the real saturation state under environmental low-temperature conditions as solubility constants of magnetite are known at high temperature (Sweeton and Baes 1970; Ziemniak et al. 1995) and extrapolations to low temperature are hazardous. In addition, recent solubility estimations on nanometre-sized minerals (Hochella 2002a, b; Wang et al. 2003) showed that at this scale, mineral solubility is several orders of magnitude different from that evaluated at the micrometre and millimetre scale.

As already stated, the variation of the chemical affinity conditions is crucial in the kinetics description and constitute the key parameter to be controlled if we want to be able to connect the formation conditions to crystal properties. For example, it is known that bacteria can form crystals with as little as 1 μM of iron in their environment, whereas magnetite cannot be formed inorganically with so little iron present (Faivre et al. 2004). Therefore, the bacteria have to use a very efficient "iron pump" to enrich the iron in their magnetosomes in order to synthesize their magnetic inclusions. In the following sections, we will therefore first present the given properties of the magnetosomes and then compare them with properties of inorganic crystals obtained at different chemical affinity.

3 Mineralogical Properties

3.1 Nature of the Crystals

Magnetotactic bacteria were discovered in 1975 (Blakemore 1975), but the nature of the intracellular mineral was determined only in 1979, when the first pure culture became available (Frankel and Blakemore 1979). It was then shown by Mössbauer spectroscopy that the iron magnetic particles essentially consisted of magnetite, with about 4% of vacancies in the octahedral sublattice or with a small admixture of a gamma-Fe_2O_3 phase (Frankel and Blakemore 1979). Later measurements, for example magnetic ones (Weiss et al. 2004), confirmed the presence of this nearly unoxidized magnetite.

Magnetotactic bacteria, however, cannot only form magnetite, but also iron–sulphur crystals. On the basis of high-resolution electron microscopy, it was proposed in two back-to-back papers published simultaneously in the same issue of *Nature* that either pyrrhotite (Farina et al. 1990) or greigite and pyrite (Mann et al. 1990) were formed. Now, it is widely accepted that in fact only greigite is formed. Remarkably, it was observed that both greigite and magnetite can coexist within the same bacterial cell (Bazylinski et al. 1993; Posfai et al., this volume). Such coexistence indicates the high degree of chemical control exerted by the cell over crystal nucleation.

The synthesis of the bacterial magnetosome seems to be a complex process that involves several discrete steps, including magnetosome vesicle formation, iron uptake by the cell, iron transport into the magnetosome vesicle and controlled Fe_3O_4 (or Fe_3S_4) biomineralization within the magnetosome vesicle. Although it is clear that the uptake, transport and mineralization steps are temporally ordered, it is unclear whether iron uptake follows vesicle formation, or if both steps occur simultaneously (Bazylinski and Frankel 2004). It is nevertheless known that vesicles are formed even under conditions where no iron is present. It is therefore thought that at least immature vesicles have to be present prior to magnetite formation. However, although the presence of magnetite was shown already in 1979 (Frankel and Blakemore 1979), the first hypothesis of a putative pathway for magnetite formation in the magnetotactic bacteria was proposed several years later. On the basis of Mössbauer spectroscopy, it was suggested that *Magnetospirillum magnetotacticum* cells precipitate magnetite via the following sequence (Frankel et al.

1983): (1) $Fe(III)_{cell\ exterior} \rightarrow Fe(II)_{cell\ interior}$, (2) low-density hydrous ferric oxide, (3) ferrihydrite and finally (4) Fe_3O_4. However, electron microscopy and Mössbauer spectroscopy experiments have not yet been complemented by biochemical analysis, and the details of the reaction pathway have remained obscure.

Abiotic magnetite can be formed experimentally via very different routes (for a list, see the annex in Thomas-Keprta et al. 2000). Following the inorganic methodology described previously (Sect. 2), Faivre et al. (2004) showed that pure magnetite can be synthesized within minutes in aqueous media at high pH (10.5), and constant ionic strength ($I = 0.01$ M), constant Fe(III)-to-Fe(II) ratio of 2, and only at high total concentration of iron. For lower [total concentration of iron, (i.e. below 30 mmol/l) other iron oxides, such as goethite, are formed and may coexist with magnetite. The relative proportion of magnetite increases with increasing total concentration of iron.

Crystalline and amorphous mineral phases containing ferric ions such as goethite are less soluble than those containing ferrous ions. The clustering of Fe(III) is thus thermodynamically favoured under basic conditions. Faivre et al. (2004) showed that the activity of the Fe(II) in solution controls the formation of magnetite. This explains the observed fast formation of ferric hydroxide. Only if the Fe(II) concentration exceeds a certain threshold, the precipitation process leading to the formation of magnetite may take place. Furthermore, when magnetite is exclusively formed, the particle size does not change as a function of the temperature or the iron activity, indicating a fast and stable reaction mechanism of nanoparticle formation (Faivre et al. 2004). It has been suggested (Jolivet et al. 1992) that an possible unstable intermediate phase, such an iron hydroxides, could be formed before the magnetite formation takes place. However, until now the unambiguous presence of this precursor could not be demonstrated.

Inorganic magnetite crystals resembling the biogenic ones can therefore also be formed at low temperature in aqueous media, but only under relatively high iron concentration and pH. This strongly suggests that the solution chemistry within the bacterial magnetosome vesicles has to be completely different from the that of the exterior growth medium and also from that of the other cellular compartments. Indeed, this magnetosome compartment can be anticipated to have a rather alkaline pH of about 9–10 compared with the physiological pH of about 7, and the total iron concentration has to exceed 10 mM to achieve supersaturation, compared with the extracellular concentrations typically between about 1 and 100 μM.

3.2
Size and Shape Factor

Magnetosome crystals of magnetite and greigite are typically from 30 to about 140 nm in size (Fig. 1) (Balazs et al. 2005; Devouard et al. 1998; Lins

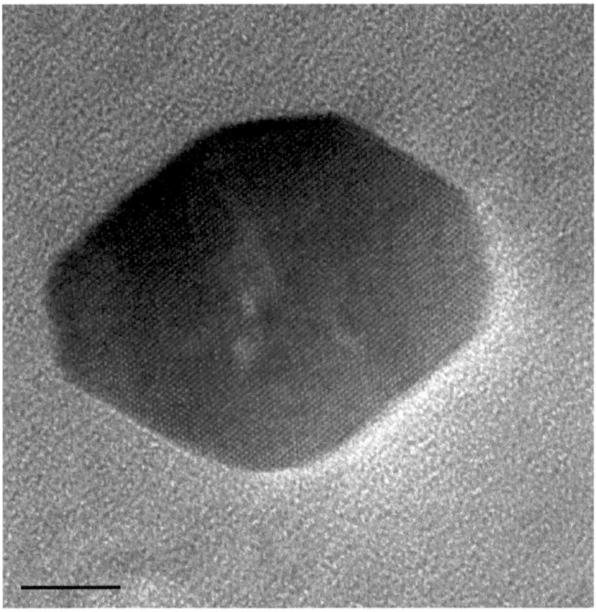

Fig. 1 Transmission electron microscopy (*TEM*) image of a typical cuboctahedral magnetosome of *Magnetospirillum magneticum* AMB-1. The *scale bar* represents 10 nm

and Farina 1998; Thomas-Keprta et al. 2000). This means that their size falls within the permanent single magnetic domain (Butler and Banerjee 1975), maximizing the efficiency of the particle as a magnetic carrier. It was therefore suggested that this relationship arises from natural selection (Kirschvink and Lowenstam 1979), as smaller crystals are observed at the end of the chains, which are predominantly superparamagnetic. Thus, they do not have a permanent magnetic moment at room temperature.

Statistical analysis of CSDs (CSDs) of magnetosomes containing magnetite from cultured strains or enriched environmental samples show narrow, asymmetrical distributions with sharp cutoffs towards larger size and with shape factors (or width-to-length ratios) consistent for a given strain or sample (Fig. 2) (Balazs et al. 2005; Devouard et al. 1998; Lins and Farina 1998; Thomas-Keprta et al. 2000).

The shape-factor analyses for magnetite crystals of *Magnetospirillum magnetotacticum* and strain MC-2 show distributions bounded by 1, with maxima around 0.85 (Devouard et al. 1998). For strain MV-1, which has elongated magnetosomes, the shape factor has a maximum frequency around 0.65. The distribution is asymmetric, with a cutoff towards the small values corresponding to the maximum elongations of the crystals (Devouard et al. 1998).

The sizes of low-temperature inorganic magnetites have also been studied or reported (Cornell and Schwertmann 1996; Devouard et al. 1998; Faivre

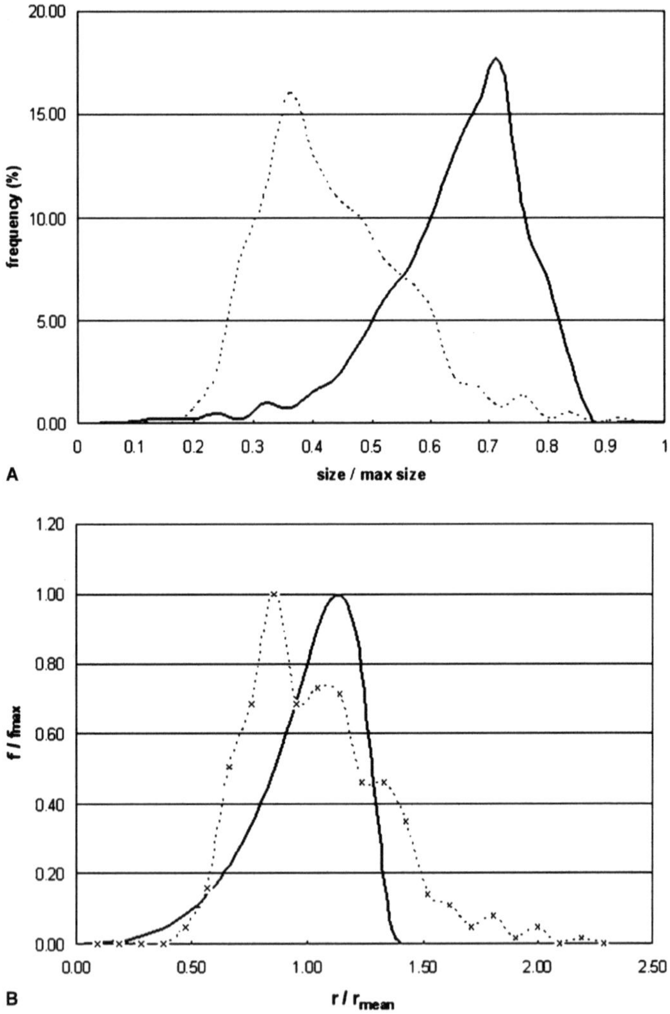

Fig. 2 **a** Typical experimental crystal size distributions (*CSDs*) from magnetotactic bacteria (*thick line*) or from inorganic synthesis (*dashed line*) and **b** experimental inorganic (*dashed line*) and theoretical Ostwald-ripened (*thick line*) CSD plotted on reduced axes (based on the theory of Eberl et al. 1998)

et al. 2004, 2005; Jolivet et al. 1992; McCartney et al. 2001; Regazzoni et al. 1981; Thomas-Keprta et al. 2000; Vayssières et al. 1998). Sizes around 10 nm are commonly reported for syntheses under constant chemical affinity conditions with log-normal CSDs and little elongation (Faivre et al. 2004, 2005). Devouard et al. (1998) used a completely different pathway for synthesizing small inorganic magnetites. In their experiments Fe(II) is oxidized at 90 °C for 60 min in a 3.33 mol/l KOH and 0.27 mol/l KNO_3 medium (Cornell and Schwertmann 1996). They observed inorganic magnetite or maghemite ranging from 20 to 120 nm, with a great majority being in the 20–80-nm interval. The sizes are much larger than in the previous studies, but the ionic strength, the nature of the ions in solution, and the strong variation of the chemical affinity of the reaction make the two synthetic pathways difficult to compare. Mann and Hannington (1987) reported a method in which magnetite, goethite or ferrihydrite was precipitated within unilamellar phosphatidylcholine vesicles of about 30 nm. Magnetite crystals precipitated from Fe(II)/Fe(III) intravesicular solutions containing NaOH, after 30 min of aging, were 2–6 nm in diameter. Again, the differences due to the presence of the vesicles make the sizes and mechanism difficult to compare with those determined in the other studies.

In general, inorganic magnetites have a log–normal CSD negatively skewed (log-normal tailing out to large crystal sizes) instead of positively for the biogenic crystals (Fig. 2). This is interesting for determining their origin but hardly gives hints about the formation processes. The CSD in fact fails to reduce to the theoretical prediction of an Ostwald-ripening type of process determined by the reduced-axes plot of Eberl (Fig. 2) (Eberl et al. 1998). So also do the CSDs of magnetosomes.

3.3 Morphology

Morphology alone could be the subject of a textbook as so many works have focused on the morphologies of magnetosomes since the first study about 20 years ago (Mann et al. 1984). Here we present a nonexhaustive list of studies, books and reviews: Baeuerlein (2000 ,2003), Bazylinski (1996), Bazylinski and Frankel (2003, 2004) Buseck et al. (2001), Clemett et al. (2002) Cornell and Schwertmann (1996, 2003), Devouard et al. (1998) Frankel and Buseck (2000) Golden et al. (2004) Lins et al. (1999, 2000, 2005), Schüler (1999, 2002), Schüler and Frankel (1999) Taylor and Barry (2004) Taylor et al. (2001) Thomas-Keprta et al. (2000, 2001, 2002) Vali et al. (1987). To make the story not too long, and as magnetite has a face-centred, inverse-spinel structure (*Fd*3*m* space group), possible morphologies for magnetite crystals are shown in Fig. 3.

However, not all morphological possibilities are observed for magnetosome crystals. The morphology therefore varies, being consistent for a given bacterial strain. The most widespread morphology seems to be the cuboctahedral one (Fig. 4), which was first discovered for *Magnetospirillum magnetotacticum* MS-1 (Mann et al. 1984). Other magnetospirilla such as *Magnetospirillum gryphiswaldense* MSR-1 or *Magnetospirillum magneticum* AMB-1 produce crystals with similar morphologies. Nonequidimensional (elongated) crystals are also reported. So-called truncated hexaoctahedral crystals (Fig. 4) are observed for MV-1 bacteria, and so-called bullet-shaped (or tooth- or arrowhead-shaped) (Fig. 4) particles are reported. Those later

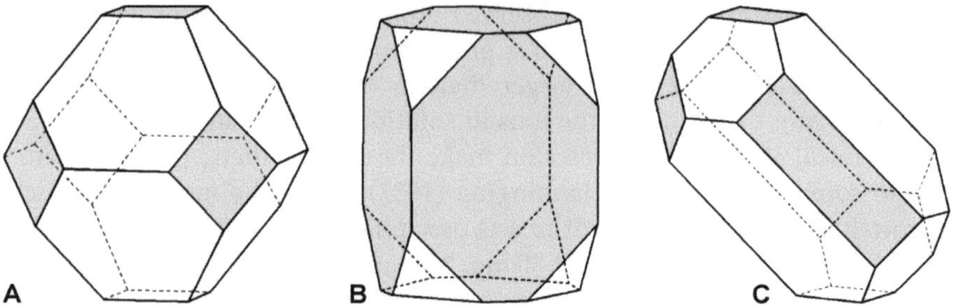

Fig. 3 Idealized possible morphologies of magnetite based on the combination of the {100} (cube, in *grey*) and {111} (octahedron in *white*) isometric forms. **a** cuboctahedron, **b** elongated, truncated cube and **c** elongated cuboctahedron. Note that the (001) face is always presented on top. Therefore, the crystal presented in **b** is elongated along the [100] direction, whereas the crystal presented in **c** is elongated along the [111] direction. For a typical MV-1 magnetosome, alignment of the magnetic moment along the [111] crystallographic direction, rather than the [100] direction, reduces the magnetostatic potential energy by about 4.5 mJ/cm^3, equivalent to about 90 kT for the MV-1 crystals (Thomas-Keprta et al. 2000). For all the possible morphologies, see Devouard et al. (1998)

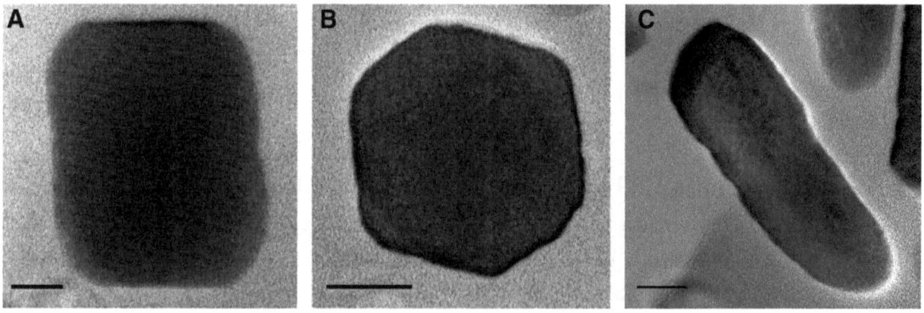

Fig. 4 Possible morphologies observed for magnetosomes based on high-resolution TEM images: **a** parallelepipedal projection of a possibly peudo-hexagonal prismatic morphology, **b** hexagonal projection of a possibly cuboctahedral crystal and **c** tooth-shaped (anisotropic) magnetosomes (the *scale bar* represents 20 nm). Note that the tooth-shaped magnetosomes are usually the larger ones and the cuboctahedral the smaller ones. (Images courtesy of A. Isambert)

elongated morphologies are important for the definition of the biogenic criterion as elongation is not commonly reported for abiogenic crystals.

For a long period it had been assumed that inorganic magnetite crystals had a spherical morphology. This was certainly due to a lack of resolution in electron microscopy. Until now, no inorganic bullet-shaped crystals or inorganic particles resembling MV-1 crystals have been reported. Such morphologies may therefore be used as a biogenic criterion. However, in the case of ALH84001 crystals, elongated cuboctahedral crystals (Golden et al. 2004) were reported, and it has been shown with high-resolution electron microscopy that inorganic magnetite can have a cuboctahedral morphology, with a little elongation (Faivre et al. 2005). In the particular case of the magnetite crystals in the meteorite, morphology cannot be used to discriminate the origin, and therefore other criteria have to be found.

3.4 Defects and Twinning

Typical imperfections observed in crystals and chains are presented in Fig. 5. In 2000, Thomas-Keprta et al. (2000) defined six criteria of biogenicity, which have to be fulfilled for a biogenic origin. Among those six criteria is the "crystallographic perfection". The authors stated that high-resolution electron microscopy studies of magnetosomes indicate they are essentially free of internal defects, with the minor exception of an occasional twin in the {111} plane, and this based on older studies (Devouard et al. 1998; Mann et al. 1988; Vali et al. 1987; Vali and Kirschvink 1989). Thomas-Keprta et al. (2000) also indicated that the lack of lattice defects acts to increase the net magnetic moment of the particle and as the [111] direction is the magnetically "easy" axis in magnetite, twins across this plane have no effect on their magnetization efficiency. However, Devouard et al. (1998) reported structural defects in magnetosomes from four strains of magnetotactic bacteria (MV-1, MC-1, MC-2 and MV-4) grown in chemically defined media. However, contrary to what Thomas-Keprta et al. suggest, twinned crystals are observed rather frequently. Indeed, as many as about 40% of MV-4 crystals are twinned, and even about 6–8% of the MV-1 crystals display twinning, which is a not negligible fraction. In a later study (Taylor et al. 2001), several imperfections were also shown, but on too few samples for meaningful statistics.

Devouard et al. (1998) also compared the structural and morphological characteristics of magnetosomes with those of synthetic magnetite with approximately the same average size and found them to have little difference

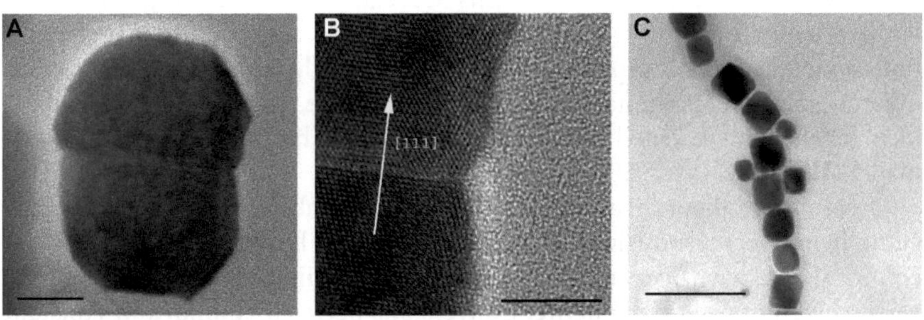

Fig. 5 Possible imperfections observed in magnetotactic bacteria. **a** crystal twinning (the *scale bar* represents 20 nm), **b** better view of the twinning, note the axis (the *scale bar* represents 10 nm) and **c** misalignment of the magnetosome chain and out of chain crystal growth (the *scale bar* represents 200 nm). (Images courtesy of A. Isambert)

in their defects. Structural anomalies in synthetic magnetite were mainly spinel-law twins with flattened twinned platelets having sparrow-tail-shaped projections when viewed along the zone axes. Irregular twins were also characterized in synthetic magnetite. As a result, crystal perfection should not be used as a biogenic criterion.

4 Isotopic Properties

A basic understanding of stable isotopic fractionations may help to explain and interpret laboratory experiments in both abiotic and biotic conditions and may be a potential tool in understanding the origin of natural samples. This may thus provide additional criteria to distinguish the origin of nano-magnetites.

The isotopic fractionation factor between two isotopes of the same element X, ${}^{\text{light}}X$ (lighter isotope) and ${}^{\text{heavy}}X$ (heavier isotope) between two phases XA and XB is usually expressed in terms of α:

$$\alpha_{\text{XA–XB}} = \left[\left({}^{\text{heavy}}X/{}^{\text{light}}X\right)_{\text{XA}} / \left({}^{\text{heavy}}X/{}^{\text{light}}X\right)_{\text{XB}}\right] . \tag{4}$$

Fractionation factors are typical very small, on the order of parts per thousand or parts per ten thousand, so it is common to use expressions that magnify the difference between α and 1.

The isotopic fractionation between two different species depends on a number of factors, including relative mass difference, the nature of the bonding environment, and the redox state. There a number of qualitative rules governing the equilibrium stable isotopic fractionations with numerous unexplained exceptions (Schauble 2004). In addition, kinetic isotopic fractionations also may control the effective isotopic fractionation values. It is commonly assumed that the fractionation factor α depends neither on the pressure nor on other constituents associated with the two given phases if these constituents do not modify the chemical state or the structure in both aqueous and solid phases. The influence of the solute composition on mineral-solution fractionation factors, for instance, termed “isotope salt effect” (O’Neil and Trusdell 1991; Horita et al. 1993) has often been underestimated (Hu and Clayton 2003).

Despite new theoretical approaches based on ab initio force-field modelling being very promising (Schauble 2004), only limited experimental, yet empirical data have been available up to now to distinguish the origin (abiotic versus bigenic) of the magnetite. Conversions between equilibrium constant and fractionation factors are more complicated, as it is often necessary to account for molecular stoichiometry and symmetry.

4.1 Oxygen Isotopes

A large number of studies have focused on oxygen isotope fractionation between magnetites and the parent aqueous solution. However, among those, very few have investigated crystals of biological origin (Mandernack et al. 1999; Zhang et al. 1997). In particular only one addressed the particles synthesized by magnetotatic bacteria (Mandernack et al. 1999). This fact is probably due to the huge number of bacteria necessary for a single analysis. There is hope that the arrival of new analytical tools such as nanoscale secondary-ion mass spectrometry and others may facilitate these investigations. With the limited data on biologically formed magnetite that we have today, it appears that the oxygen isotope fractionation factor from bacterial magnetite produced intracellularly (Mandernack et al. 1999) and extracellularly (Zhang et al. 1997) decreases as a function of the temperature, following the function (T in Kelvin):

$$1000 \ln \alpha_{\text{biogenic}} = 7.9(10^5/T^2) - 7.64\,. \qquad (5)$$

Equation 5 is derived from a simple linear regression without error propagation. In Fig. 6 we report the biotic and the abiotic magnetite–water oxygen isotope fractionation factor (α_{abiotic}) as a function of the temperature. It appears that in the range between 5 and 69 °C, the abiotic oxygen isotope fractionation factor increases by about 2.5‰ as a 10-times increase of the temperature according to the following equation (T in Kelvin) has been parametrized (Faivre and Zuddas 2006):

$$10^3 \ln \alpha_{\text{abiotic}} =- 0.55(\pm 0.13)(10^6/T^2) + 5.64(\pm 1.42)\,, \qquad (6)$$

with a correlation coefficient $R^2 = 0.88$.

In a recent study (Cole et al. 2004), various experimental results on inorganic magnetite–water fractionation for oxygen isotopes over a large range of high temperatures were reviewed. According to these data, when the temperature exceeds 300 °C, the slope of $10^3 \ln \alpha_{\text{m-w}}$ vs. $10^6/T^2$ is negative. A minimum value and then a positive slope at lower temperatures allow the high-temperature data to be rendered consistent the with data from biogenic magnetite of Mandernack et al. (1999) and Zhang et al. (1997). The temperature corresponding to these minimum values is, however, not very well defined. Other data indicate that the minimum for the fractionation curve is just below 300 °C (Blattner et al. 1983; Ripperdan et al. 1998), whereas according to Becker and Clayton, it is below 200 °C. Finally, the minimum can also be observed below 100 °C (Rowe et al. 1994). This great discrepancy is problematic.

Extrapolation of these high-temperature data to lower-temperature fractionation leads therefore to $1000 \ln \alpha_{\text{abiotic}}$ values that differ by about 15‰ (as reported by Ripperdan et al. 1998) at 0 °C. This may be explained by the fact

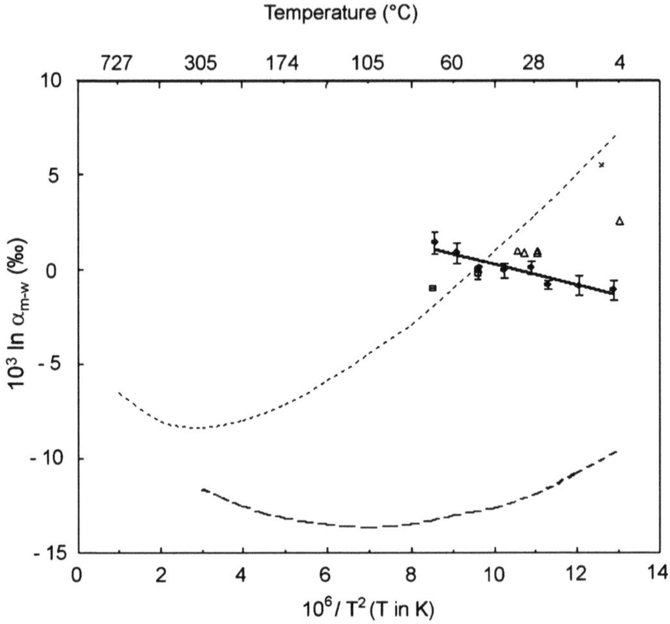

Fig. 6 Oxygen isotope fractionation factor between magnetite and water ($10^3 \ln \alpha_{m\text{-}w}$) as a function of temperature ($10^6/T^2$, T in Kelvin). *Circles*: pure inorganic magnetite from Faivre and Zuddas (2006). *Triangles*: magnetite formed by magnetotactic bacteria (intracellular crystals) (Mandernack et al. 1999). *Squares*: magnetite formed by thermophilic iron-reducing bacteria (Zhang et al. 1997). *Cross*: magnetite from chiton tooth (O'Neil and Clayton 1964). The *dotted curve* is from the compilation of data by Ripperdan et al. (1998) (Becker and Clayton 1976; Blattner et al. 1983; Fortier et al. 1995; Zhang et al. 1997). The *dashed curve* is also from Ripperdan et al. (1998) after Rowe et al. (1994) with higher-temperature extension calculated using data from Clayton and Kiefer (1991) and Richet et al. (1977). These two curves represent the extreme behaviour of the magnetite–water fractionation at low temperature (more than about 15‰ difference between 0 and 50 °C)

that those fractionations are based on high-temperature experiments and that they cannot necessarily be extrapolated at low temperatures following a linear law. This has already been observed for the calcite–water system, where high-temperature data (O'Neil et al. 1969) do not extrapolate to low temperature (Epstein et al. 1953), perhaps owing to differing isotopic reaction mechanisms. This leads to the general difficulty of quantitatively interpreting the oxygen isotopes data from minerals in natural systems where parent fluids and the physicochemical conditions of mineral formation are not identified.

We notice that the formation of magnetite nanoparticles at low temperature by bacteria is supposed to happen at high local pH and under high supersaturation conditions (Baeuerlein 2000; Faivre et al. 2004), and thus might possibly not lead to oxygen isotope equilibrium between magnetite and water. This would not be the first example of disequilibrium oxygen isotope variations. Indeed, Dickson (1991) has shown that oxygen isotope fractionation could vary between different crystallographic phases of the same synchronous growth surface of calcite.

As previously shown, the results obtained by Faivre and Zuddas (2006) are also different from the trend defined by results on biogenic magnetites (Fig. 6) (Mandernack et al. 1999). This may mean either that the biogenic magnetites are close to oxygen isotope equilibrium, or that they alternatively also suffer from a kinetics effect, which however can be distinguished from the abiotic situation. Previous studies of abiotic (Faivre et al. 2004; Jolivet et al. 1992) and biotic (Frankel et al. 1983) precipitations of magnetite have shown that the formation of magnetite nanocrystals generally occurs via an unstable ferric oxide intermediate. In bacterial intracellular magnetites, the magnetosome formation is achieved by a mineralization process with control over the accumulation and deposition of magnetite particles resulting from a complex biochemical process involving phospholipids and proteins (Baeuerlein 2003). One of the differences between biotic and abiotic reaction pathways appears to be the nature of the ferric iron oxide precursor. This difference might have an effect on the oxygen isotope fractionation factor between abiotic and biotic systems since variations in bonding can be correlated with variations in oxygen fractionation.

Isotopic properties can also be inherited from precursors and in these cases variations in fractionations could be attributed to the formation or transformation of the ferric oxide intermediate. Thus, the abiotic/biotic difference in oxygen fractionation with temperature could be explained by a difference in the nature of the iron oxide intermediate and therefore by a difference in the mechanism, such as the interaction between the oxygen ions around the ferric ions. The isotopic signature of the final product might be recorded in this step and might not change during the final addition of ferrous ions to form magnetite.

Crystal size might also influence the oxygen isotope ratio in metamorphic magnetite (Sharp 1991; Valley and Graham 1993). It has also been shown that differences in solubility could be observed at the nanoscale when compared with the microcale and larger scales (Hochella 2002a, b; Michard 1989; Wang et al. 2003). Thus, it seems that oxygen isotope compositions of nanocrystals and macrocrystals do significantly differ and that size could also have an effect on the oxygen isotope signature of magnetite.

4.2 Iron Isotopes

As in the case of oxygen isotopes, recent advance in analytical techniques and specifically in inductively coupled plasma mass spectrometry have made it feasible to detect variations in iron isotope ratios (^{56}Fe to ^{54}Fe) as low as 0.1‰ in samples as small as 1 ng of iron. Again, the study by Mandernack et al. (1999) is the only report on iron isotope fractionations during magnetite formation by magnetotactic bacteria. In this study, no detectable fractionation was observed compared with the situation for the growth medium. As not many studies have been performed yet, we can just compare these results with those obtained with other microorganisms to see if this absence of fractionation is a common feature (Fig. 7).

Iron isotope fractionation produced by diverse Fe(II)-oxidizing anaerobic phototropic bacteria results in poorly crystalline hydrous ferric oxide products that have ^{56}Fe-to-^{54}Fe ratios that are about 1.5‰ higher than that of the aqueous Fe(II) electron donor (Croal et al. 2004). Experiments with dissimilatory Fe-reducing bacteria of the species *Shewanella algae* grown on a ferrihydrite substrate indicate that the delta^{56}Fe of Fe(II) in solution is isotopically lighter than the ferrihydrite substrate by 1.3‰ (Beard et al. 1999).

At the same time, similar fractionations have also been observed for abiotic processes (Anbar et al. 2000; Bullen et al. 2001), therefore making this

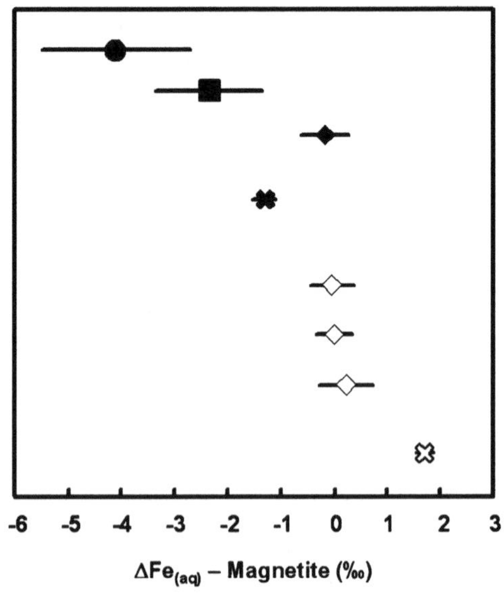

Fig. 7 Comparison of isotopic fractionations between Fe in solution [ferrous (*in black*) and ferric (*in white*) species] and magnetite. All fractionation factors were calculated relative to ^{56}Fe-to-^{54}Fe ratios. *Error bars* reflect reported uncertainties. The data are from predictions based on spectroscopic data (*circle*) (Polyakov and Mineev 2000; Schauble et al. 2001), natural samples (*square*) (Johnson et al. 2003), experimental studies of biogenic magnetite formation from magnetotactic bacteria (*diamonds*) (Mandernack et al. 1999) and from dissimilatory iron-reducing bacteria (*crosses*) (Johnson et al. 2005). This last result, measured for $Fe(II)_{aq}$ magnetite has been converted to an $Fe(III)_{aq}$ magnetite fractionation factor using the $Fe(III)_{aq}$–$Fe(II)_{aq}$ fractionation factor reported (Johnson et al. 2002; Welch et al. 2003), for comparison with the $Fe(III)_{aq}$ magnetite fractionation factors measured for magnetotactic bacteria by Mandernack et al. (1999)

less meaningful as a biosignature. This application is nevertheless still being developed, with particular attention being paid to fractionation mechanisms. Understanding such mechanisms could in fact provide new insights into the environmental biogeochemistry of Fe.

5
Conditions of Formation and the Criterion of Biogenicity

The purpose of this chapter is to provide a concise description of mineralogical and isotopic properties of magnetite nanocrystals as a criterion of biogenicity and to identify processes possibly responsible for their formation under biogenic and eventually abiotic conditions. This debate has been stimulated by the possible biogenic origin of nanosized magnetite crystal founded in the Martian meteorite ALH84001. Nanocrystals of magnetite can be found in present sediments and sedimentary rocks and can be generated by magnetic bacteria as well as by abiotic processes.

Recent experimental developments have shown that magnetite nanocrystals can be primarily formed under high saturation conditions and that the key kinetics parameter is the activity of Fe(II). In particular, the biogenic production of magnetite nanocrystals should correspond to aqueous media with high pH and solution saturation.

Criteria of biogenicity have been defined in the case of crystals having similar size when magnetic properties are not sufficient to differentiate biogenic from abiogenic particles. Cuboctahedral morphology is observed in crystals generated by magnetotactic bacteria, in the Martian meteorite ALH84001 (of unknown origin) and in abiotic nanocrystals formed under constant chemical affinity conditions. Therefore, morphology alone cannot be used as a biogenicity criterion. However, opposite asymmetries of the CSDs from biogenic and from abiotic nanocrystals of synthetic magnetites have been observed. The analysis of CSDs should thus provide a criterion to distinguish between biocontrolled mineralization and inorganic mineralization. In the case of a mixed sample, the situation is naturally much more complex, and the unambiguous differentiation more difficult or not necessarily possible (Arató et al. 2005). Therefore, the addition of another property is required. Fractionation factors of oxygen isotopes during magnetite precipitation at different temperature allow the differentiation between biogenic and abiotic magnetite outside the limit of a crossing region in the range 30–40 °C (Faivre and Zuddas 2006). A combination of oxygen isotope properties and CSDs is this a potential powerful tool for a more accurate determination of a reliable biogenicity criterion.

Acknowledgements We thank A. Isambert for Figs. 4 and 5, N. Menguy for his help with Figs. 1, 4 and 5 and O. Lopez for his help with Fig. 6. Valuable discussions, chronologically, with R. Hellmann, F. Guyot, N. Menguy, P. Agrinier, A. Isambert, A. Komeili, D. Newman, T. Bullen, D. Schüler, and members of his group, D. Bazylinski, B. Matzanke and R. Dunin-Borkowski helped us in understanding the different aspects presented in this chapter. D.F. acknowledges support from a Marie Curie fellowship from the European Union.

References

Anbar AD, Roe JE, Barling J, Nealson KH (2000) Nonbiological fractionation of iron isotopes. Science 288:126–128

Arató B, Szanyi Z, Flies CB, Schüler D, Frankel RB, Buseck PR, Pósfai M (2005) Crystal-size and shape distributions of magnetite from uncultured magnetotactic bacteria as a potential biomarker. Am Mineral 90:1233–1241

Baeuerlein E (2000) Biomineralization. Wiley, Weinheim

Baeuerlein E (2003) Biomineralization of unicellular organisms: an unusual membrane biochemistry for the production of inorganic nano- and microstructures. Angew Chem Int Ed Engl 42:614–641

Bahaj AS, Croudace IW, James PAB, Moeschler FD, Warwick PE (1998a) Continuous radionuclide recovery from wastewater using magnetotactic bacteria. J Magn Magn Mater 184:241–244

Bahaj AS, James PAB, Moeschler FD (1998b) Low magnetic-field separation system for metal-loaded magnetotactic bacteria. J Magn Magn Mater 177–181:1453–1454

Balazs A, Szanyi Z, Flies CB, Schueler D, Frankel RB, Buseck PR, Posfai M (2005) Crystal-size and shape distributions of magnetite from uncultured magnetotactic bacteria as a potential biomarker. Am Mineral 90:1233–1241

Bazylinski DA (1996) Controlled biomineralization of magnetic minerals by magnetotactic bacteria. Chem Geol 132:191–198

Bazylinski DA, Frankel RB (2003) Biologically controlled mineralization in prokaryotes. Rev Mineral Geochem 54:217–248

Bazylinski DA, Frankel RB (2004) Magnetosome formation in prokaryotes. Nat Rev 2:217–230

Bazylinski DA, Heywood BR, Mann S, Frankel RB (1993) Fe_3O_4 and Fe_3S_4 in a bacterium. Nature 366:218

Beard BL, Johnson CM, Cox L, Sun H, Nealson KH, Aguilar C (1999) Iron isotope biosignatures. Science 285:1889–1892

Becker RH, Clayton RN (1976) Oxygen isotope study of a Precambrian banded iron-formation. Hamersley Range. Western Australia. Geochim. Cosmochim Acta 40:1153–1165

Blakemore RP (1975) Magnetotactic bacteria. Science 190:377–379

Blattner P, Braithwaite WR, Glover RB (1983) New evidence on magnetite oxygen isotope geothermometers at 175° and 112 °C in Wairakei steam pipeline (New Zealand). Isotope Geosci 1:195–204

Bullen TD, White AF, Childs CW, Vivit DV, Schultz MS (2001) Demonstration of significant abiotic iron isotope fractionation in nature. Geology 29:699–702

Buseck PR, Dunin-Borkowski RE, Devouard B, Frankel RB, McCartney MR, Midgley PA, Posfai M, Weyland M (2001) Magnetite morphology and life on Mars. Proc Natl Acad Sci USA 98:13490–13495

Butler RF, Banerjee S (1975) Theoretical single domain grain size range in magnetite and titanomagnetite. J Geophys Res 80:4049–4058

Clayton RN, Kiefer SW (1991) Oxygen isotopic thermometer calibration. In: Taylor HP Jr, O'Neal JR, Kaplan IR (eds) Stable isotope geochemistry: a tribute to Samuel Epstein, vol 3. Geochemical Society Special Publication, p 3–10

Clemett SJ, Thomas-Keprta KL, Shimmin J, Morphew M, McIntosh JR, Bazylinski DA, Kirschvink JL, Wentworth SJ, McKay DS, Vali H, Gibson EK Jr, Romanek CS (2002) Crystal morphology of MV-1 magnetite. Am Mineral 87:1727–1730

Cole DR, Horita J, Polyakov VB, Valley JW, Spicuza MJ, Coffey DW (2004) An experimental and theoretical determination of oxygen isotope fractionation in the system magnetite-H_2O from 300 to 800 °C. Geochim Cosmochim Acta 68:3569–3585

Cornell RM, Schwertmann U (1996) The iron oxides: structure, properties, reactions, occurrence and uses. Wiley, new York

Cornell RM, Schwertmann U (2003) The iron oxides (structure, properties, reactions, occurrences and uses). Wiley, New York

Courtillot V, Hulot G, Alexandrescu M, le Mouël J-L, Kirschvink JL (1997) Sensitivity and evolution of sea-turtle magnetoreception: observations, modelling and constraints from geomagnetic secular variation. Terr Nova 9:203–207

Croal LR, Johnson CM, Beard BL, Newman D (2004) Iron isotope fractionation by Fe(II)-oxidizing photoautotrophic bacteria. Geochim Cosmochim Acta 68:1227–1242

Deutschlander ME, Borland SC, Phillips JB (1999) Extraocular magnetic compass in newts. Nature 400:324–325

Devouard B, Posfai M, Hua X, Bazylinski DA, Frankel RB, Buseck PB (1998) Magnetite from magnetotactic bacteria: size distributions and twinning. Am Mineral 83:1387–1398

Dickson JAD (1991) Disequilibrium carbon and oxygen isotope variations in natural calcite. Nature 353:842–844

Dobson J, Grassi P (1996) Magnetic properties of human hippocampal tissue—evaluation of artefact and contamination sources. Brain Res Bull 39:255–259

Dunn JR, Fuller M, Zoeger J, Dobson J, Heller F, Hammann J, Caine E, Moskowitz BM (1995) Magnetic material in the human hippocampus. Brain Res Bull 36:149–153

Eberl DD, Drits VA, Srodon J (1998) Deducing growth mechanisms for minerals from the shapes of crystal size distributions. Am J Sci 298:499–533

Epstein S, Bushsbaum R, Lowenstam HA, Urey HC (1953) Revised carbonate-water temperature scale. GSA Bull 64:1315–1326

Faivre D, Agrinier P, Menguy N, Zuddas P, Pachana K, Gloter A, Laval J-Y, Guyot F (2004) Mineralogical and isotopic properties of inorganic nanocrystalline magnetites. Geochim Cosmochim Acta 68:4395–4403

Faivre D, Menguy N, Guyot F, Lopez O, Zuddas P (2005) Morphology of nanomagnetite crystals: implications for formation conditions. Am Mineral 90:1793–1800

Faivre D, Zuddas P (2006) An integrated approach for determining the origin of magnetite nanoparticles. Earth Planet Sci Lett 243(1–2):53–60

Farina M, Esquivel DMS, Lins de Barros H (1990) Magnetic iron-sulphur crystals from a magnetotactic microorganism. Nature 343:256–258

Fortier SM, Cole DR, Wesolowski DJ, Riciputi LR, Paterson BA, Valley JW, Horita J (1995) Determination of the magnetite-water equilibrium oxygen isotope fractionation factor at 350 °C: a comparison of ion micrprobe and laser fluorination techniques. Geochim Cosmochim Acta 59:3871–3875

Frankel RB, Bazylinski DA (2003) Biologically induced mineralization by bacteria. Rev Mineral Geochem 54:95–114

Frankel RB, Blakemore R (1979) Magnetite in freshwater magnetotactic bacteria. Science 203:1355–1356

Frankel RB, Buseck PB (2000) Magnetite biomineralization and ancient life on Mars. Curr Opin Chem Biol 4:171–176
Frankel RB, Papaefthymiou GC, Blakemore RP, O'Brian W (1983) Fe_3O_4 Precipitation in magnetotactic bacteria. Biochim. Biophys Acta 763:147–159
Golden DC, Ming DW, Morris RV, Brearley AJ, Lauer HV Jr, Treiman AH, Zolensky ME, Schwandt CS, Lofgren GE, McKay GA (2004) Evidence for exclusively inorganic formation of magnetite in Martian meteorite ALH84001. Am Mineral 89:681–695
Hochella MF Jr (2002a) Nanoscience and technology: the next revolution in the Earth sciences. Earth Planet Sci Lett 203:593–605
Hochella MF Jr (2002b) There's plenty of room at the bottom: nanoscience in geochemistry. Geochim Cosmochim Acta 66:735–743
Hu G, Clayton RN (2003) Oxygen isotope salt effect at high pressure an high temperature and the calibration of oxygen isotope geothermometers. Geochim Cosmochim Acta 67:3227–3246
Johnson CM, Beard BL, Beukes NJ, Klein C, O'Leary JM (2003) Ancient geochemical cycling in the Earth as inferred from Fe isotope studies of banded iron formations from the Transvaal Craton. Contrib Mineral Petrol 144:523–547
Johnson CM, Roden EE, Welch SA, Beard BL (2005) Experimental constraints on Fe isotope fractionation during magnetite and Fe carbonate formation coupled to dissimilatory hydrous ferric oxide reduction. Geochim Cosmochim Acta 69:963–993
Johnson CM, Skulan JL, Beard BL, Sun H, Nealson KH, Braterman PS (2002) Isotopic fractionation between Fe(III) and Fe(II) in aqueous solutions. Earth Planet Sci Lett 195:141–153
Jolivet JP, Belleville P, Tronc E, Livage J (1992) Influence of Fe(II) on the formation of the spinel iron oxide in alkaline medium. Clays Clay Mineral 40:531–539
Kirschvink JL, Lowenstam HA (1979) Mineralization and magnetization of chiton teeth: paleomagnetic, sedimentilogic and biologic implications of organic magnetite. Earth Planet Sci Lett 44:193–204
Lasaga AC (1998) Kinetic theory in the earth sciences. Princeton University Press, Princeton
Lins U, Farina M (1998) Magnetosome size distribution in uncultured rod-shaped bacteria as determined by electron microscopy and electron spectroscopic imaging. Microsc Res Tech 42:459–464
Lins U, Kachar B, Farina M (1999) Imaging faces of shadowed magnetite (Fe_3O_4) crystals from magnetotactic bacteria with energy-filtering transmission electron microscopy. Microsc Res Tech 46:319–324
Lins U, Freitas F, Keim CN, Farina M (2000) Electron spectroscopic imaging of magnetotactic bacteria: magnetosome morphology and diversity. Microsc Microanal 6:463–470
Lins U, McCartney MR, Farina M, Frankel RB, Buseck PR (2005) Habits of magnetosome crystals in coccoid magnetotactic bacteria. Appl Environ Microbiol 71:4902–4905
Lohmann KJ, Johnsen S (2000) The neurobiology of magnetoreception in vertebrate animals. Trends Neurosci 23:153–159
Lowe CR (2000) Nanobiotechnology: the fabrication and applications of chemical and biological nanostructures. Curr Opin Structur Biol 10:428–434
Maher BA (1998) Magnetite biomineralization in termites. Proc R Soc Lond Ser B 265:733–737
Mandernack KW, Bazylinski DA, Shanks WC, Bullen TD (1999) Oxygen and Iron Isotope Studies of Magnetite Produced by Magnetotactic Bacteria. Science 285:1892–1896
Mann S, Frankel RB, Blakemore RP (1984) Structure, morphology and crystal growth of bacterial magnetite. Nature 310:405–407

Mann S, Sparks NHC, Frankel RB, Bazlinski DA, Jannasch HW (1990) Biomineralization of ferrimagnetic greigite (Fe_3S_4) and iron pyrite (FeS_2) in a magnetotactic bacterium. Nature 343:258–261

Mann S, Sparks NHC, Walker MM, Kirschvink JL (1988) Ultrastructure, morphology and organization of biogenic magnetite from sockeye salmon, Oncorhynchus nerka: implications for magnetroreception. J Exp Biol 140:35–49

McCartney MR, Lins U, Farina M, Buseck PR, Frankel RB (2001) Magnetic microstructure of bacterial magnetite by electron holography. Eur J Mineral 13:685–689

McKay DS, Gibson EK Jr, Thomas-Keprta KL, Vali H, Romanek CS, Clemett SJ, Chilier XDF, Maechling CR, Zare RN (1996) Search for past life on Mars: possible relic biogenic in Martian meteorite ALH84001. Science 273:924–930

Michard G (1989) Equilibres chimiques dans les eaux naturelles. Publisud, Paris

O'Neil JR, Trusdell AH (1991) Oxygen isotope fractionation studies of solute-water interactions. In: Taylor JHP, O'Neil JR, Kaplan IR (eds) Stable isotope geochemistry: a tribute to Samuel Epstein. The Geochemical Society, San Antonio, Lancaster Press, Inc., p 17–25

O'Neil JR, Clayton RN (1964) Oxygen isotope geothermometry. In: Craig H (ed) Isotopic and cosmic chemistry. North Holland, Amsterdam, pp 157–168

O'Neil JR, Clayton RN, Mayeda TK (1969) Oxygen isotope fractionation in divalent metal carbonates. J Chem Phys 51:5547–5558

Polyakov VB, Mineev SD (2000) The use of Mössbauer spectroscopy in stable isotope geochemistry. Geochim Cosmochim Acta 64:849–865

Regazzoni AE, Urrutia GA, Blesa MA, Maroto AJG (1981) Some observations on the composition and morphology of synthetic magnetites obtained by different routes. J Inorg Nucl Chem 43:1489–1493

Richet P, Bottinga Y, Javoy M (1977) A review of hydrogen, carbon, notrogen, oxygen, sulphur and chlorine stable isotope fractionation among gaseous molecules. Ann Rev Earth Planet Sci 5:65–110

Ripperdan RL, Riciputi LR, Cole DR, Elmore RD, Banerjee S, Engel MH (1998) Oxygen isotope ratios in authigenic magnetites from the Belden Formation, Colorado. J Geophys Res 103:21015–21023

Rowe MW, Clayton RN, Mayeda TK (1994) Oxygen isotopes in separated components of CI and CM meteorites. Geochim Cosmochim Acta 58:5341–5347

Safarik I, Safarikova M (2002) Magnetic nanoparticles and biosciences. Monatsh Chem 133:737–759

Sarikaya M, Tamerler C, Jen AK-Y, Schulten K, Baneyx F (2003) Molecular biomimetics: nanotechnology through biology. Nat Mater 2:577–585

Schauble EA (2004) Applying stable isotope fractionation theory to new systems. Rev Mineral Geochem 55:65–111

Schauble EA, Rossman GR, Taylor HPJ (2001) Theoretical estimates of equilibrium Fe-isotope fractionations from vibrational spectroscopy. Geochim Cosmochim Acta 65:2487–2497

Schüler D (1999) Formation of magnetosomes in magnetotactic bacteria. J Mol Microbiol Biotechnol 1:79–86

Schüler D (2002) The biomineralisation of magnetosomes in Magnetospirillum gryphiswaldense. Int Microbiol 5:209–214

Schüler D, Frankel RB (1999) Bacterial magnetosomes: microbiology, biomineralization and biotechnological applications. Appl Microbiol Biotechnol 52:464–473

Sharp ZD (1991) Determination of oxygen diffusion rates in magnetite from natural isotopic variations. Geology 19:653–656

Sweeton FH, Baes CF Jr (1970) The solubility of magnetite and hydrolysis of ferrous ion in aqueous solutions at elevated temperatures. J Chem Thermodyn 2:479–500

Tartaj P, Morales MD, Veintemillas-Verdaguer S, Gonzalez-Carreno T, Serna CJ (2003) The preparation of magnetic nanoparticles for applications in biomedicine. J Phys D Appl Phys 36:R182–R197

Taylor AP, Barry JC (2004) Magnetosomal matrix: ultrafine structure may template biomineralization of magnetosomes. J Microsc 213:180–197

Taylor AP, Barry JC, Webb RI (2001) Structural and morphological anomalies in magnetosomes: possible biogenic origin for magnetite in ALH84001. J Microsc 201:84–106

Thomas-Keprta KL, Bazlinski DA, Kirschvink JL, Clemett SJ, McKay DS, Wentworth SJ, Vali H, Gibson EK Jr, Romanek CS (2000) Elongated prismatic crystals in AL84001 carbonate globules: potential Martian magnetofossils. Geochim Cosmochim Acta 64:4049–4081

Thomas-Keprta KL, Clemett SJ, Bazlinski DA, Kirschvink JL, McKay DS, Wentworth SJ, Vali H, Gibson EK Jr, McKay MF, Romanek CS (2001) Truncates hexa-octahedral magnetite crystals in ALH84001: presumptive biosignature. Proc Natl Acad Sci USA 98:2164–2169

Thomas-Keprta KL, Clemett SJ, Bazylinski DA, Kirschvink JL, McKay DS, Wentworth SJ, Vali H, Gibson EK Jr, Romanek CS (2002) Magnetofossils from ancient Mars: a robust biosignature in the Martian meteorite ALH84001. Appl Environ Microbiol 68:3663–3672

Vali H, Forster O, Amarantidid G, Petersen H (1987) Magnetotactic bacteria and their magnetofossils in sediments. Earth Planet Sci Lett 86:389–400

Vali H, Kirschvink JL (1989) Magnetofossil dissolution in a paleomagnetically unstable deep-sea sediment. Nature 339:203–206

Valley JW, Graham CM (1993) Cryptic grain-scale heterogeneity of oxygen isotope ratios in metamorphic magnetite. Science 259:1729–1733

Vayssières L, Chanéac C, Tronc E, Jolivet JP (1998) Size tailoring or magnetite particles formed by aqueous precipitation: an example of thermodynamic stability of nanometric oxide particles. J Colloid Interface Sci 205:205–212

Wang Y, Bryan C, Xu H, Gao H (2003) Nanogeochemistry: Geochemical reactions and mass transfers in nanopores. Geology 31:387–390

Weiss BP, Kim SS, Kirschvink JL, Kopp RE, Sankaran M, Kobayashi A, Komeili A (2004) Ferromagnetic resonance and low-temperature magnetic tests for biogenic magnetite. Earth Planet Sci Lett 224:73–89

Welch SA, Beard BL, Johnson CM, Braterman PS (2003) Kinetic and equilibrium Fe isotope fractionation between aqueous Fe(II) and Fe(III). Geochim Cosmochim Acta 67:4231–4250

Wiltschko W, Munro U, Wiltschko R, Kirschvink JL (2002) Magnetite-based magnetoreception in birds: the effect of a biasing field and a pulse on migratory behavior. J Exp Biol 205:3031–3037

Winklhofer M, Holtkamp-Rötzler E, Hanzlik M, Fleissner G, Petersen N (2001) Clusters of superparamagnetic magnetite particles in the upper-beak skin of homing pigeons: evidence of a magnetoreceptor? Eur J Mineral 13:659–669

Zhang C, Liu S, Phelps TJ, Cole DR, Horita J, Fortier SM, Elless M, Valley JW (1997) Physiochemical, mineralogical, and isotopic characterization of magnetite-rich iron oxides formed by thermophilic iron-reducing bacteria. Geochim. Cosmochim Acta 61:4621–4632

Ziemniak SE, Jones ME, Combs KES (1995) Magnetite solubility and phase stability in alkaline media at elevated temperatures. J Solution Chem 24:837–877

Microbiol Monogr (3)
D. Schüler: Magnetoreception and Magnetosomes in Bacteria
DOI 10.1007/7171_044/Published online: 30 August 2006

Characterization of Bacterial Magnetic Nanostructures Using High-Resolution Transmission Electron Microscopy and Off-Axis Electron Holography

Mihály Pósfai[1] (✉) · Takeshi Kasama[2,3] · Rafal E. Dunin-Borkowski[2,3]

[1]Department of Earth and Environmental Sciences, Pannon University, POB 158, 8200 Veszprém, Hungary
posfaim@almos.vein.hu

[2]Frontier Research System, The Institute of Physical and Chemical Research, Hatoyama, 350-0395 Saitama, Japan

[3]Department of Materials Science and Metallurgy, University of Cambridge, Pembroke Street, Cambridge CB2 3QZ, UK

Abstract Magnetotactic bacteria can be regarded as model systems for studying the structural, chemical, and magnetic properties of arrangements of ferrimagnetic iron oxide and sulfide nanocrystals. The aim of the present chapter is to show how the size, shape, crystal structure, crystallographic orientation, and spatial arrangement of bacterial magnetite (Fe_3O_4) and greigite (Fe_3S_4) crystals affect their magnetic properties. We present recent results obtained using transmission electron microscopy (TEM) techniques, including high-resolution TEM imaging and off-axis electron holography.

1 Introduction

Magnetotactic bacteria contain intracellular ferrimagnetic crystals that are typically 30–120 nm in size and are arranged in a linear chain configuration. The magnetic properties of the bacterial iron oxide and sulfide nanocrystals depend on several factors, including crystal structure and habit, the direction of the crystal's elongation, and the orientation and spacing of each crystal with respect to its neighbors. It is necessary to study these properties in order to obtain an insight into the magnetic behavior of such magnetic particles.

The transmission electron microscope is a powerful tool for characterizing the physical and chemical properties of materials at high spatial resolution, and has been used for the study of bacterial magnetosomes since the discovery of magnetotactic bacteria (Blakemore 1975). Advanced transmission electron microscopy (TEM) techniques, including off-axis electron holography (EH) (Tonomura 1992; Völkl et al. 1998) and electron tomography (ET), for applications in the physical sciences (Midgley et al. 2001) have become widely available in the past decade. The combination of these new techniques and conventional high-resolution TEM (HRTEM) and selected-area electron diffraction (SAED) now allows the relationships between the structures, three-dimensional morphologies, configurations, and magnetic properties of magnetosomes to be studied in detail.

In this chapter, we briefly describe the principles of HRTEM, EH, and ET. The structural and magnetic properties of magnetite and sulfide magnetosomes are then discussed, with an emphasis on the interpretation of recent EH results. The quantitative nature of EH is highlighted.

2 Techniques

2.1 High-Resolution Transmission Electron Microscopy

In HRTEM, an image is formed by using a large objective aperture to cause one or more beams that have been diffracted by a specimen to interfere with the transmitted beam. In modern generations of transmission electron microscopes, the image resolution is better than 0.2 nm, and individual atomic columns can be resolved in many crystalline inorganic materials. If the objective lens defocus and image astigmatism and incident beam alignment are optimized, then the resulting image can be interpreted directly in terms of the projected crystal potential. With care, crystal lattice spacings, angles between lattice planes, and defects can be identified from such images. As the electron dose used to acquire a single image is typically approximately 500–2000 electrons per square angstrom, it is not usually possible to examine organic materials directly under such intense imaging conditions.

2.2
Off-Axis Electron Holography

EH allows the phase shift of a high-energy electron wave that has passed through a specimen to be recovered. The phase shift is, in turn, sensitive to the in-plane component of the magnetic induction and the electrostatic potential in the specimen. The TEM mode of off-axis EH is illustrated schematically in the form of a ray diagram in Fig. 1. The specimen is examined using highly coherent illumination from a field-emission gun electron source, with the region of interest positioned so that it covers approximately half the field of view. A (typically positive) voltage is applied to an electron biprism, which is located close to a conjugate image plane in the microscope, to overlap the electron wave that has passed through a vacuum (or through a thin region of support film) with a part of the same electron wave that has passed through the region of interest on the specimen, to form a hologram in a slightly defocused image plane. When magnetic materials are examined, holograms are normally recorded with the conventional microscope objective lens turned off, as its strong magnetic field would saturate the magnetization in the specimen in the electron beam direction. A Lorentz lens (a high-strength minilens located below the lower objective pole piece) can then be used to record

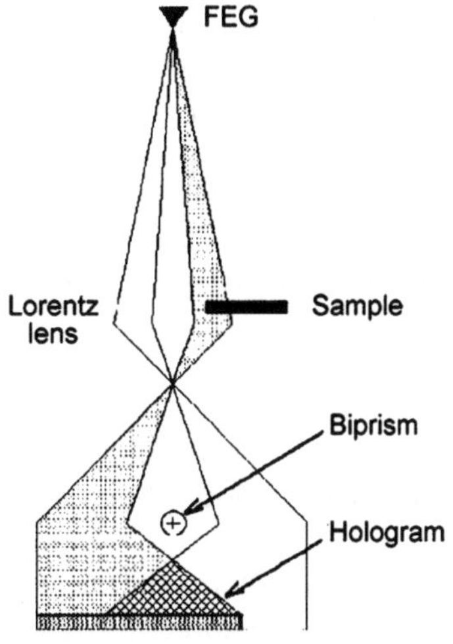

Fig. 1 Ray diagram illustrating the components of a transmission electron microscope that are required for off-axis electron holography (*EH*) of nanoscale magnetic materials. See text for details. *FEG* field-emission gun

holograms at high magnification with the specimen in a magnetic-field-free environment. The objective lens can also be excited slightly and the specimen tilted to apply known magnetic fields to follow magnetization processes in situ in the transmission electron microscope. When magnetic nanostructures are examined, the magnetic contribution to the phase shift recorded from a hologram must usually be separated from the dominant mean inner potential (MIP) contribution. This separation can be achieved by recording two holograms that differ only in the (opposite) directions of magnetization in the specimen. The magnetic contribution is then calculated by taking half of the difference between the phases of the holograms (Dunin-Borkowski et al. 2004). All of the magnetic induction maps presented in this chapter show the ferrimagnetic crystals in their remanent state, since the holograms were recorded in zero magnetic field.

2.3 Electron Tomography

In the life sciences, ET has been used to image complex nanoscale structures in three dimensions since the late 1960s (DeRosier and Klug 1968). In general, tomography requires a series of images (projections) of an object to be recorded at successive specimen tilt angles, using a signal that is related monotonically to the projected specimen thickness. Bright-field images, in which mass-thickness contrast dominates, are suitable for applications in the life sciences. However, in the physical sciences, crystalline objects result in the presence of diffraction and Fresnel contrast, and an incoherent form of imaging is more suitable. Images are then typically recorded using high-angle annular dark-field imaging using a scanned probe or energy-filtered transmission electron microscope. As many images as possible are recorded over as large a range of tilt angles as possible. After acquisition of the tilt series, each image is back-projected along the original tilt angle. The overlap of all projections then leads to a reconstruction of the three-dimensional object. Artifacts are minimized by employing an iterative routine that constrains the final reconstruction to match the original projections (Frank 1992).

3 The Structures and Magnetic Properties of Magnetite Magnetosomes

The structures and morphologies of magnetite magnetosomes have been studied since the discovery of magnetotactic bacteria (Balkwill et al. 1980; Towe and Moench 1981; Matsuda et al. 1983). Most studies found essentially no defects in the magnetite crystals (Mann et al. 1984; Sparks et al.

1990; Meldrum et al. 1993a, b). More structural disorder, primarily spinel-type twinning, was reported subsequently (Devouard et al. 1998), whereas other defects, which were interpreted as screw dislocations, were described in magnetite from uncultured magnetotactic bacteria (Taylor et al. 2001). Nevertheless, most bacterial magnetite crystals are free of extended structural defects.

The crystal habits of magnetite magnetosomes are species-specific, and have been described in terms of categories that include cubooctahedral, prismatic and arrowhead-shaped (Mann et al. 1984; Vali and Kirschvink 1990; Bazylinski and Frankel 2004). The sizes and shapes of magnetite crystals were found to be affected only slightly by variations in culturing conditions (Meldrum et al. 1993a, b). When they are elongated, bacterial magnetite crystals typically have elongation directions that are parallel to [111] (Mann et al. 1984; Sparks et al. 1990; Lins et al. 2005); however, highly elongated magnetite magnetosomes with elongation directions that are parallel to [112] (Mann et al. 1987; Taylor et al. 2001), [100] (Iida and Akai 1996; Taylor et al. 2001; Kasama et al. 2006a; Pósfai et al. 2006), and [110] (Taylor and Barry 2004) have also been described.

The magnetic dipole moments of magnetosome chains were calculated by Frankel (1984). On the basis of their sizes, the individual magnetosomes were assumed to be single magnetic domains. If the individual moments were assumed to be aligned parallel to each other and to the chain direction, then an ordered magnetite chain was calculated to have an overall magnetic dipole moment approximately equal to the sum of the individual magnetosome magnetic moments. Rock magnetism measurements of intact cells and separated chains were in agreement with these calculations, and indicated that both the dominant single-domain character and the chain configuration of the magnetite particles could be detected using bulk magnetic methods (Moskowitz et al. 1989, 1993; Weiss et al. 2004). Magnetic force microscopy was used to measure the magnetic dipole moment of a single cell (Proksch et al. 1995). Because of its high spatial resolution and sensitivity, EH takes the study of the magnetic properties of bacterial magnetosomes forward significantly, as described later.

Since crystal shape is important for determining the magnetic behavior of nanocrystals, it is convenient to describe the structural and magnetic characteristics of magnetite magnetosomes according to the three morphological groups that we have already described. The structures of all three types of crystals have been studied using HRTEM and SAED, and the magnetic microstructures of cubooctahedral and prismatic magnetosomes have been studied using EH. In the following sections, we discuss the structural and magnetic properties of cubooctahedral and prismatic magnetite magnetosomes. EH studies of arrowhead-shaped crystals, which are in progress, will be described elsewhere.

3.1
Cubooctahedral Magnetite Crystals

Magnetotactic spirilla such as *Magnetospirillum magnetotacticum* and *Magnetospirillum gryphiswaldense* produce magnetite crystals that are usually termed cubooctahedral magnetite (Bazylinski and Frankel 2000). Their morphologies, which are approximately equidimensional, can be described most simply in terms of a combination of the {111} and {100} forms (the octahedron and the cube, respectively). Some bacterial strains contain magnetite crystals that appear to be almost perfect octahedra, whereas other strains produce nanocrystals that are combinations of the octahedron, the cube, and the dodecahedron {110} (Devouard et al. 1998; Arató et al. 2005). For simplicity, all of these variations of roughly equidimensional magnetite morphologies are referred to here by the term cubooctahedral.

The precise identification of all of the faces of a nanocrystal is not straightforward, and has only been carried out in a few cases. As discussed by Buseck et al. (2001), the projected outlines of magnetosomes in bright-field TEM images provide ambiguous information about their three-dimensional shapes. In particular, it is difficult to identify whether a straight segment of the crystal outline is a projection of a face that is parallel to the electron beam or that of an edge shared by two faces (Lins et al. 2005). This difficulty is exacerbated by the fact that some parts of a crystal outline may appear rounded or rough. In addition, the details of such contrast features may vary with objective lens defocus. The ambiguity of crystal shape determination from two-dimensional projections was illustrated by a debate on the potential biogenic origin of magnetite crystals in the Martian meteorite ALH84001 (Thomas-Keprta et al. 2000; Buseck et al. 2001; Clemett et al. 2002; Golden et al. 2004), the central issue of which was whether the bacterial magnetite crystals have unique morphological characteristics that distinguish them from crystals that formed inorganically.

Since the contrast in HRTEM images is sensitive to the thickness of a crystal, it is possible to obtain some information about the three-dimensional crystal morphology by observing the contrast variation in experimental images. Careful examination of the HRTEM image of a magnetite magnetosome in Fig. 2a reveals that the contrast shows a characteristic variation from the edges to the center of the crystal. Regions having different contrast are marked by different shades of gray in Fig. 2b, with each shade representing regions of similar crystal thickness projected in the direction of the electron beam. Although care is required with the interpretation of such images because the exit surface of the crystal also varies, the crystal appears to be much thinner at regions marked A than at those marked B, even though both regions are near the outline of the crystal's projected image. These observations suggest that the surfaces that are parallel to (01–1) and (0–11) planes are faces that are aligned parallel to the incident electron beam, whereas at the points

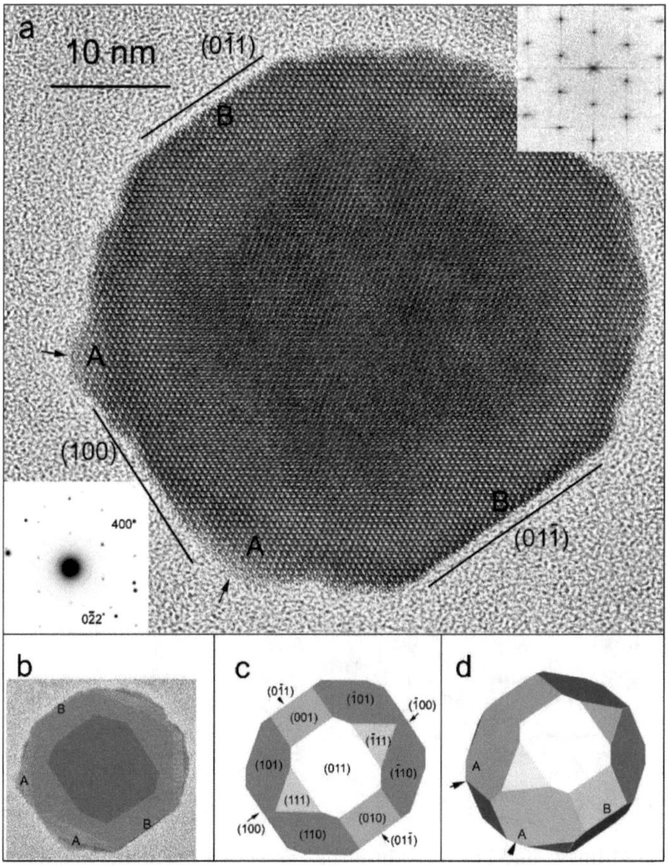

Fig. 2 **a** High-resolution transmission electron microscopy (*HRTEM*) image of a structurally perfect cubooctahedral magnetite crystal from *Magnetospirillum gryphiswaldense*, viewed along [011]. The *insets* in the *lower-left corner* and the *upper-right corner* are an experimental selected-area electron diffraction (*SAED*) pattern and a Fourier transform of the image, respectively. The labels *A* and *B* and the *short arrows* are discussed in the text. **b** "Thickness map" of the magnetosome in **a**, inferred from the variations in the HRTEM image contrast. **c, d** Assumed morphological model for the magnetite crystal in **a**, displayed in two slightly different orientations

that are marked by short arrows in Fig. 2a several edges meet, resulting in regions that are thinner in projection. A possible morphological model for this cubooctahedral magnetite crystal is presented in Fig. 2c. In order to illustrate the crystal morphologies at the regions marked A and B, the same model is shown from a slightly different direction in Fig. 2d.

Another example of the determination of crystal thickness from an HRTEM image is shown in Fig. 3. This small magnetite magnetosome is from a cell of a mutant of *Magnetospirillum gryphiswaldense* that synthesizes

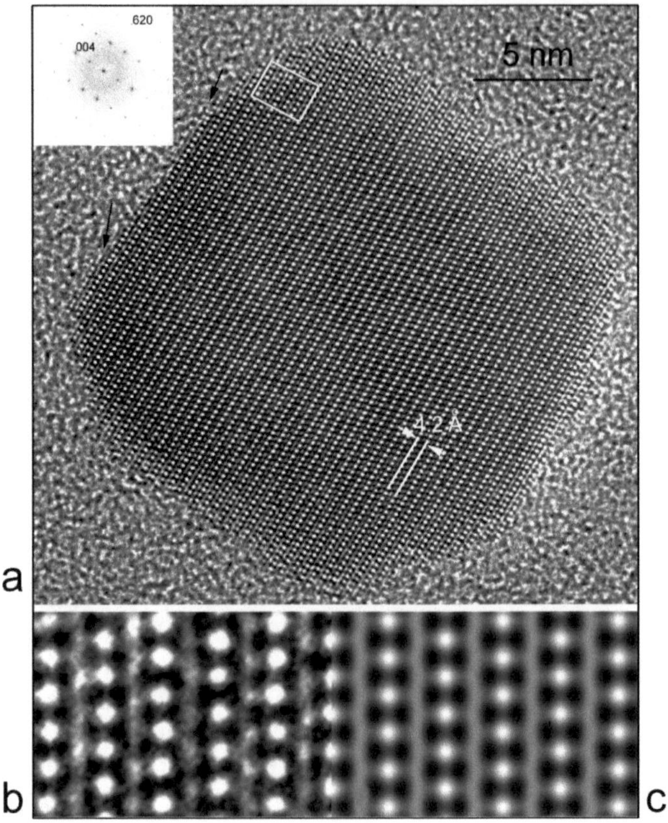

Fig. 3 **a** HRTEM image and Fourier transform (*inset*) of a magnetite magnetosome from a mutant of *Magnetospirillum gryphiswaldense*, viewed along [1–30]. **b** A magnified image of the boxed region in **a.** **c** Calculated HRTEM image, simulated using the parameters 400-kV accelerating voltage, C_s = 1 mm, – 74-nm objective lens defocus, 6-nm^{-1} objective aperture, 4-nm crystal thickness, no crystal or beam tilt. The *arrows* in **a** mark terminating structural planes on the crystal surface

slightly smaller magnetosome crystals. By comparing the contrast in experimental and simulated HRTEM images, the approximate thickness of the crystal can be determined. The boxed region in Fig. 3a is shown enlarged in Fig. 3b, and a corresponding calculated image is shown in Fig. 3c. Although it is notoriously difficult to match calculated HRTEM images reliably to experimental images, the simulated and experimental images in Fig. 3 show a good match of contrast, suggesting that the thickness value (4 nm) that was used for the calculation is approximately correct. The contrast varies little from the edge to the center of the particle, suggesting that this magnetosome is very thin and may have a tabular morphology. On the edge of the crystal, terminating structural planes (arrowed in Fig. 3a) appear to be edge-on views of atomic-scale steps on the surface of the crystal.

Cubooctahedral magnetite magnetosomes typically do not contain extended defects other than twin boundaries (Fig. 4). Twins are of spinel type, as also observed in inorganically formed magnetite. Spinel twins are related to each other by a twofold axis that is parallel to [111]. The twin boundary can lie along a single (111) plane, although in many bacterial magnetite crystals

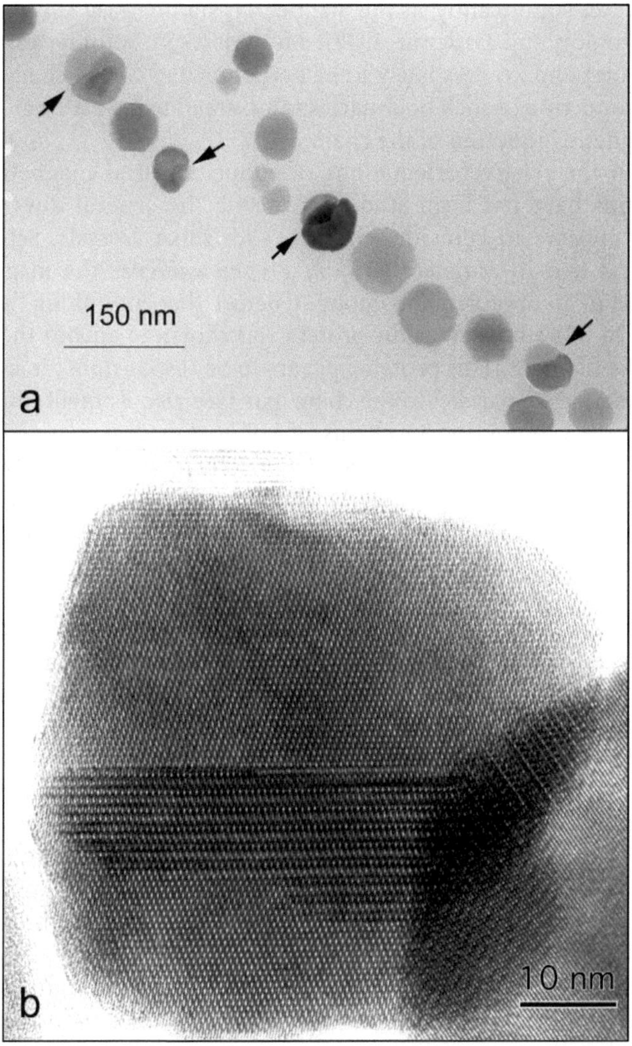

Fig. 4 **a** Bright-field TEM image of a chain fragment of magnetite magnetosomes from *Magnetospirillum gryphiswaldense* (wild-type), showing twinned crystals (*arrowed*). **b** HRTEM image of a twinned magnetite crystal from *Magnetospirillum magnetotacticum* strain MS-1. The two twin member crystals overlap, producing a Moiré effect. (**b** From Devouard et al. 1998)

the two parts of the twin meet at an irregular surface (Fig. 4b). The proportion of twinned magnetosomes is strain-specific (Devouard et al. 1998). Whereas in some strains fewer than 1% of all magnetite crystals are twinned, a strain containing a large proportion (approximately 70%) of twinned crystals has also been observed (D. Bazylinski, private communication). With rare exceptions, twin boundaries occur perpendicular to the chain axis; thus, they are not expected to affect the direction of the magnetic induction within the magnetosomes significantly, as the magnetocrystalline easy axis of magnetite is [111] (Dunlop and Özdemir 1997). Multiple twin boundaries tend to be more irregular and are less likely to be perpendicular to the chain axis; however, the proportion of such boundaries is too small to have a significant effect on the magnetic induction of the chain.

Although the relative orientations of cubooctahedral magnetite crystals within chains have not been studied in detail, the general direction of the chain axis appears to coincide with [111] for most crystals. Scheffel et al. (2005) found that in *Magnetospirillum gryphiswaldense* the magnetosomes are attached to an organic filamentous structure that runs along the long axis of the cell. In this study, an acidic protein was shown to anchor the magnetosomes to the filament. This protein appears to be responsible for setting [111] in each crystal to be parallel to the chain axis (see also Komeili, this volume).

The magnetic induction in chains of cubooctahedral magnetite crystals has been studied in *Magnetospirillum magnetotacticum* strain MS-1 using EH (Dunin-Borkowski et al. 1998, 2001). Figure 5 shows a magnetic induction map obtained from EH images of a magnetosome chain that can be regarded as representative for MS-1. The magnetite crystals are about 45 nm in size. The direction of the magnetic phase contours shown in Fig. 5 represents the direction of the local magnetic induction (projected in the electron beam direction), and the density of the contour lines provides a measure of the strength of the local in-plane induction. The same amount of magnetic flux is enclosed between any two adjacent contours.

In Fig. 5, the crystal size along the chain is not uniform. The sizes of the magnetosomes gradually decrease on the right side of the chain, whereas the corresponding change on the left side of the chain is more abrupt. This variation in crystal size affects the strength of the local magnetic induction. The contour map shows a larger number of closely spaced contours at the left end of the chain, where larger crystals are present. The crystals at the right end of the chain are below approximately 20 nm in size, and thus would be expected to be superparamagnetic if they were isolated. However, the direction of the magnetic contours is confined to be parallel to the chain axis in these crystals, indicating that they are stable single magnetic domains at room temperature. This behavior likely results from magnetic interactions between adjacent nanocrystals in the linear chain configuration.

Two smaller crystals are present at the position marked by the white arrow in Fig. 5, creating a defect in the chain structure and resulting in slightly

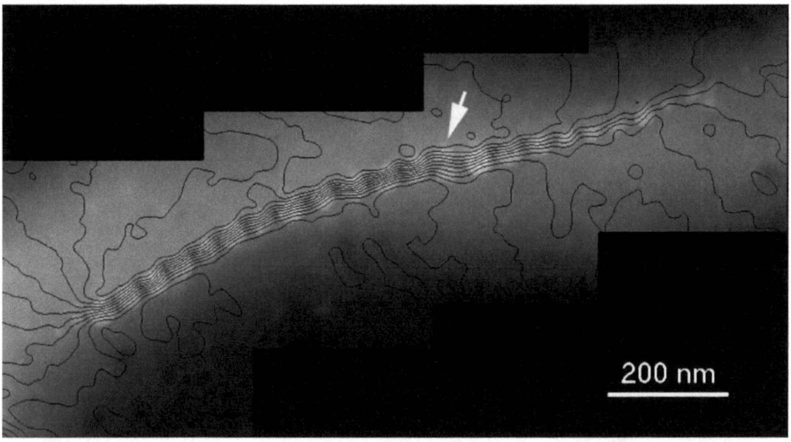

Fig. 5 Magnetic induction map recorded using EH from a chain of magnetite magnetosomes in a cell of *Magnetospirillum magnetotacticum* strain MS-1. The mean inner potential contribution to the phase shift was removed from the measured signal before creating the contour map. The *white arrow* marks a defect in the chain that consists of two small crystals. The spacing of the contours is 0.064 rad. (From Dunin-Borkowski et al. 1998, 2001)

poorer confinement of the magnetic phase contours. In a similar cell, slight waviness of the contour lines was caused by a small displacement of the magnetosomes from the axis of the magnetosome chain (Dunin-Borkowski et al. 1998, 2001). The contour lines within these magnetosomes were observed to bend to follow the positions of the crystals, irrespective of their crystallographic orientations. Thus, the positions of the magnetite crystals dictate the local direction of the magnetic induction. Since the crystals are roughly equidimensional, neither shape nor magnetocrystalline anisotropy has a significant effect on the magnetic induction for chains of cubooctahedral magnetite magnetosomes.

3.2 Elongated Prismatic Magnetite Crystals

Magnetite magnetosomes with elongated "prismatic" habits typically occur in magnetotactic vibrios and cocci, collected from both freshwater and marine environments. In many types of cells, these crystals form well-organized single or double chains. Twinning occurs in prismatic magnetite magnetosomes (Devouard et al. 1998), but other types of structural defect have not been observed.

The magnetite crystals from a single cell of a freshwater magnetotactic coccus that are shown in Fig. 6a and b are free of extended defects. Each of the numbered crystals in the double chain shown in Fig. 6a was aligned to either

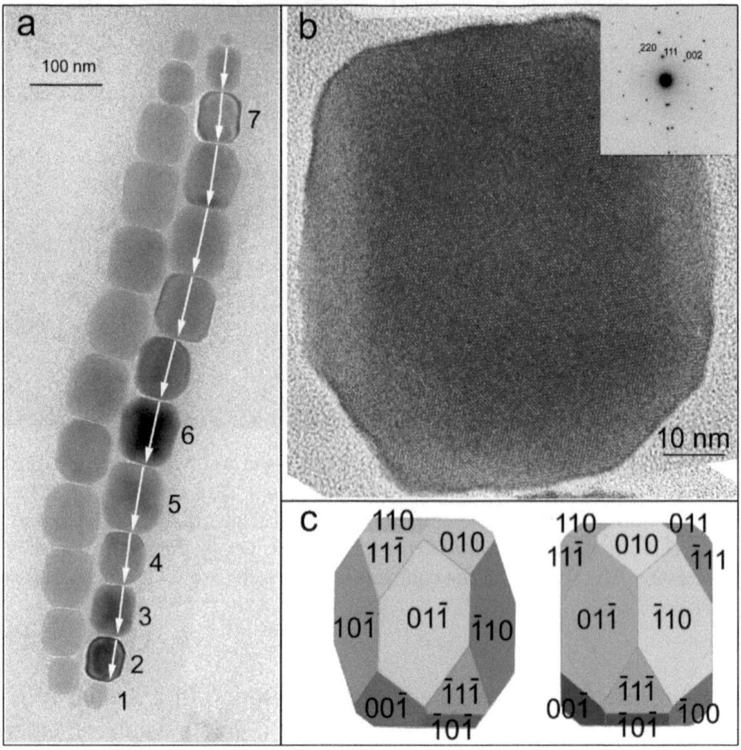

Fig. 6 **a** Bright-field TEM image of a double chain of magnetite magnetosomes from a single cell of a freshwater magnetotactic coccus. The numbered crystals were found to have their [111] directions parallel to the magnetosome chain axis to within 3° (except for crystal 1 at the end of the chain, which was misaligned with respect to the chain axis). **b** HRTEM image and SAED pattern of crystal 4 from **a**, viewed along [- 110]. **c** Model for the assumed morphology of the magnetite crystals in **a**, shown in the two directions ([011] and [121]) in which HRTEM images were recorded. (**a**, **b** From Simpson et al. 2005)

a [110] or a [112] axis, in order to determine the relative orientations of the crystals with respect to one another (Simpson et al. 2005). Significantly, the [111] axes of almost all of the crystals were found to be aligned parallel to the chain axis to within a few degrees, whereas their crystallographic directions perpendicular to the chain axis were random. Similar arrangements were found in other cells that contained single chains of magnetite magnetosomes with prismatic habits (Lins et al. 2005; I. Dódony, private communication). Assuming that the relative orientations of the crystals are not affected by air-drying of the cells on TEM grids, these results suggest that biological control of the orientations of magnetosomes is stricter in setting the [111] magnetic easy axis to be parallel to the chain axis than in controlling the orientations of the crystals perpendicular to this direction.

A tentative model for the typical morphology of the magnetosomes shown in Fig. 6a is illustrated for two orientations of the crystal in Fig. 6c. The inferred morphology is formed from a combination of the {111}, {110}, and {100} forms. Characteristically for prismatic magnetite magnetosomes, the largest faces are those six faces of {110} that are parallel to the [111] axis of elongation.

A magnetic induction map obtained from the double magnetite chain shown in Fig. 6a using EH is presented in Fig. 7. In each crystal, the direction of the recorded magnetic induction is parallel to the direction of elongation of the crystal. Since this direction coincides with [111] and the chain axis, the effects of shape and magnetocrystalline anisotropy and interactions between neighboring crystals reinforce each other. In other chains of prismatic magnetite magnetosomes, magnetic phase contours have also been found to be parallel to the elongation axes of the individual crystals (Figs. 8, 9). As a consequence of the precise alignment of the elongation axes of the magnetosomes with the chain axis, as well as the proximity of adjacent crystals, the magnetic phase contours are highly parallel to each other and confined within the magnetosomes. In Fig. 9, the two chains within a single cell show little magnetic interaction with each other and appear to be largely magnetically independent. In contrast, the double chain shown in Figs. 6 and 7 can be regarded as similar to a single bar magnet, with the magnetic flux returning outside the combined double chain.

One of the most studied magnetotactic strains is MV-1. Cells of this strain also contain elongated, prismatic magnetite crystals that form a single chain in each cell (Fig. 8). Each magnetosome is a single magnetic domain. However, the gaps between the magnetosomes are larger than in the double chain shown in Figs. 6 and 7. As a result, the separation of the magnetic contour

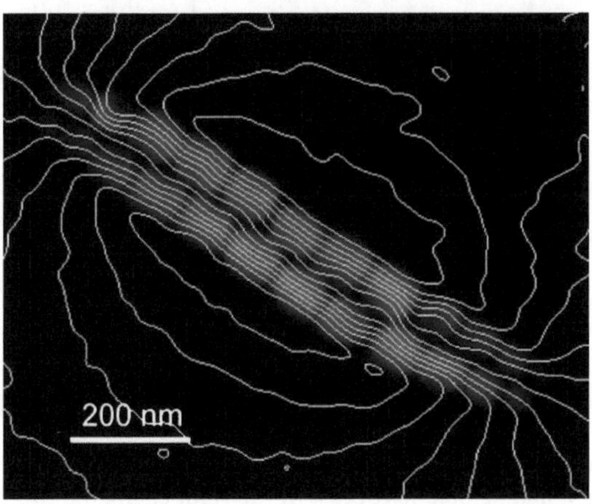

Fig. 7 Magnetic induction map recorded using EH from the double chain of magnetite magnetosomes shown in Fig. 6a. The spacing of the contours is 0.3 rad

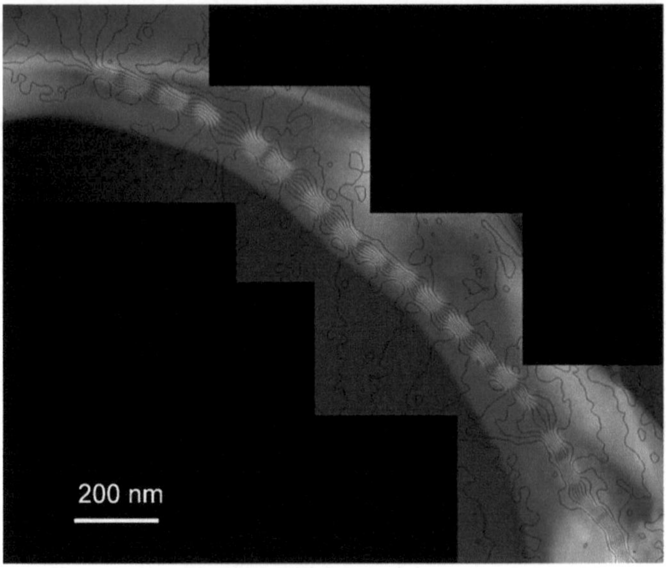

Fig. 8 Magnetic induction map recorded using EH from a magnetite magnetosome chain from bacterial strain MV-1. The phase contour spacing is 0.064 rad. (From Dunin-Borkowski et al. 1998, 2001)

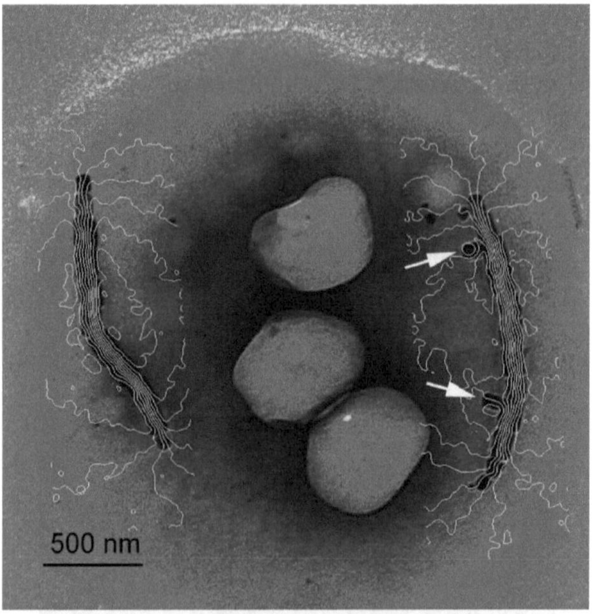

Fig. 9 Magnetic induction map recorded using EH from two single chains of magnetite crystals in a cell of a magnetotactic coccus. In the right chain, the two *arrowed* magnetosomes are away from the primary chain axis, with their axis of elongation perpendicular to the chain. (From Dunin-Borkowski et al. 2004)

lines increases in the gaps between the crystals. Even if a crystal is misplaced from the chain axis (for example, the fourth magnetosome from the left in Fig. 8), the direction of the contour lines inside the crystal is parallel to its elongation axis, suggesting that shape anisotropy plays a dominant role in the elongated magnetite crystals in MV-1.

In the cell shown in Fig. 9, two single chains of prismatic magnetite magnetosomes are positioned at opposite sides of the cell. Hanzlik et al. (1996) suggested that such chain arrangements may balance such a cell magnetically as it swims parallel to the direction of an external magnetic field. A noteworthy feature of the magnetosome chain on the right of Fig. 9 is that two magnetite crystals are in irregular positions perpendicular to the chain axis. The fact that the magnetic phase contours inside these two magnetosomes remain parallel to their elongation axes suggests that for these crystals shape anisotropy dominates over the effect of interactions with other particles in the chain.

The results presented in Figs. 6–9 show that, in general, in ordered chains that consist of well-aligned, elongated magnetite magnetosomes, the effects of magnetocrystalline and shape anisotropy, as well as magnetic interactions between adjacent particles, reinforce each other to produce a magnetic moment that is strictly parallel to the elongation direction of each crystal and to the chain axis.

3.3 Unusually Large Magnetite Magnetosomes, and Scattered or Clustered Crystals

Unusually large magnetosomes were described from a magnetotactic coccus from the Itaipu Lagoon in Brazil (Farina et al. 1994; Spring et al. 1998). The cells of this organism contain magnetite crystals that are approximately 200 nm in length and thus would not be expected to be in a single-domain magnetic state (Dunlop and Özdemir 1997). Nevertheless, magnetic induction maps obtained from EH measurements by McCartney et al. (2001) reveal that the large magnetite crystals are essentially single magnetic domains, so long as they are in a linear chain configuration (Fig. 10a). The magnetic contour lines fringe out at the edges of the last crystals in the chain, creating flowerlike patterns, but inside the crystals the closely spaced contour lines follow a path that is parallel to the elongation direction of each crystal and to the chain axis. Interestingly, a broken chain configuration results in a more complicated magnetic microstructure, with domain walls and possibly vortex states similar to those reported by Hÿtch et al. (2003) appearing inside the large crystals (Fig. 10b). It therefore appears that for crystals of this size interparticle interactions constrain the direction of the magnetic field and are responsible for the single-domain state of these large magnetite crystals within linear magnetosome chains.

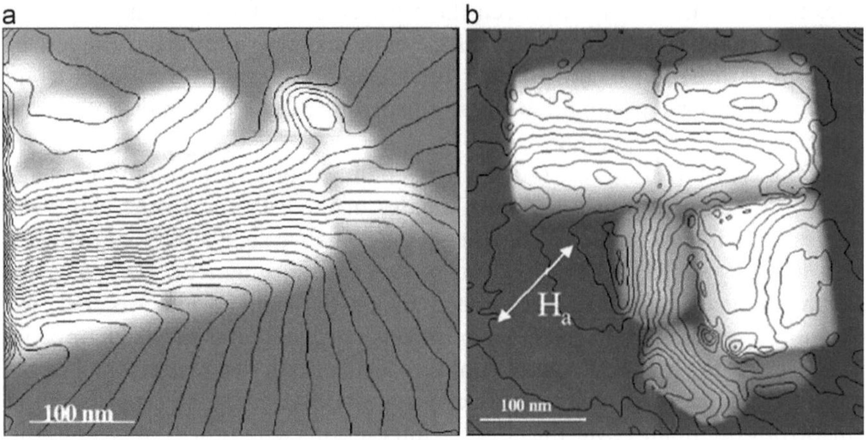

Fig. 10 Magnetic induction maps recorded using EH from large magnetite magnetosomes from magnetotactic cells that were collected in the Itaipu Lagoon, Brazil. **a** When the large magnetosomes are in a chain configuration, magnetic interactions result in parallel magnetic contour lines within the crystals. Smaller magnetosomes from other types of cells cling to the larger particles. **b** The large magnetosomes are no longer in a single-domain state in broken chains, and exhibit a more complicated magnetic microstructure. The *double-headed arrow* labeled H_a indicates the direction of the applied magnetic field before recording the remanent magnetic state. The contour spacing is 0.5 rad. (From McCartney et al. 2001)

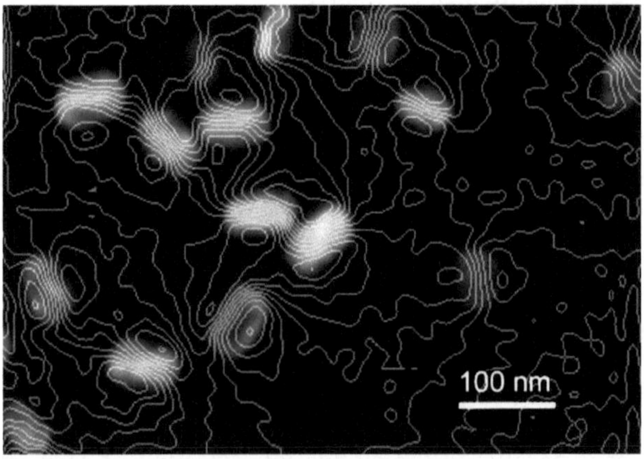

Fig. 11 Magnetic induction map recorded using EH from scattered magnetite crystals in a single cell of a freshwater coccus. All of the particles are magnetic single domains. The phase contour spacing is 0.125 rad

The profound effect of the orientations and separations of magnetite nanocrystals on their magnetic induction is illustrated by the magnetic induction map shown in Fig. 11. In this freshwater, spherical cell, a disordered arrangement of prismatic magnetite magnetosomes, which may have resulted in part from air-drying of the cell, results in highly variable directions of the magnetic field inside the crystals. However, as discussed before, in most cases the magnetic contour lines remain parallel to the elongation axis of each crystal, as dictated by shape anisotropy.

4
The Crystal Structures and Magnetic Properties of Iron Sulfide Magnetosomes

Iron sulfide magnetosomes contain a larger variety of inorganic phases than iron oxide magnetosomes. Whereas the only magnetosome iron oxide mineral that has been identified is magnetite, the inorganic phase of iron sulfide magnetosomes can comprise either greigite (Fe_3S_4) (Mann et al. 1990; Farina et al. 1990; Heywood et al. 1990) or mackinawite (FeS) (Pósfai et al. 1998a, b). In addition, a third structure, sphalerite-type cubic FeS, was identified tentatively (Pósfai et al. 1998a). It appears that nonmagnetic mackinawite is initially precipitated, before converting into ferrimagnetic greigite. The time that is required for this solid-state phase transition in living cells is unknown, but is presumably within the limits set by the requirement that the majority of the magnetosomes in a cell must always be magnetic. The mackinawite to greigite transition was observed to take place over 10 days in a dehydrated cell on a TEM grid that was stored in air (Pósfai et al. 1998a). The transition may be faster in living cells in their natural sulfide-rich environment.

Mackinawite and greigite have the same cubic-close-packed sulfur substructure. The orientation relationship between mackinawite-like and greigite-like structural elements in crystals that appear to be in a transitional state between the two minerals (Fig. 12) suggests that the conversion of mackinawite takes place with the sulfur substructure remaining intact. Only the iron atoms change their crystallographic positions, one quarter of them being lost from each crystal. The fate of the iron that is released from the magnetosomes remains unknown.

Sulfide magnetosomes typically show heterogeneous, patchy contrast in HRTEM images. This feature likely results from defects that remain in the greigite structure as a result of the solid-state transformation from mackinawite. The greigite structure appears to have nucleated in several places within the crystal shown in Fig. 12. When such greigite nuclei grow and merge, antiphase domain boundaries may remain in the structure. It is also possible that some of the contrast variation seen in HRTEM images is caused by the uneven thickness of the sulfide magnetosomes (Fig. 13b, c). Many greigite magnetosomes appear to be aggregates of several smaller crystals, each in the same crystallographic orientation, producing a single-crystal-like

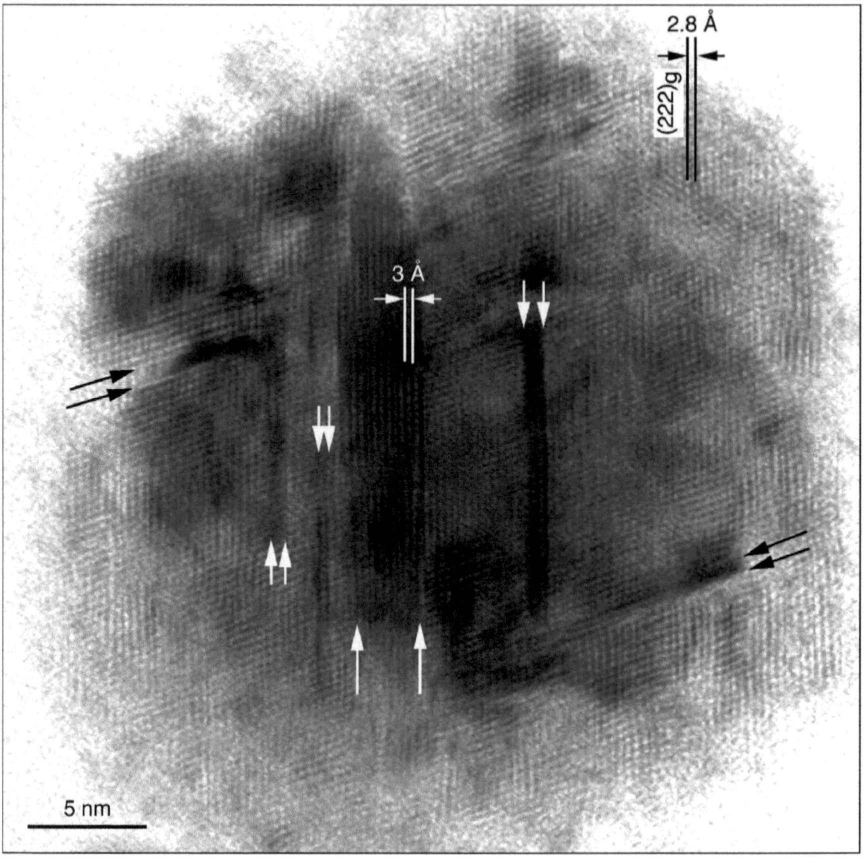

Fig. 12 HRTEM image of an iron sulfide magnetosome from a multicellular magnetotactic aggregate that contains both mackinawite and greigite structural elements, and is in a transitional state between the two structures. The spacing of lattice fringes in the bands between the *double white arrows* and the *double black arrows* is 3 Å, consistent with $d(011)$ of mackinawite, whereas in the rest of the crystal the 2.8-Å spacing is consistent with $d(222)$ of greigite. (From Pósfai et al. 1998a)

aggregate (Kasama et al. 2006a). Interestingly, even though greigite and magnetite are isostructural, spinel-type twins have not been observed in greigite magnetosomes.

Since sulfide-producing magnetotactic bacteria are not yet available in pure culture, and since the number of cells separated magnetically from samples collected from the environment is small, bulk magnetic measurements have not yet been performed on iron sulfide magnetosomes. However, measurements of the magnetic properties of individual sulfide magnetosomes and their chains have recently been performed using EH (Kasama et al. 2006a).

Iron sulfide magnetosomes have been observed both in organisms that are usually referred to as magnetotactic multicellular aggregates or prokaryotes (MMA or MMP) (see also Keim et al., this volume), and in various types of rod-shaped cells (Fig. 13a) (Heywood et al. 1990; Bazylinski et al. 1993; Simmons et al. 2004). In any of these cell types, the sulfide magnetosomes generally form chains that are not as well ordered as is typical in magnetite-bearing organisms. In most studies of sulfide-producing bacteria, the cells were not fixed but left to dry on TEM grids (Mann et al. 1990; Pósfai et al. 1998a, b; Kasama et al. 2006a), raising the concern that some of the disorder in the orientations and positions of the sulfide magnetosomes may be an artifact of the dehydration of the cell. However, even fixed sulfide-producing cells contain fairly disorganized magnetosome chains (Keim et al. 2004), suggesting that the random orientations of the crystallographic and elongation directions of the sulfide magnetosomes within a chain may be typical features of sulfide-producing magnetotactic bacteria.

The disorganized arrangement of sulfide magnetosomes has a strong influence on the magnetic properties of the chain, as illustrated in Fig. 14. This cell is at the point of cell division and contains a multiple chain of magnetosomes that is divided approximately equally between the two daughter cells. Magnetic contour lines derived from EH are superimposed on the image and show that the direction of the magnetic field follows a meandering path along the chain. Highly variable directions of the magnetic moments in individual magnetosomes are apparent in the enlarged inset that shows the central part of the cell and the division plane. Many of the magnetosomes appear

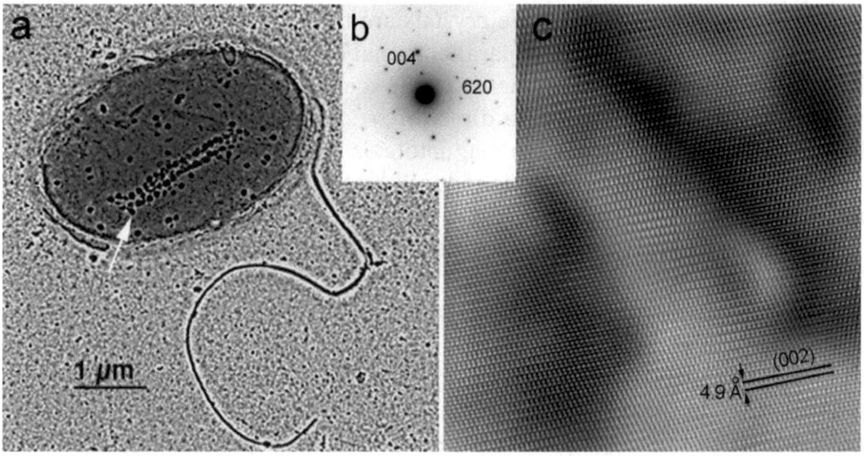

Fig. 13 **a** Bright-field TEM image of a rod-shaped cell that contains a double chain of iron sulfide magnetosomes. **b** SAED pattern and **c** HRTEM image of the greigite magnetosome marked by an *arrow* in **a**. The inhomogeneous contrast in **c** is likely caused by variations in the thickness of the crystal. (From Kasama et al. 2006b)

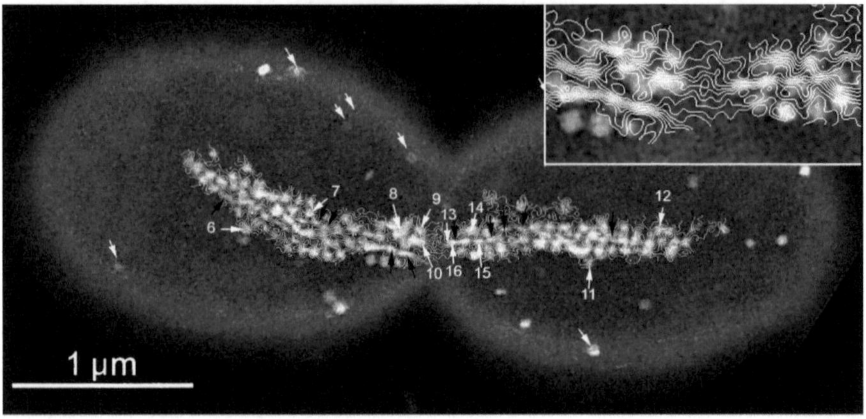

Fig. 14 Magnetic induction map recorded using EH from a dividing cell that contains iron sulfide magnetosomes. The structures of the numbered crystals were identified and found to be consistent with greigite. The *inset* shows the central part of the cell at higher magnification. The magnetic phase contours are wavy, and their density within individual magnetosomes is highly variable. The *black arrows* mark elongated magnetite crystals. *White arrows without numbers* point to magnetosome-sized, iron-free crystals that contain sulfur and oxygen. The magnetic phase contour spacing is 0.098 rad

to be only weakly magnetic or not magnetic at all. In order to determine whether this magnetic behavior is related to the mineral phase of the sulfide magnetosomes, SAED patterns were obtained from each of the numbered crystals in Fig. 14. Despite the fact that these particles include both apparently strongly and weakly magnetic crystals, all of them were found to be greigite. The apparently nonmagnetic crystals are therefore probably magnetized along a direction that is almost parallel to that of the electron beam. Since EH is sensitive only to the components of the magnetic induction that are perpendicular to the electron beam direction, such crystals may then appear to be nonmagnetic. With regard to magnetotaxis, it is a disadvantage for the cell that its total magnetic moment is reduced by the random orientations of the magnetosomes.

In general, in Fig. 14 the contour lines inside the magnetosomes are parallel to the elongation direction of each crystal. On the basis of SAED patterns obtained from the numbered crystals in Fig. 14, there is no preferred crystallographic direction for the elongation of the sulfide magnetosomes. This observation suggests that magnetocrystalline anisotropy plays little role in determining the direction of elongation of the magnetosomes. The structural defects and the aggregated nature of some of the greigite magnetosomes may affect their magnetic properties. However, as for the magnetite crystals described before, it appears that the magnetic moments of the greigite magnetosomes are influenced most strongly by the shapes and the relative positions of the crystals.

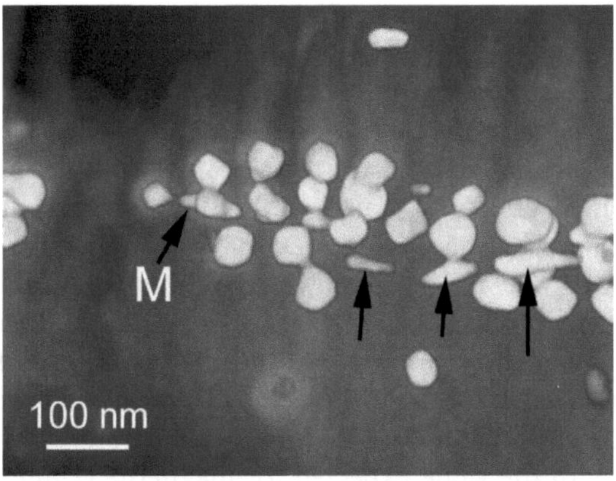

Fig. 15 Visualization of a three-dimensional electron tomographic reconstruction of the shapes of the magnetosomes within the right half of the boxed region in Fig. 14, derived from a tilt series of high-angle annular dark-field images. The *arrows* mark highly elongated iron oxide crystals. The other magnetosomes are greigite

In addition to the sulfide magnetosomes, magnetite particles occur in the multiple chain shown in Fig. 14. The particles that are marked by dark arrows are highly elongated and contain dense magnetic contours. Quantitative measurements of their magnetic induction (as discussed later), compositional data obtained using energy-filtered images, and an SAED pattern obtained from the particle marked M in Fig. 15 are all consistent with the elongated particles being magnetite. Magnetotactic cells that contained both iron sulfide and iron oxide magnetosomes were described by Bazylinski et al. (1995). In the case of the cell presented in Fig. 14, results from ET (Fig. 15) confirm that the greigite magnetosomes are oriented randomly in the multiple chain, whereas the highly elongated magnetite particles are parallel to the chain axis. The magnetite particles therefore provide the cell with a strongly magnetic "backbone".

5 Quantitative Measurements Using Electron Holography

Using EH, we can measure the magnetic properties of both individual magnetosomes and entire cells quantitatively. As mentioned before, holographic phase images contain information about both the magnetic induction $\boldsymbol{B}$ and the MIP V_0 of the sample. In situ magnetization reversal experiments allow the magnetic contribution to the phase shift to be separated from the MIP contribution. If a region of the sample is chosen in which demagnetizing fields are negligible, such as a crystal close to the center of a chain of closely spaced magnetosomes, then the magnetic and MIP contributions to the phase shift can be used to measure the ratio V_0/B. The equations that describe the details of this approach are given elsewhere (Simpson et al. 2005).

The ratio V_0/B was measured experimentally from cells that contained closely spaced prismatic magnetite crystals. The results are plotted in the form of a histogram in Fig. 16. The scatter in the measured values results from the combined effects of noise, diffraction contrast, and residual demagnetizing fields. The magnetic induction can be determined from such measurements of V_0/B if V_0 is known. Experimentally, the MIP of magnetite has been measured as 17 V (Harrison et al. 2002). Despite the scatter in the measurements shown in Fig. 16, and the possibility of systematic errors, the use of this value for the MIP results in a mean value for the measured magnetic induction of 0.58 ± 0.02 T at room temperature, which is in excellent agreement with the value expected for magnetite (Dunlop and Özdemir 1997). This agreement confirms the chemical purity and magnetic homogeneity of the biogenic crystals.

Quantitative measurements of the magnetic properties of the greigite and magnetite magnetosomes that are visible within the inset in Fig. 14 reveal a relationship between the morphologies, compositions, and magnetic inductions of the magnetosomes in this multiple chain, as shown in Fig. 17. As mentioned before, the sulfide particles in this chain are roughly equidimensional, whereas the oxide particles have elongated shapes. The measured magnetic inductions in the sulfide particles (Fig. 17a) are approximately consistent with values expected from the literature for greigite (approximately 0.16 T) (Dunlop and Özdemir 1997). The magnetic inductions in the elongated oxide particles (Fig. 17b) range from 0.1 to 0.6 T, suggesting that these particles are either magnetite (0.60 T) or maghemite (0.48 T). Since several of the oxide particles show induction values larger than 0.48 T, it is likely that they are magnetite. Many values in Fig. 17 are below the expected 0.16 and 0.60 T for greigite and magnetite, respectively. Several factors may contribute

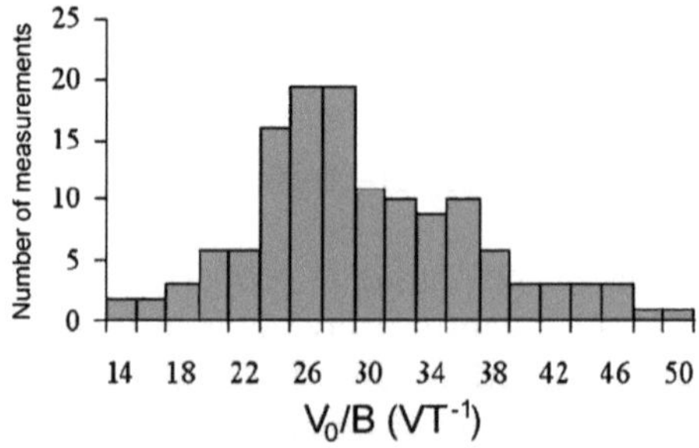

Fig. 16 The V_0/B ratio measured from a large number of magnetite magnetosomes in cells that contained two double chains of elongated prismatic crystals (such as the double chain shown in Fig. 6). The mean value of V_0/B is 29.5 ± 0.8 V T^{-1}. (From Simpson et al. 2005)

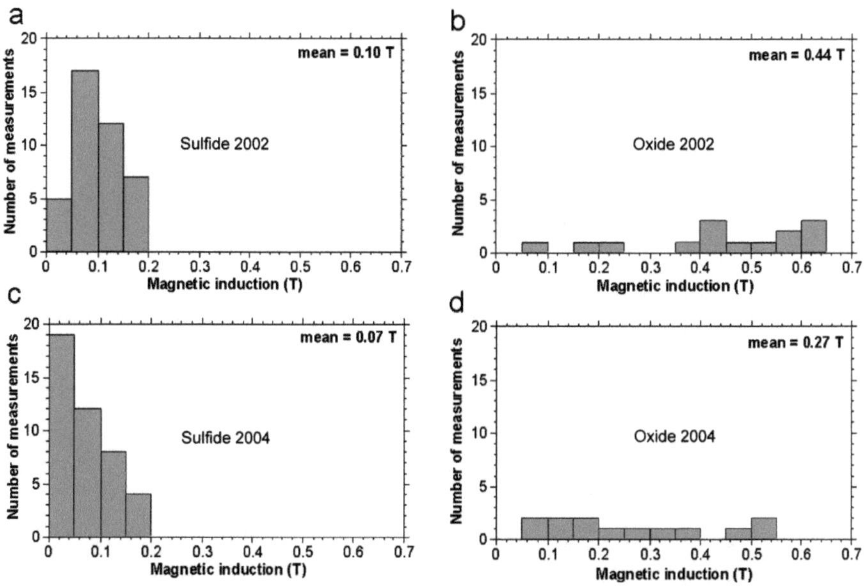

Fig. 17 The measured magnetic induction of a large number of sulfide and oxide particles in the multiple magnetosome chain shown in Fig. 14, measured both in 2002 and in 2004. Over the 2 years of sample storage, both sulfide and oxide magnetosomes partially oxidized and their magnetic induction decreased. (From Kasama et al. 2006a)

to the apparent reduction in the magnetic induction, including magnetization directions that are not perpendicular to the electron beam direction, the difficulty of measuring the physical dimensions of the nanocrystals precisely, and partial oxidation of the magnetosome minerals. The measurements also provide an opportunity to assess the effects of oxidation of the magnetosomes on their magnetic induction. Figure 17c and d shows that, when the sample was stored in air for 2 years, the magnetic induction of both the sulfide and the oxide magnetosomes decreased.

Using EH, we can measure the magnetic moments of entire cells of magnetotactic bacteria. The procedure, which is described in detail elsewhere (Dunin-Borkowski et al. 2001, 2004), involves measuring the area under the gradient of the magnetic contribution to the phase shift, evaluated in a direction perpendicular to that of the desired magnetic moment. Magnetic moments measured using this approach from single, double, and multiple chains containing cubooctahedral and prismatic magnetite and greigite magnetosomes are listed in Table 1. From Table 1, it is apparent that both the magnetic moment and the magnetic moment per unit length are very similar for all of the chains.

Table 1 Magnetic moments of individual cells measured using EH

Strain	Magnetosome mineral	Average length of each magnetosome (nm)	Number of magnetosomes in chain	Magnetic moment (Am^2)	Length of chain (μm)	Magnetic moment per unit length ($Am^2\ \mu m^{-1}$)	Reference
MS-1 (Fig. 5)	Magnetite (single chain)	~ 45	22	5×10^{-16}	1.2	4.2×10^{-16}	Dunin-Borkowski et al. (1998, 2001)
MV-1 (Fig. 8)	Magnetite (single chain)	~ 60	15	7×10^{-16}	1.6	4.4×10^{-16}	Dunin-Borkowski et al. (1998, 2001)
Uncultured coccus (Figs. 6, 7)	Magnetite (double chain)	~ 80	25	1.7×10^{-15}	0.95	1.8×10^{-15}	Simpson et al. (2005)
Uncultured, rod-shaped cell (Fig. 13)	Greigite and possibly magnetite (double chain)	~ 60	57	9.0×10^{-16}	2.19	4.1×10^{-16}	Kasama et al. (2006a,b)
Uncultured, dividing, rod-shaped cell (Fig. 14)	Greigite and magnetite (multiple chain)	~ 60 (greigite) ~ 80 (magnetite)	~ 155	1.8×10^{-15}	2.94	6.1×10^{-16}	Kasama et al. (2006a)

6 Summary

In this chapter, the application of HRTEM and off-axis EH to the characterization of the physical and magnetic properties of ferrimagnetic iron oxide and iron sulfide crystals in magnetotactic bacteria has been described. It has been shown that HRTEM images can be used to provide important information about the sizes, shapes, orientations, spatial arrangements, crystal structures, and defect structures of such crystals, and that this information can now be complemented by three-dimensional information about nanoparticle morphologies obtained using the evolving technique of ET. In contrast, off-axis EH can be used to provide quantitative images of the strength and direction of the local magnetic induction, projected in the electron beam direction, inside and outside each crystal, with a spatial resolution of approximately 5 nm. As a result, this technique can be used to assess the magnetic states of individual biogenic crystals both at remanence and during reversal, as well as to measure the strengths of magnetic interactions between neighboring crystals and parameters such as coercivities and magnetic moments.

The results that have been presented in this chapter show that, in general, the ferrimagnetic crystals in magnetite-containing bacteria are crystallographically perfect, morphologically similar to each other, and oriented with their [111] crystallographic axes parallel to the directions of the chains that contain them. Magnetically, the crystals are single domains that are oriented parallel to each other and exhibit little flux leakage. Crystals that are small enough to be superparamagnetic at room temperature or large enough to contain several magnetic domains if they were isolated are constrained to be single domains by magnetostatic interactions with neighboring crystals in the chain. Shape anisotropy appears to be the most important factor controlling the magnetic microstructure of each crystal, followed by interparticle interactions, with magnetocrystalline anisotropy being least important. Similar conclusions are presented for iron sulfide containing cells, the primary difference being the more scattered arrangement of the crystals, resulting in poorer confinement and greater variability in the directions of the field lines in the cell. Significantly, despite the different crystal structures, orientations, arrangements, and magnetic properties of the ferrimagnetic crystals that are formed by different strains of bacteria, EH measurements show that the magnetic moments of different bacterial cells are remarkably consistent with each other, satisfying the requirement for magnetotaxis that a large proportion of the forward motion of each cell should be in the direction of the external magnetic field.

The electron microscopy techniques that have been described here are being developed actively, offering the prospect of improved structural and magnetic information about magnetotactic bacteria in the future. In particular, it may be possible to map both the morphologies and the magnetic vector fields of entire cells in three dimensions with nanometer spatial resolution by combining EH with ET. In future experiments, it will also be important to optimize sample preparation techniques so that the internal arrangements of cells are always preserved faithfully for electron microscopy.

Acknowledgements We thank Ed Simpson and Anna Finlayson for contributions to the EH measurements, and Ryan Chong for ET. This chapter benefited greatly from ongoing collaborations and discussions with Richard Frankel, Dirk Schüler, Dennis Bazylinski, Peter Buseck, and István Dódony. We thank Jian-Min Zuo and Jim Mabon for use of their WebEMAPS HRTEM image simulation software, and Werner Kaminsky for making the WinXMorph crystal morphology software available. R.D.B. acknowledges the Royal Society for support and M.P. acknowledges support from the Hungarian Science Fund (OTKA-T030186).

References

Arató B, Szányi Z, Flies C, Schüler D, Frankel RB, Buseck PR, Pósfai M (2005) Crystal-size and shape distributions of magnetite from uncultured magnetotactic bacteria as a potential biomarker. Am Mineral 90:1233–1241

Balkwill DL, Maratea D, Blakemore RP (1980) Ultrastructure of a magnetic spirillum. J Bacteriol 141:1399–1408

Bazylinski DA, Frankel RB (2000) Magnetic iron oxide and iron sulfide minerals within microorganisms. In: Baeuerlein E (ed) Biomineralization. Wiley, Weinheim, pp 25–46

Bazylinski DA, Frankel RB (2004) Magnetosome formation in prokaryotes. Nat Rev Microbiol 2:217–230

Bazylinski DA, Garratt-Reed AJ, Abedi A, Frankel RB (1993) Copper association with iron sulfide magnetosomes in a magnetotactic bacterium. Arch Microbiol 160:35–42

Bazylinski DA, Frankel RB, Heywood BR, Mann S, King JW, Donaghay PL, Hanson AK (1995) Controlled biomineralization of magnetite (Fe_3O_4) and greigite (Fe_3S_4) in a magnetotactic bacterium. Appl Environ Microbiol 61:3232–3239

Blakemore RP (1975) Magnetotactic bacteria. Science 190:377–379

Buseck PR, Dunin-Borkowski RE, Devouard B, Frankel RB, McCartney MR, Midgley PA, Posfai M, Weyland M (2001) Magnetite morphology and life on Mars. Proc Natl Acad Sci USA 98:13490–13495

Clemett SJ, Thomas-Keprta KL, Shimmin J, Morphew M, McIntosh JR, Bazylinski DA, Kirschvink JL, Wentworth SJ, McKay DS, Vali H, Gibson EK, Romanek CS (2002) Crystal morphology of MV-1 magnetite. Am Mineral 87:1727–1730

DeRosier DJ, Klug A (1968) Reconstruction of three dimensional structures from electron micrographs. Nature 217:130–134

Devouard B, Pósfai M, Xin H, Bazylinski DA, Frankel RB, Buseck PR (1998) Magnetite from magnetotactic bacteria: size distributions and twinning. Am Mineral 83:1387–1398

Diaz-Ricci JC, Kirschvink JL (1992) Magnetic domain state and coercivity predictions for biogenic greigite (Fe_3S_4): a comparison of theory with magnetosome observations. J Geophys Res 97:17309–17315

Dunin-Borkowski RE, McCartney MR, Frankel RB, Bazylinski DA, Pósfai M, Buseck PR (1998) Magnetic microstructure of magnetotactic bacteria by electron holography. Science 282:1868–1870

Dunin-Borkowski RE, McCartney MR, Posfai M, Frankel RB, Bazylinski DA, Buseck PR (2001) Off-axis electron holography of magnetotactic bacteria: magnetic microstructure of strains MV-1 and MS-1. Eur J Mineral 13:671–684

Dunin-Borkowski RE, McCartney MR, Smith DJ (2004) Electron holography of nanostructured materials. In: Nalwa HS (ed) Encyclopedia of nanoscience and nanotechnology. American Scientific, pp 41–100

Dunlop DJ, Özdemir Ö (1997) Rock magnetism: fundamentals and frontiers. Cambridge University Press, Cambridge

Keim CN, Lopes Martins J, Lins de Barros HGP, Lins U, Farina M (2006) The Magnetotactic Multicellular Organism, a multicellular prokaryote from hypersaline environments. In: Schüler D (ed) Magnetoreception and magnetosomes in bacteria. Springer, Berlin Heidelberg New York

Farina M, Esquivel DMS, Lins de Barros HGP (1990) Magnetic iron-sulphur crystals from a magnetotactic microorganism. Nature 343:256–258

Farina M, Kachar B, Lins U, Broderick R, Lins de Barros HGP (1994) The observation of large magnetite crystals from magnetotactic bacteria by electron and atomic force microscopy. J Microsc 173:1–8

Frank J (ed) (1992) Electron tomography: three-dimensional imaging with the transmission electron microscope. Springer, Berlin Heidelberg New York

Frankel RB (1984) Magnetic guidance of organisms. Annu Rev Biophys Bioeng 13:85–103

Golden DC, Ming DW, Morris RV, Brearley A, Lauer HV, Treiman AH, Zolensky ME, Schwandt CS, Lofgren GE, McKay GA (2004) Evidence for exclusively inorganic formation of magnetite in Martian meteorite ALH84001. Am Mineral 89:681–695

Hanzlik M, Winklhofer M, Petersen N (1996) Spatial arrangement of chains of magnetosomes in magnetotactic bacteria. Earth Planet Sci Lett 145:125–134

Harrison RJ, Dunin-Borkowski RE, Putnis A (2002) Direct imaging of nanoscale magnetic interactions in minerals. Proc Natl Acad Sci USA 99:16556–16561

Heywood BR, Bazylinski DA, Garratt-Reed A, Mann S, Frankel RB (1990) Controlled biosynthesis of greigite (Fe3S4) in magnetotactic bacteria. Naturwissenschaften 77:536–538

Hÿtch MJ, Dunin-Borkowski RE, Scheinfein MR, Moulin J, Duhamel C, Mazaleyrat F, Champion Y (2003) Vortex flux channeling in magnetic nanoparticle chains. Phys Rev Lett 91:257207

Iida A, Akai J (1996) TEM study on magnetotactic bacteria and contained magnetite grains as biogenic minerals, mainly from Hokuriku-Niigata region, Japan. Sci Rep Niigata Univ Ser E 11:43–66

Kasama T, Pósfai M, Chong RKK, Finlayson AP, Buseck PR, Frankel RB, Dunin-Borkowski RE (2006a) Magnetic properties, microstructure, composition and morphology of greigite nanocrystals in magnetotactic bacteria from electron holography and tomography. Am Mineral 91:1216–1229

Kasama T, Pósfai M, Chong RKK, Finlayson AP, Dunin-Borkowski RE, Frankel RB (2006b) Magnetic microstructure of iron sulfide crystals in magnetotactic bacteria from off-axis electron holography. Physica B (in press)

Keim CN, Abreu F, Lins U, de Barros HL, Farina M (2004) Cell organization and ultrastructure of a magnetotactic multicellular organism. J Struct Biol 145:254–262

Komeili A (2006) Cell biology and magnetosome protein targeting in MTB. In: Schüler D (ed) Magnetoreception and magnetosomes in bacteria. Springer, Berlin Heidelberg New York

Lins U, McCartney MR, Farina M, Frankel RB, Buseck PR (2005) Habits of magnetosome crystals in coccoid magnetotactic bacteria. Appl Environ Microbiol 71:4902–4905

Mann S, Moench TT, Williams RJP (1984) A high resolution electron microscopic investigation of bacterial magnetite. Implications for crystal growth. Proc R Soc Lond Ser B 221:385–393

Mann S, Sparks NHC, Blakemore RP (1987) Structure, morphology and crystal growth of anisotropic magnetite crystals in magnetotactic bacteria. Proc R Soc Lond Ser B 231:477–487

Mann S, Sparks NHC, Frankel RB, Bazylinski DA, Jannasch HW (1990) Biomineralization of ferrimagnetic greigite (Fe3S4) and iron pyrite (FeS2) in a magnetotactic bacterium. Nature 343:258–261

Matsuda T, Endo J, Osakabe N, Tonomura A (1983) Morphology and structure of biogenic magnetite particles. Nature 302:411–412

McCartney MR, Lins U, Farina M, Buseck PR, Frankel RB (2001) Magnetic microstructure of bacterial magnetite by electron holography. Eur J Mineral 13:685–689

Meldrum FC, Mann S, Heywood BR, Frankel RB, Bazylinski DA (1993a) Electron microscopy study of magnetosomes in a cultured coccoid magnetotactic bacterium. Proc R Soc Lond Ser B 251:231–236

Meldrum FC, Mann S, Heywood BR, Frankel RB, Bazylinski DA (1993b) Electron microscopy study of magnetosomes in two cultured vibrioid magnetotactic bacteria. Proc R Soc Lond Ser B 251:237–242

Midgley PA, Weyland M, Thomas JM, Johnson BFG (2001) Z-contrast tomography: A technique in 3-dimensional nanostructural analysis based on Rutherford scattering. Chem Commun 10:907–908

Moskowitz BM, Frankel RB, Bazylinski DA (1993) Rock magnetic criteria for the detection of biogenic magnetite. Earth Planet Sci Lett 120:283–300

Moskowitz BM, Frankel RB, Bazylinski DA, Jannasch HW, Lovley DR (1989) A comparison of magnetite particles produced anaerobically by magnetotactic and dissimilatory iron-reducing bacteria. Geophys Res Lett 16:665–668

Pósfai M, Buseck PR, Bazylinski DA, Frankel RB (1998a) Reaction sequence of iron sulfide minerals in bacteria and their use as biomarkers. Science 280:880–883

Pósfai M, Buseck PR, Bazylinski DA, Frankel RB (1998b) Iron sulfides from magnetotactic bacteria: Structure, composition, and phase transitions. Amer Mineral 83:1469–1481

Pósfai M, Moskowitz B, Arató B, Schüler D, Flies C, Bazylinski DA, Frankel RB (2006) Properties of intracellular magnetite crystals produced by *Desulfovibrio magneticus* strain RS-1. Earth Planet Sci Lett (in press)

Proksch R, Schäffer TE, Moskowitz BM, Dahlberg ED, Bazylinski DA, Frankel RB (1995) Magnetic force microscopy of the submicron magnetic assembly in a magnetotactic bacterium. Appl Phys Lett 66:2582–2584

Scheffel A, Gruska M, Faivre D, Linaroudis A, Plitzko JM, Schüler D (2005) An acidic protein aligns magnetosomes along a filamentous structure in magnetotactic bacteria. Nature 440:110–114

Simmons SL, Sievert SM, Frankel RB, Bazylinski DA, Edwards KJ (2004) Spatiotemporal distribution of marine magnetotactic bacteria in a seasonally stratified coastal salt pond. Appl Environ Microbiol 70:6230–6239

Simpson ET, Kasama T, Pósfai M, Buseck PR, Harrison RJ, Dunin-Borkowski RE (2005) Magnetic induction mapping of magnetite chains in magnetotactic bacteria at room temperature and close to the Verwey transition using electron holography. J Phys Conf Ser 17:108–121

Sparks NHC, Mann S, Bazylinski DA, Lovley DR, Jannasch HW, Frankel RB (1990) Structure and morphology of magnetite anaerobically-produced by a marine magnetotactic bacterium and a dissimilatory iron-reducing bacterium. Earth Planet Sci Lett 98:14–22

Spring S, Lins U, Amann R, Schleifer K-H, Ferreira LCS, Esquivel DMS, Farina M (1998) Phylogenetic affiliation and ultrastructure of uncultured magnetic bacteria with unusually large magnetosomes. Arch Microbiol 169:136–147

Taylor AP, Barry JC (2004) Magnetosomal matrix: ultrafine structure may template biomineralization of magnetosomes. J Microsc 213:180–197

Taylor AP, Barry JC, Webb RI (2001) Structural and morphological anomalies in magnetosomes: possible biogenic origin for magnetite in ALH84001. J Microsc 201:84–106

Thomas-Keprta KL, Bazylinski DA, Kirschvink JL, Clemett SJ, McKay DS, Wentworth SJ, Vali H, Gibson EK Jr, Romanek CS (2000) Elongated prismatic magnetite crystals in ALH84001 carbonate globules: potential Martian magnetofossils. Geochim Cosmochim Acta 64:4049–4081

Tonomura A (1992) Electron-holographic interference microscopy. Adv Phys 41:59–103

Towe KM, Moench TT (1981) Electron optical characterization of bacterial magnetite. Earth Planet Sci Lett 52:213–220

Vali H, Kirschvink JL (1990) Observation of magnetosome organization, surface structure and iron biomineralization of undescribed magnetic bacteria: Evolutionary speculations. In: Blakemore RP, Frankel RB (eds) Iron biominerals. Plenum, New York, pp 97–115

Völkl E, Allard LF, Joy DC (eds) (1998) Introduction to electron holography. Plenum, New York

Weiss BP, Kim SS, Kirschvink JL, Kopp RE, Sankaran M, Kobayashi A, Komeili A (2004) Ferromagnetic resonance and low-temperature magnetic tests for biogenic magnetite. Earth Planet Sci Lett 224:73–89

Microbiol Monogr (3)
D. Schüler: Magnetoreception and Magnetosomes in Bacteria
DOI 10.1007/7171_045/Published online: 8 September 2006

Molecular Bioengineering of Bacterial Magnetic Particles for Biotechnological Applications

Tadashi Matsunaga (✉) · Atsushi Arakaki

Department of Biotechnology, Tokyo University of Agriculture and Technology, 2-24-16, Naka-cho, Koganei-shi, 184-8588 Tokyo, Japan
tmatsuna@cc.tuat.ac.jp

Abstract Magnetic particles are currently one of the most significant materials in the industrial sector, where they have been widely used for biotechnological and biomedical applications. Much effort has been devoted to the preparation of nanosized magnetite particles, and there is great interest with reference to their well-controlled size and shape. Magnetotactic bacteria synthesize uniform nanosized magnetite particles, the magnetosomes (also referred to as "bacterial magnetic particles", BMPs), which are enveloped by organic lipid membrane. BMPs offer high technological potential since they can be conveniently manipulated with magnetic force. The thin organic membrane enveloping the individual BMPs confers high and even dispersion in aqueous solutions compared to artificial magnetites, making them ideal biotechnological materials. Based on the molecular studies of BMP biomineralization from genome analysis, proteomics, and transcriptomics of *Magnetospirillum magneticum* AMB-1, methods for the construction of functional BMPs were established through genetic engineering. They are applicable to high-sensitivity immunoassay, DNA detection, drug screening, and cell separation. Furthermore, high-throughput immunoassay, single-nucleotide polymorphism discrimination, and DNA recovery systems have been developed to use these functionalized BMPs. The nanosized fine magnetic particles offer vast potential in new nanotechniques.

1 Introduction

Magnetotactic bacteria synthesize intracellular nanosized magnetic particles of iron oxide and/or iron sulfides known as magnetosomes, which are aligned in chains and individually covered with a stable organic membrane consisting mainly of lipids and proteins (Gorby et al. 1988; Nakamura et al. 1991). Magnetosomes are easily separated and purified from disrupted magnetotactic bacteria by magnetic separation. Magnetosomes are also referred to as bacterial magnetic particles (BMPs) to distinguish them from artificial magnetic particles (AMPs). BMPs can be easily dispersed in aqueous solutions because they are bounded by organic membranes, unlike AMPs (Nakamura et al. 1991). An individual BMP has a single magnetic domain of magnetite yielding a superior magnetic property (Bazylinski et al. 1994; Matsunaga and Kamiya 1987; Bazylinski et al. 1994). On the basis of the above advantageous properties, BMPs have been utilized in immunoassays (Nakamura et al. 1991; Matsunaga et al. 1996b; Tanaka and Matsunaga 2000; Tanaka et al. 2004b), DNA recovery (Yoza et al. 2002, 2003a), DNA discrimination (Takeyama et al. 2000; Nakayama et al. 2003; Tanaka et al. 2003; Maruyama et al. 2004), cell separation (Kuhara et al. 2004), and as drug carriers (Matsunaga et al. 1997). BMPs have vast potential for various technological applications and the molecular mechanism of their formation is of particular interest. This review describes the development of functional BMPs and their biotechnological applications based on current fundamental studies of these unique biomaterials.

2 Molecular Study of Magnetosome Formation in *Magnetospirillum magneticum* AMB-1

The abundant occurrence of magnetotactic bacteria has been observed in various aquatic environments. We have isolated two facultative anaerobic spirilla, *Magnetospirillum magneticum* AMB-1 (ATCC 700264) (Matsunaga et al. 1991) and MGT-1 (FERM P-16617) (Matsunaga et al. 1990), and an obligate anaerobe, *Desulfovibrio magneticus* RS-1 (ATCC 700980) (Sakaguchi et al. 1993). *M. magneticum* AMB-1 (Fig. 1a) and MGT-1 are nitrate-reducing bacteria which belong to α-Proteobacteria. *D. magneticus* RS-1 (Fig. 1b) is a dis-

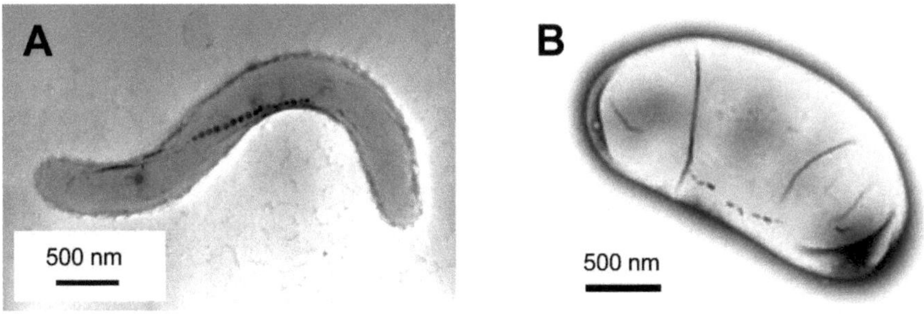

Fig. 1 Transmission electron micrographs of magnetotactic bacteria. **a** *Magnetospirillum magneticum* AMB-1, and **b** *Desulfovibrio magneticus* RS-1

similatory sulfate-reducing bacterium which belongs to δ-Proteobacteria and produces unique, irregular, bullet-shaped BMPs (Kawaguchi et al. 1995; Sakaguchi et al. 2002). *M. magneticum* AMB-1 and MGT-1 are capable of growing under both microaerobic and aerobic conditions in liquid or solid media, hence they are amenable to genetic manipulations (Matsunaga et al. 1992). This facultative anaerobic characteristic of the bacteria has enabled us to generate nonmagnetic mutant strains by transposon mutagenesis (Nakamura et al. 1995a; Wahyudi et al. 2001, 2003). We have subsequently developed electroporation (Okamura et al. 2003) and conjugation techniques (Matsunaga et al. 1992) specifically designed for AMB-1 and MGT-1 strains, providing valuable information on genes involved in BMP formation as well as protein display on BMPs based on gene fusion techniques.

In order to elucidate the mechanism of BMP formation in magnetotactic bacteria, the genome, proteome, and transcriptome profiles were analyzed. Clarification of magnetite biomineralization pathways contributes to the biotechnological applications in a broad range of research disciplines. Recent efforts on the basic studies in *M. magneticum* AMB-1 were introduced.

2.1 Proteome Analysis of Magnetosome

Proteins expressed on the BMP membrane play a direct role in the BMP biosynthesis process. The thin section of the *M. magneticum* AMB-1 cell revealed the presence of intracellular membranous structures surrounding the magnetite crystals (Fig. 2). Five dominant proteins in the BMP membrane of *M. magneticum* AMB-1 were identified by SDS-PAGE (Okamura et al. 2000). One of these is a 24-kDa protein designated as Mms24 (identical with MamA) (Komeili et al. 2004). This protein was observed in other *Magnetospirillum* species and corresponds to Mam22 in *M. magnetotacticum* MS-1 (Okuda et al. 1996) and MamA in *M. gryphiswaldense* MSR-1 (Grünberg et al. 2001). A 16-kDa protein (Mms16) possessing GTPase activity was found to be the

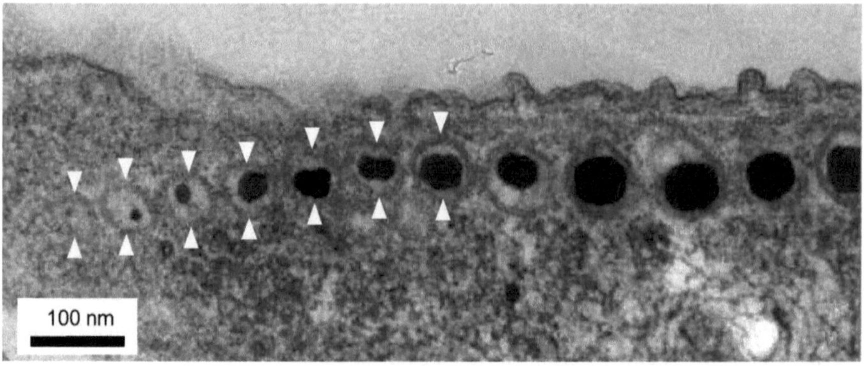

Fig. 2 Thin section of *M. magneticum* AMB-1. Intracellular membranous structures surrounding magnetite crystals are indicated with *arrow heads*

most abundantly expressed of the five proteins isolated from the BMP membrane of *M. magneticum* AMB-1 (Okamura et al. 2001). Mms16 shows similar properties to eukaryotic small GTPases that control priming and trafficking of budding vesicles. This finding led to the hypothesis that the BMP membrane is derived from the cytoplasmic membrane through an invagination process. Schultheiss et al. reported that Mms16 in strain MSR-1 is a poly(3-hydroxybutyrate) (PHB) granule-bound protein (ApdA), as concluded from the localization analysis and hydrolysis activity of PHB granules in vitro, and phenotypic analysis of an *mms16* mutant (Schultheiss et al. 2005). Mms16 in strain MSR-1 does not possess a P-loop structure based on a homology search, and indicated no GTPase activity. In contrast, Mms16 in strain AMB-1 has a P-loop (ATP/GTP binding motif) and clearly shows GTPase activity. Consequently, it is most probable that these Mms16 proteins in *M. magneticum* AMB-1 and *M. gryphiswaldense* MSR-1 may be different proteins with different functions.

By further analysis, four novel proteins tightly bound to the magnetite crystal, designated Mms5, 6, MamD (Mms7), and MamC (Mms13), were identified (Arakaki et al. 2003). These proteins show the common amphiphilic features of containing hydrophobic N-terminal and hydrophilic C-terminal regions. The N-terminal regions in Mms5, 6, and MamD possess a common leucine and glycine repetitive sequence, LGLGLGLGAWGPX(L/I)(L/V) GX(V/A)GXAGA (Fig. 3). A further gene homolog (ORF1) was identified from AMB-1 genome encoding a protein with similar motif to this protein group. Similar sequences were reported in silk fibroin like molluscan shell framework protein (Sudo et al. 1997) and collagen like fish otolith structure protein (Murayama et al. 2002), which are considered to form a β-sheet and produce a self-assembled framework structure for carbonate crystal formation. Although the LG-rich regions in BMP proteins are rather short to form an intermolecular β-sheet structure, this region may participate in the

Fig. 3 Sequence alignment and comparison of N-terminal regions of Mms5, Mms6, Mms7, and ORF1. Highlighted sequences indicate conserved residues. The numbers represent amino acid positions in each protein

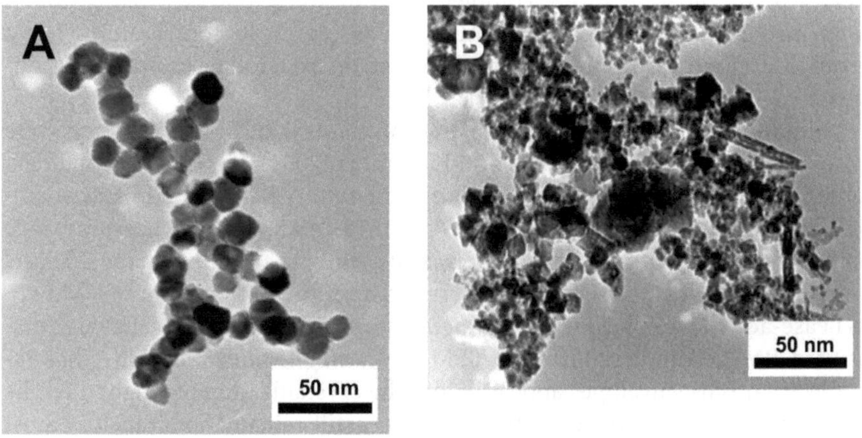

Fig. 4 Electron micrographs of magnetic particles synthesized in the **a** presence and **b** absence of Mms6 protein

self-assembling of proteins for magnetite crystal formation. The C-terminal regions in Mms6 contain carboxyl and hydroxyl groups which are found as iron binding sites. The magnetite particles synthesized in the presence of Mms6 in vitro showed a cuboidal morphology similar to strain AMB-1 BMPs with sizes ranging from 20 to 30 nm in diameter (Fig. 4) (Arakaki et al. 2003). These results suggest that Mms6 binds iron ions to initiate magnetite crystal formation and/or regulates morphology by producing a self-assembled framework structure. Crucial proteins for BMP alignment in the cell were recently reported. MamK is a homolog of bacterial actin-like protein, and is proposed to form filaments to establish the specific localization of BMPs through the proper subcellular targeting (Komeili et al. 2006). MamJ is an acidic protein associated with the filamentous structure and directs the assembly of the BMP chain (Scheffel et al. 2006; Komeili 2006, in this volume).

Several genes involved in BMP formation were also isolated through transposon mutagenesis. The *magA* gene encodes a proton-driving $H^+/Fe(II)$

antiporter protein (Nakamura et al. 1995a). Intracellular localization of the MagA protein was examined using a MagA-luciferase fusion protein and indicated that this protein is localized in both the cytoplasmic and BMP membranes (Nakamura et al. 1995b). Interestingly, MagA topology is inversely oriented between the cytoplasmic and the BMP membrane. MagA appears to function for iron efflux in the former and iron influx in the latter. A tungsten-containing aldehyde ferredoxin oxidoreductase (AOR) which functions for aldehyde oxidation was also isolated (Wahyudi et al. 2003). AOR was found to be expressed under microaerobic conditions and localized in the cytoplasm of AMB-1. Transmission electron microscopy (TEM) of the mutant revealed that no BMPs were completely synthesized, but PHB-like granules were persistently produced. The AOR is considered to contribute to ferric iron reduction during BMP synthesis under microaerobic respiration (Calugay et al. 2004; Wahyudi et al. 2003).

Although alternative models for the mechanism of BMP formation were also proposed by Schüler (2004) and Fukumori (2004), the experimental data obtained from our studies with *M. magneticum* AMB-1 allowed us to postulate that BMP formation is comprised of three major stages. The first stage entails the invagination of the cytoplasmic membrane primed by the GTPase, and the vesicle formed serves as the precursor of the BMP membrane. Mms16 GTPase-activity inhibition in *M. magneticum* AMB-1 yielded cells with disrupted BMP chain arrangement. TEM of *M. magnetotacticum* MS-1 has revealed that BMP envelopes appear before the crystallization of magnetite. The mechanism of envelope formation, however, still remains unclear. It is most probable that magnetotactic bacteria have similar mechanisms of vesicle formation to most eukaryotes and that a specific GTPase mediates the priming of the invagination (for a detailed discussion, see the chapter by A. Komeili, in this volume). The second stage of BMP formation entails the accumulation of ferrous ions into the vesicles by the BMP transmembrane transporter MagA. External iron is internalized by transport proteins and siderophores. The internal iron is controlled strictly by an oxidation–reduction system. In the final stage, Mms6 triggers magnetite crystal nucleation and/or regulates morphology. Various proteins associated with the BMP membrane could play functional roles in magnetite generation. These include the accumulation of supersaturating iron concentrations, maintenance of reductive conditions, and the oxidation of iron to induce mineralization, or the partial reduction and dehydration of ferrihydrite to magnetite.

2.2
Genome Analysis of *M. magneticum* AMB-1

Parallel to the protein analysis, the entire genome of *M. magneticum* AMB-1 was sequenced (Fig. 5) (Matsunaga et al. 2005). The genome consists of a single circular chromosome of 4 967 148 bp. The genes were annotated with an

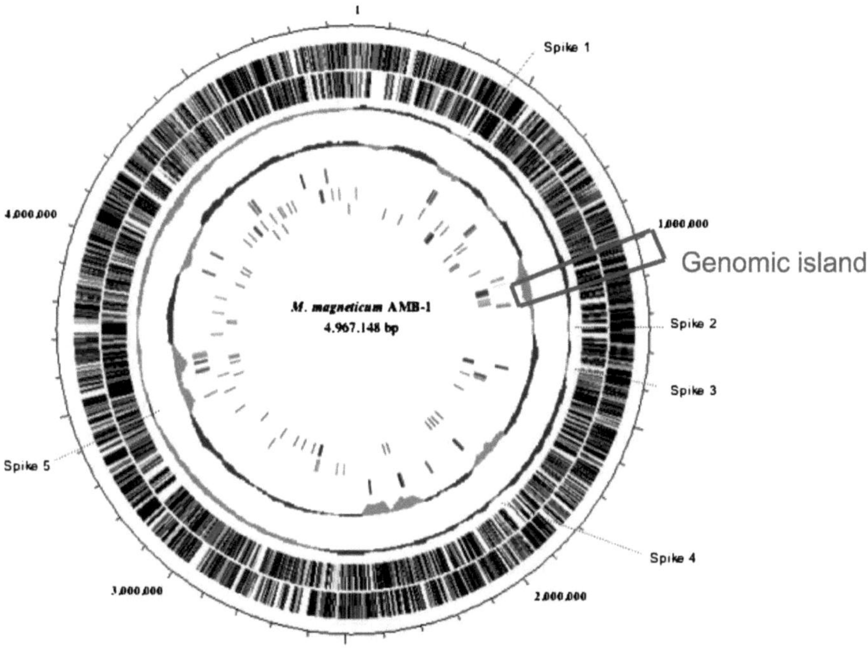

Fig. 5 The *outer and first inner circles* represent predicted ORFs on the plus and minus strands, respectively (*salmon*: translation, ribosomal structure, and biogenesis; *light blue*: transcription; *cyan*: DNA replication, recombination, and repair; *turquoise*: cell division; *deep pink*: posttranslational modification, protein turnover, and chaperones; *olive drab*: cell envelope biogenesis; *purple*: cell motility and secretion; *forest green*: inorganic ion transport and metabolism; *magenta*: signal transduction; *red*: energy production; *sienna*: carbohydrate transport and metabolism; *yellow*: amino acid transport; *orange*: nucleotide transport and metabolism; *gold*: coenzyme transport and metabolism; *dark blue*: lipid metabolism; *blue*: secondary metabolites, transport, and catabolism; *gray*: general function prediction only; *black*: function unclassified or unknown). The *second inner circle* represents GC skew: *purple* indicates > 0, *orange* indicates < 0. The *third inner circle* further represents GC content: *purple* indicates higher than average, *orange* indicates less than average. The *fourth inner circle* represents insertion sequence (IS) elements (*black*: ISmag1; *orange*: ISmag2; *pink*: ISmag3; *purple*: ISmag4; *blue*: ISmag5; *red*: ISmag6; *light blue*: ISmag7; *light green*: ISmag8; *green*: ISmag9; *brown*: ISmag10; *yellow*: ISmag11). The *fifth and sixth inner circles* indicate the genes encoding magnetosome membrane proteins and loci of genes disrupted by transposon, respectively

E value 1×10^{-1}; however, almost half of the 4559 open reading frames (ORFs) remained unknown with respect to functional prediction. Remarkable numbers of sensor and response domains were identified. The genome of AMB-1 includes many vestiges of past exogenous gene transfers, such as insertion sequence (IS) elements, integrases, and large regions containing phage-coding genes. The IS-concentrated regions have a lower GC content than the average of the whole genome, suggestive of gene transfer from other bacteria or phages. The inversions of GC content, or spikes, are observed mainly in IS insertion positions. During the genome analysis, we obtained a spontaneous nonmagnetic mutant which lacks a 98-kb genomic region. The region has the characteristics of a genomic island: low GC content, location between two 1.1-kb repetitive sequences, and the presence of an integrase in the flanking region of the first repetitive sequence. The region was identified as a "magnetosome island" because nonmagnetic spontaneous mutants were obtained which lack this genomic region (Schübbe et al. 2003; Jogler and Schüler 2006, in this volume). The spontaneous deletion of this 98-kb genomic region and the circular form after excision from the chromosome were detected by polymerase chain reaction (PCR) amplification (Fukuda et al. 2006). The island seems to be derived from lateral gene transfer considering the typical structure of a genomic island and the dynamics after excision from the chromosome. It is also postulated that an ancestral magnetotactic bacterium came into contact with another bacterium harboring the

Table 1 Clusters of orthologous group categories of proteins annotated in *M. magneticum* AMB-1 genome

Categorized number of genes	2290
Information storage and processing	
Translation, ribosomal structure, and biogenesis	142
Transcription	148
DNA replication, recombination, and repair	139
Cellular process	
Cell division and chromosome partitioning	27
Cell envelope biogenesis, outer membrane	186
Cell motility and secretion, protein turnover, chaperones	83
Posttranslational modification, protein turnover, chaperones	117
Inorganic ion transport and metabolism	171
Signal transduction	205
Metabolism	
Energy production and conversion	226
Amino acid transport and metabolism	234
Nucleotide transport and metabolism	51
Carbohydrate transport and metabolism	123
Coenzyme metabolism	109
Lipid metabolism	113
Secondary metabolite biosynthesis, transport, and catabolism	66
Poorly characterized	
General function prediction only	315
Function unknown	168
Uncategorized number of genes	2269
Total number of ORFs	4559

genomic island, and the resulting generation eventually diverged to AMB-1, MS-1, MC-1, and MSR-1.

To further identify the genes involved in BMP formation within the 2269 ORFs found in the whole genome of *M. magneticum* AMB-1, transposon mutagenesis and BMP proteomic analyses were performed (Table 1) (Matsunaga et al. 2005). Except for the magnetosome island, BMP-related genes are scattered throughout the genome. Five proteins identified in the BMP proteome profiles were encoded within the magnetosome island. Other housekeeping genes were also found within the magnetosome island, suggesting that they may be necessary for complete BMP formation. Although gene predictions were confined to annotated genes, the predictions were also observed in other bacteria. Perhaps the machinery for iron uptake or other molecules for BMP formation might be simple, but they may be strictly regulated in a complex manner.

2.3
Transcriptome Analysis of Iron Uptake Genes

Magnetotactic bacteria take up approximately 100 times more iron than non-magnetic bacteria to synthesize intracellular BMPs. To identify global gene expression profiles, especially those of iron uptake genes under different iron concentrations, transcriptome analysis through microarrays imprinted with DNA from the whole genome sequence of AMB-1 was investigated (Suzuki et al. 2006). Global gene expression profiling of AMB-1 yielded five gene expression profile patterns. Three ferrous and two ferric iron uptake systems encoded by 16 genes were identified from the genome sequence. Our study was focused on these genes for expression analysis. Two distinct patterns were observed in relation to BMP synthesis and gene expression (Table 2). High-affinity ferrous iron transport genes were upregulated in BMP-forming cells grown under iron-rich conditions, and ferric iron transport genes were downregulated under these conditions. In high iron conditions wherein magnetite synthesis occurs, the ferrous iron uptake system was dominant while ferric iron or siderophore transport genes were expressed after the high initial iron concentrations were taken up by the magnetic cells. This is consistent with our previous report on the siderophore production of *M. magneticum* AMB-1 grown in media containing different initial iron ions (Calugay et al. 2003). The initial high concentration of iron was rapidly assimilated from the medium within only 4 h after inoculation, reaching levels comparable to iron-deficient cultures and thereby triggering siderophore production. A concentration of at least 6 μM ferric iron was required to initiate the siderophore production. Furthermore, enhancement of BMP production was observed when ferrous iron was utilized as iron source (see Sect. 3.4) (Yang et al. 2001a). These results suggest that ferrous iron uptake systems mainly serve as iron supply lines for magnetite formation.

Table 2 Expression of genes involved in ferrous and ferric ion transport and nitrogen respiration in *M. magneticus* AMB-1 at various initial iron concentrations

Gene	Gene identification	Fold change at iron concentration (μM) 20	40	80	100	150	200	250	300
Fe^{2+} transporter									
ftr1	amb0937	**4.08**	**4.73**	**2.69**	1.16	1.19	1.10	1.23	1.22
ftr1	amb1681	**2.30**	**1.84**	**1.95**	**1.78**	**1.80**	**1.94**	**1.94**	**1.96**
tpd	amb0939	**4.20**	**4.30**	**3.14**	0.97	0.96	1.07	1.09	1.10
tpd	amb0940	**4.11**	**3.76**	**2.61**	1.03	1.05	0.98	0.95	0.97
tpd	amb4411	**5.78**	**3.56**	**2.01**	**2.32**	**2.34**	**2.44**	**2.41**	**2.43**
feoA	amb1022	**3.74**	**3.73**	**3.10**	**2.25**	**2.26**	**2.23**	**2.23**	**2.24**
feoB	amb1024	**2.32**	**3.21**	**2.52**	**2.48**	**1.52**	**2.12**	**2.08**	**2.06**
feoA	amb2730	**1.48**	**2.27**	**1.72**	**1.77**	**1.81**	**1.59**	**1.61**	**1.59**
feoB	amb2731	1.06	**1.57**	**2.17**	**1.97**	**1.98**	**2.17**	**2.16**	**2.17**
Fe^{3+} transporter									
cirA	amb0540	1.33	**1.72**	**2.09**	**2.07**	**2.11**	**2.47**	1.21	1.49
cirA	amb0846	**1.55**	**2.48**	**2.33**	**1.90**	**1.88**	**2.23**	1.21	1.23
fepC	amb4338	1.48	**1.48**	**1.77**	**1.72**	**1.76**	**2.10**	**2.12**	**2.10**
tonB	amb3212	*0.22*	*0.54*	*0.48*	*0.59*	*0.57*	*0.13*	*0.11*	*0.13*
tonB	amb3546	*0.52*	*0.51*	*0.52*	*0.55*	*0.53*	*0.64*	*0.62*	*0.64*
fepA	amb3547	*0.68*	*0.72*	*0.66*	*0.61*	*0.60*	*0.60*	*0.62*	*0.63*
exbB	amb3548	*0.57*	*0.70*	*0.65*	*0.63*	*0.65*	*0.62*	*0.69*	*0.61*
tonB	amb3549	*0.39*	*0.35*	*0.33*	*0.50*	*0.51*	*0.15*	*0.15*	*0.16*
tolQ	amb3550	*0.37*	*0.38*	*0.02*	*0.54*	*0.58*	*0.62*	*0.68*	*0.56*
exbD	amb3551	*0.14*	*0.11*	*0.41*	*0.40*	*0.43*	*0.63*	*0.60*	*0.59*
exbD	amb3552	*0.56*	*0.28*	*0.67*	*0.60*	*0.58*	*0.70*	*0.75*	*0.77*
napA	amb2690	**2.74**	**2.79**	**2.99**	**2.76**	**2.75**	**2.89**	**2.85**	**2.86**
napB	amb2687	**2.21**	**2.30**	**2.27**	**2.55**	**2.59**	**2.27**	**2.29**	**2.27**
napC	amb2686	**2.20**	**2.30**	**2.30**	**2.63**	**2.64**	**2.31**	**2.30**	**2.31**
napC	amb3335	1.40	**1.66**	**1.75**	**1.75**	**1.79**	**2.07**	**2.11**	**2.09**

Highlighted in **bold**, upregulated; *italic*, downregulated

Our results indicate that despite the unusual high-iron requirement of *M. magneticum* AMB-1, it utilizes robust but simple iron uptake systems similar to those of other gram-negative bacteria. It is also shown that AMB-1 possesses unique regulation systems for the iron uptake.

3
Development of Functional Magnetic Particles by Bioengineering

Assembling functional molecules on nanoparticles is important for various biotechnological applications. The amount and stability of assembled proteins are strongly dependent on the method used. This technique requires the assembly of sufficient functional proteins to perform specific bioassays on matrices which provide stable environments for the attached proteins. The presence of organic membranes surrounding BMPs enables us to assemble biological molecules on their surfaces. Proteins and oligonucleotides can be attached covalently to solid supports that prevent the desorption of these molecules during assay. Functional proteins have also been directly expressed onto BMPs by a protein display technique with membrane-integrated or -attached proteins, which guided target proteins onto the surface of BMPs. This technique permits preservation of protein activity and simple preparation of functional protein–BMP complexes. The techniques for functional molecules assembling on BMPs are introduced in the following sections.

3.1
Immobilization of Biomolecules on Magnetic Particles

BMPs are covered with a lipid bilayer membrane that mainly consists of phospholipids (comprising 58–65% of the total lipids), 50% of which is phosphatidylethanolamine (Frankel et al. 1998; Nakamura and Matsunaga 1993). The outward-pointing amine groups from phospholipids enable the immobilization of functional molecules on the BMP surface through cross-linking reaction. In order to immobilize antibodies and oligonucleotides, various cross-linkers were utilized. Homofunctional cross-linkers, which contain two aldehydes or *N*-hydroxysuccinimide (NHS) esters, reacted with amines on BMPs and antibodies or oligonucleotides (Matsunaga et al. 2003; Tanaka et al. 2004b). A heterofunctional cross-linker, succinimidyl 3-(2-pyridyldithio)propionate (SPDP), which contains one NHS ester, was used for the thiolation of BMPs (Nakamura and Matsunaga 1993; Nakamura et al. 1993; Matsunaga et al. 1996b). Dithiothreitol was reacted with antibodies for the reaction. Two heterofunctional cross-linkers were used for the preservation of antibody specificity after chemical conjugation. Succinimidyl 4-(*N*-maleimidomethyl) cyclohexane-1-carboxylate (SMCC), which contains NHS ester, and maleimide were reacted with antibodies (Matsunaga et al. 1996b). Streptavidin immobilized on BMPs is prepared by the same principle. Sulfo-NHS-LC-LC-biotin was used to modify the magnetic particles with biotin (Amemiya et al. 2005). The biotin was then utilized for the immobilization of streptavidin. The streptavidin-modified BMPs were designed to immobilize biotin-modified antibodies.

3.2
Protein Display on Magnetic Particles by Gene Fusion

Functional proteins and random peptide libraries have been displayed on the surface of bacteria, bacteriophages, viruses, and yeast, allowing the manipulation of diverse molecules (Clackson et al. 1991; Lee et al. 2000). The technique is often used for the identification of therapeutics, affinity purification, and imaging agents. Protein display on BMPs was utilized by the development of a fusion technique involving anchor proteins isolated from magnetotactic bacteria. Several proteins identified from the proteome analysis of the BMP membrane were used to display functional molecules on BMPs. The variety of the BMP membrane proteins enabled us to design functional molecules retaining their native activities. Our reports on the techniques for functional protein display on BMPs based on molecular engineering are described in the following sections.

3.2.1 Anchor Proteins for Display

Protein displays were designed using anchor molecules to introduce foreign proteins on the surface of BMPs. The MagA (47 kDa) (Nakamura et al. 1995b; Matsunaga et al. 1999, 2000, 2002), Mms16 (16 kDa) (Yoshino et al. 2004, 2005), and Mms13 (MamC) (13 kDa) (Yoshino and Matsunaga 2005, 2006) proteins were utilized as anchor molecules (Fig. 6). The fusion proteins usually become highly dependent on the structural properties of the inserted foreign protein domain, since the protein will be more constrained when inserted into a permissive site of an anchor molecule. Selections of an anchor protein and fusion sites are very crucial for protein display (Yoshino and Matsunaga 2006, Yoshino et al. 2004). It is also known that expression properties were sometimes different, even though the same vector construct and promoter were employed.

MagA is a transmembrane protein identified from an *M. magneticum* AMB-1 mutant strain generated by transposon mutagenesis (Nakamura et al. 1995a). This was used as an anchor molecule for expression of foreign proteins on the surface of BMPs. Soluble proteins, such as luciferase (Nakamura et al. 1995b), protein A (Matsunaga et al. 1999), and acetate kinase (Matsunaga et al. 2000) were expressed on BMPs. However, MagA maintains a large hydrophobic domain making it unsuitable for assembling membrane proteins. Mms16 was found to be a smaller protein abundantly expressed in the BMP membrane (Okamura et al. 2001). This is tightly anchored in the membrane by myristoylation. This characteristic was considered to be useful for expression of large transmembrane proteins, since it has less effect on the expression of foreign protein by dimensional obstruction. For example, G-protein coupled receptors (GPCRs) are seven transmembrane proteins involved in signal transduction cascades contributing to many biological activities. Mms16 was therefore selected as an anchor for assembly of a GPCR and the efficient expression on BMPs was confirmed (Yoshino et al. 2004). A newly found integral BMP membrane protein, Mms13 (MamC), was also investigated for its efficiency to serve as an anchor molecule to display functional

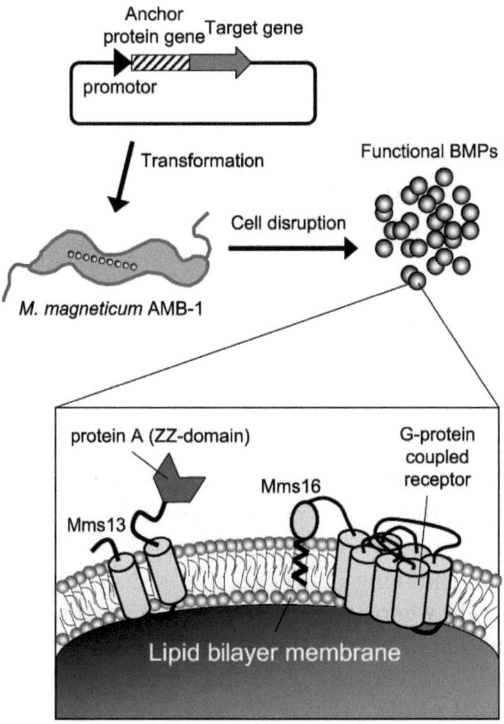

Fig. 6 Schematic diagram for preparation of functional BMPs for protein display

proteins onto BMPs (Arakaki et al. 2003). The anchoring properties of Mms13 were confirmed by luciferase fusion studies. The C terminus of Mms13 was shown to be expressed on the surface of BMPs (Yoshino and Matsunaga 2006). Moreover, Mms13 was found to be bound tightly on the magnetite surface permitting stable localization of a large protein like luciferase (61 kDa) on BMPs. Consequently, the luminescence intensity obtained from BMPs using Mms13 as an anchor molecule was 400 to 1000 times higher than that from Mms16 or MagA. Based on these investigations, an efficient technique for protein display on BMPs was developed.

3.2.2
Isolation of Active Promoters for Gene Expression

To establish highly expressed protein display on BMPs, strong promoters were identified based on the AMB-1 genome and proteome databases (Yoshino and Matsunaga 2005). Upstream DNA sequences of ORFs that code for highly expressed proteins may contain strong promoters. Therefore, highly expressed proteins in AMB-1 were isolated. Several proteins that were highly expressed on the cell membrane and also localized on the BMP membrane were selected. The N-terminal amino acids were determined based on the protein database of AMB-1. The sequences prior to start codons (ATG) of identified ORFs were analyzed using a computer program and the promoter regions were predicted. To evaluate the activity of each promoter, the plasmids used in a luciferase–reporter gene assay were constructed. Consequently, the luminescence intensity obtained using the *msp3* promoter showed approximately 400 times higher activity than the *magA* promoter previously used.

3.2.3 Designing of a Shuttle Vector

A 3.7-kb cryptic plasmid designated pMGT found in *M. magneticum* MGT-1 was characterized and used for the development of an improved expression system in strain AMB-1 through the construction of a shuttle vector, pUMG (Okamura et al. 2003). An electroporation method for magnetotactic bacteria that uses the cryptic plasmid was also developed. Recombinant plasmid pUMG was constructed by ligating pUC19 and pMGT. The copy number of pUMG containing the replicon of pMGT in AMB-1 was approximately 39 copies/cell, whereas the vector pRK415 yielded only 3 copies/cell. In addition, evaluation of pUMG as an expression vector through a reporter–luciferase expression system exhibited five times higher activity than those from cells harboring pKML previously used. These results are useful in improving the mass production of BMPs and protein A displayed on BMPs in batch-fed cultures, as we have previously described.

3.3 In Vitro Protein Integration for Protein Display

As an alternative method for protein display on BMPs, in vitro reconstitution of fusion protein with integral membrane protein or peptides into lipid vesicles was investigated. Artificial integration of useful protein is a powerful method for the highly efficient assembling of soluble proteins on BMPs. Electrostatic and hydrophobic interactions are considered to be the driving forces for spontaneous insertion of proteins and peptides into lipid membranes (Knol et al. 1998). MagA-luciferase fusion proteins prepared from recombinant *Escherichia coli* membranes were integrated by sonication (Matsunaga et al. 2002). A maximum luminescence was obtained which was 18 times higher compared with recombinant luciferase-MagA displayed on BMPs generated using gene fusion techniques.

Furthermore, an antimicrobial peptide, temporin L, and its derivative were also employed as anchor peptides and immobilized streptavidin on BMPs (Fig. 7) (Tanaka et al. 2004b). Temporin L is a cationic linear peptide with varied structure, length, and orientation (Rinaldi et al. 2002). This forms an amphiphilic structure with the hydrophobic part organized in a helix when

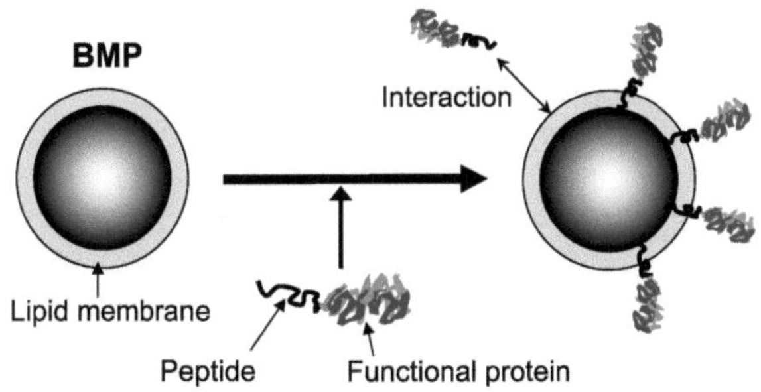

Fig. 7 In vitro assembling of proteins on BMP using a transmembrane peptide

associated with membrane. This spontaneous integration mechanism was applied for protein assembly onto BMPs. The C terminus of temporin L was incorporated into a BMP membrane, and the N terminus was located on the BMP membrane surface. The conjugation of biotin to the N terminus of temporin L leads to binding of streptavidin on a BMP membrane. Improved efficiency of integration of functional proteins is obtained by selecting the most suitable anchor molecule under optimized conditions.

3.4
Mass Production of Functional Magnetic Particles

Mass production of BMPs was investigated in order to obtain functional BMPs of biotechnological use. In our previous work, a setup for the large-scale cultivation of 1000 L was prepared and yielded approximately 2.6 mg BMPs per liter of culture (Matsunaga et al. 1990). To enhance the productivity, the growth conditions for mass production of luciferase-BMPs by a recombinant AMB-1 were investigated in a pH-regulated fed-batch culture (Matsunaga et al. 2000). Recombinant cells harboring plasmids with *magA-luc* gene fusion were cultured. The culture conditions including iron source, nutrients, and reducing agents in the medium were investigated (Matsunaga et al. 1996a). Fed-batch culture of AMB-1 harboring expression of a stably maintained plasmid in the cells for recombinant protein was carried out in a 10-L fermentor under microaerobic conditions. The addition of fresh nutrients was feedback-controlled as a function of the pH of the culture. The yield of BMPs was optimized by adjusting the rate of ferric iron addition. Feeding ferric quinate at 15.4 μg/min yielded 7.5 μg/L BMPs (Yang et al. 2001b). The transformants yielded an average luminescence of 0.41 kcounts/min/μg BMPs. Stable expression of *magA-luc* fusion gene was indicated. BMP formation in *M. gryphiswaldense* was also investigated by using an oxygen-controlled fermentor. Although the value of the amount of BMPs is obtained from calculation, a significant BMP yield (6.3 mg/L/day) was reported (Heyen and Schüler).

Table 3 Effect of various iron sources on BMP production in fed-batch culture. Iron solutions (31 mM, 16 mL) were added every 12 h four times after the cell concentration reached 1×10^7 cells/mL. Cells were cultured for 144 h

Iron sources	BMP production (mg/L)
Ferric quinate	4.5 ± 0.4
Ferric malate	4.7 ± 0.3
Ferric citrate	7.2 ± 0.5
Ferric gallate	11.4 ± 0.6
Ferrous sulfate	11.8 ± 0.8

Enrichment of growth medium with L-cysteine, yeast extract, and polypeptone enhanced both bacterial growth and BMP production (Yang et al. 2001a). The presence of L-cysteine in the medium was useful for induction of cell growth. Strictly anaerobic conditions led to a prolonged lag phase and limited the final cell density. As iron sources, ferrous sulfate and ferric gallate dramatically enhanced BMP yield as compared to ferric quinate, an iron chelate conventionally used. The optimized conditions increased cell density to 0.59 ± 0.03 g cell dry weight/L and BMP production to 14.8 ± 0.5 g cell dry weight/L in fed-batch culture for 4 days (Table 3). The optimized culture conditions are applicable to the production of other functional BMPs.

4 Application of Functional Magnetic Particles

The use of functional magnetic particles in bioassays enables the separation of bound and free analytes by applying a magnetic field. Because BMPs disperse evenly throughout the reaction mixture, their use facilitated rapid reaction kinetics without the need for continuous mixing or shaking, enabling the coupling of antibodies (Nakamura et al. 1991). Furthermore, nanometer-sized BMPs provide a large surface area for reactions. These characteristics have enabled the use of BMPs as stable platforms for antibodies that are used in sensitive immunoassays, high-performance DNA/mRNA recovery (Sode et al. 1993; Yoza et al. 2002), and DNA discrimination analysis (Nakayama et al. 2003). Fully automated systems were also developed for precise, rapid, and high-throughput processing of these assays (Maruyama et al. 2004; Tanaka and Matsunaga 2000; Yoza et al. 2003b). Applications of BMPs developed by the authors are introduced in this section.

4.1
Immunoassay and Receptor Binding Assay

Competitive chemiluminescence enzyme immunoassays using antibodies immobilized onto BMPs was developed for the rapid and sensitive detection of small molecules, such as environmental pollutants, hormones, and toxic detergents (Matsunaga et al. 2003; Tanaka et al. 2004b). Xenoestrogens, such as alkylphenol ethoxylates (APEs), bisphenol A (BPA), and linear alkylbenzenesulfonates (LASs) were detectable using monoclonal antibodies immobilized onto BMPs based on the competitive reaction of xenoestrogens. The entire procedure was completed in 15 min while typical plate methods took more than 2.5 h. This method gave a wider range and lower detection limit than enzyme-linked immunosorbent assay (ELISA), in which the same antibodies were used for detection. Furthermore, the detection limits of LASs and BPA were of similar levels compared with those obtained through gas chromatography–mass spectrometry (GC-MS). These experiments suggest that BMP-based immunoassay systems have superior advantages due to the high sensitivity and rapid measurement of samples.

The Z domain of protein A in *Staphylococcus aureus* has the binding ability to the Fc region of immunoglobulin G (IgG) (Lowenadler et al. 1987). An assembly technique of the Z domain on the BMP surface by gene fusion using a protein A-*magA* hybrid gene was constructed (Matsunaga et al. 1999). The antibody was accurately oriented on the BMP due to its association with protein A, unlike in immobilization by chemical conjugation. When a chemiluminescence sandwich immunoassay was carried out using antibody–protein A–BMP complexes and BMPs chemically immobilized with antibodies, the antigen binding activity per microgram of antibodies on antibody–protein A–BMP complexes was two times higher than that for antibody–BMP conjugates prepared by chemical immobilization (Tanaka and Matsunaga 2000). Human insulin concentrations in blood serum were measured by a fully automated sandwich immunoassay using antibody–protein A–BMP complexes and alkaline phosphatase-conjugated antibodies as primary and secondary immunoreactants, respectively (Fig. 8). Dose–response curves were obtained from the luminescence intensity for human insulin concentrations. A detection limit of 2 μU/mL and a linear correlation between signal and concentration were apparent over the range of 19–254 μU/mL (Tanaka and Matsunaga 2000).

BMPs displaying receptors were developed and utilized for receptor binding assays. G-protein coupled receptors (GPCRs) represent one of the most predominant families of transmembrane proteins and are a prime target for drug discovery (Mirzabekov et al. 2000). Various challenges to ligand screening of GPCRs have been undertaken; however, these proteins are generally expressed at very low levels in the cell and are extremely hydrophobic, rendering the analysis of ligand interaction very difficult. Also pu-

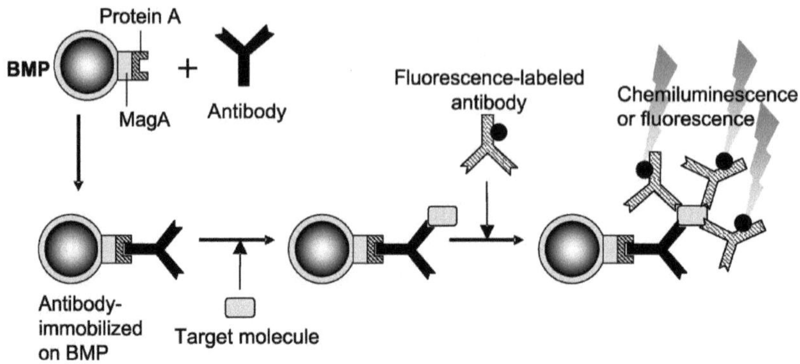

Fig. 8 Schematic diagram of chemiluminescence and fluorescence sandwich immunoassay on BMPs

rification of GPCRs from cells is frequently time-consuming and typically results in loss of native conformation. In BMPs displaying D1 dopamine receptor (D1R), GPCR was used as a model of the ligand binding assay (Yoshino et al. 2004). Efficient assembling of D1R into the lipid membrane of nanosized BMPs was accomplished, and this was used for a competitive dopamine binding assay. This system conveniently refines the native conformation of GPCRs without the need for detergent solubilization, purification, and reconstitution after cell disruption. This novel system provides advantages for studying various membrane proteins, which are usually difficult to assay.

4.2 Cell Separation

Immunomagnetic particles have been used preferentially in target cell separation from leukocytes for in vitro diagnosis, because of the more rapid and simple methodology compared with cell sorting using a flow cytometer (Parra et al. 1997). Magnetic bead technology allows the simple, rapid, and efficient enrichment of target cell populations. In general, nanosized magnetic particles, rather than microsized beads, are preferred for cell separation because separated cells using the latter are subsequently useful for flow cytometric analysis. Microsized magnetic particles, on the other hand, have inhibitory effects on cell growth and differentiation after magnetic separation. However, commercially available nanoparticles are superparamagnetic and require the use of specially designed magnetic columns for cell separation to produce a high magnetic field gradient. Because BMPs consist of ferromagnetic iron oxide, they are easily separated from cell suspensions using a permanent magnet with no special column. We have therefore applied BMPs to develop highly efficient magnetic cell separation (Kuhara et al.

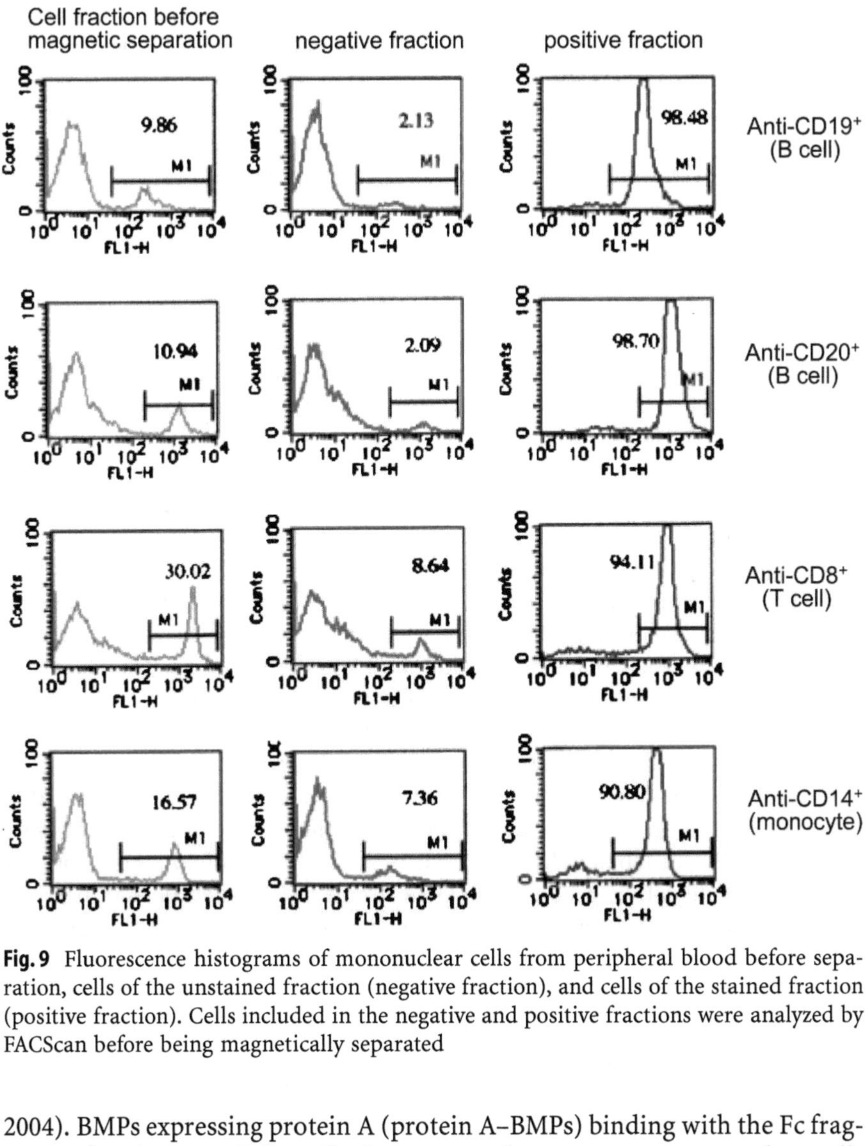

Fig. 9 Fluorescence histograms of mononuclear cells from peripheral blood before separation, cells of the unstained fraction (negative fraction), and cells of the stained fraction (positive fraction). Cells included in the negative and positive fractions were analyzed by FACScan before being magnetically separated

2004). BMPs expressing protein A (protein A–BMPs) binding with the Fc fragment of anti-mouse IgG antibody were used to separate mononuclear cells from peripheral blood. The procedure of positive selection involves incubation of mononuclear cells and mouse monoclonal antibodies against different cell surface antigens (CD8, CD14, CD19, CD20) prior to treatment with protein A–BMP binding with rabbit anti-mouse IgG secondary antibodies. The average purities of the separated mononuclear cells of $CD19^+$ and $CD20^+$ were 97.5 and 97.6%, respectively (Fig. 9). Furthermore, more than 95% of the recovery ratio was obtained for these cells. Stem cell marker (CD34) positive cells were also separated using protein A–BMP binding with antibody.

CD34 is the best-identified surface antigen expressed on hematopoietic stem cells which are known as rare cells. The separated stem cells using BMPs maintained their capability of colony formation as hematopoietic stem cells without inhibition of the proliferation and differentiation abilities. Specific separation of target cells using BMPs was achieved by simple magnetic separation from cell suspensions using a permanent magnet with no special magnetic column.

4.3 Single-Nucleotide Polymorphism Discrimination by Automated System

Single-nucleotide polymorphisms (SNPs) are the most common form of human genetic polymorphism and serve as markers for identifying genes responsible for diseases such as cancer, hypertension, and diabetes induced by environmental factors (Nyren et al. 1997). High-throughput and accurate multiple assay systems are therefore required for SNP analysis. For DNA detection employing sub-microparticles and fluorescent dyes, the wavelength from light scattering caused by the small particles (1000–100 nm) overlaps the fluorescent dye emissions, resulting in low sensitivity (Nakayama et al. 2003; Tanaka et al. 2003). This phenomenon is called Mie scattering of nanometer-size particles in solution, and has disturbed accurate SNP detection in a BMP system when using fluorescent dye-labeled probes such as fluorescein isothiocycanate (FITC) or cyanine dyes Cy3 and Cy5, which show a Stokes shift with narrow wavelength. To avoid this interference of fluorescence signal, we developed a protocol based on DNA thermal dissociation curve analysis for an automated system with BMPs (Maruyama et al. 2004). Biotin-labeled PCR products in ALDH2, two allele-specific probes (Cy3-labeled detection probe for ALDH2∗1 and Cy5-labeled detection probe for ALDH2∗2), and streptavidin-immobilized BMPs (SA–BMPs) were simultaneously mixed. The mixture was denatured at 70 °C and cooled slowly to 25 °C, allowing the DNA duplex to form. The DNA–BMP complex was then heated to 58 °C, a temperature determined by dissociation curve analysis, and a single-base mismatched detection probe was removed at the same temperature under precise control. The fluorescence signal from the detection probe was liberated into the supernatant from completely matched duplex DNA–BMP complex by heating to 80 °C and measured. Based on this principle, three genotypes of SNP in ALDH2 gene categorized into the homozygous (ALDH2∗1/∗1), heterozygous (ALDH2∗1/∗2), and mutant (ALDH2∗2/∗2) were successfully discriminated using the automated detection system with BMPs (Fig. 10).

An automated SNP detection processor for BMP-based SNP discrimination was designed based on the protocol for DNA thermal dissociation curve analysis (Fig. 11) (Maruyama et al. 2004). The designed processor is equipped with a 96-way automated pipetter which collects and dispenses fluids as it

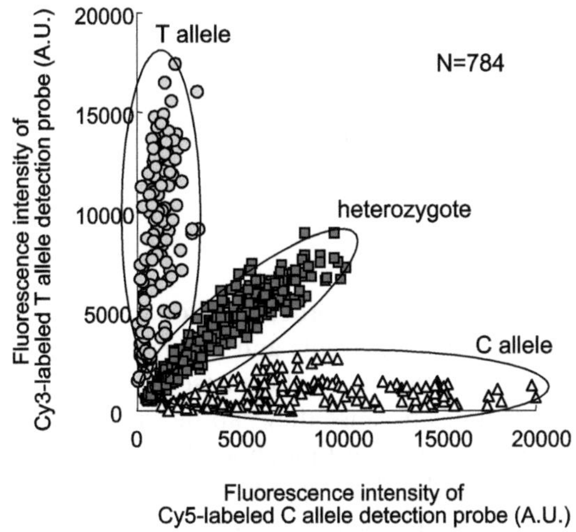

Fig. 10 SNP detection in TGF-β1 gene using an automated SNP detection system with BMPs

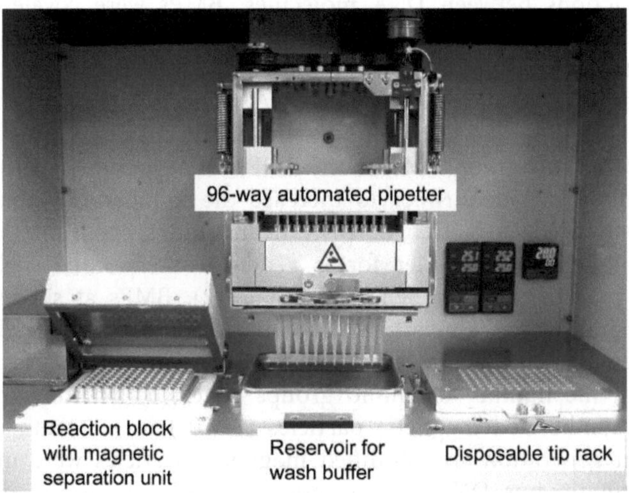

Fig. 11 Photograph of an automated SNP detection processor for BMP-based SNP discrimination

moves in the vertical and horizontal directions. The platform contains a disposable tip rack, a reagent vessel, and a reaction station with magnetic separation unit for a 96-well microtiter plate. One pole of the magnet allows the application of a magnetic field for one well. Eight poles of the magnet were aligned on an iron rod and 12 rods were set on the rear side of the microtiter plate to apply magnetic fields to the 96 wells. The applied magnetic field is switched on and off by rotating the rods 180°. The reaction station is combined with a heat block regulated in the range of 4–99 °C, configured to perform the hybridization step. This system permits simultaneous SNP discrimination of 96 samples per assay.

4.4 DNA Extraction

DNA extraction is a fundamental procedure for molecular biological techniques, but the use of organic solvents and intensive labor is often involved. The use of magnetic particles as a solid-phase adsorbent is well suited for DNA extraction techniques because they can be easily manipulated through simple application of a magnet. Although artificially synthesized iron oxide has the capacity to adsorb DNA, aggregation is prominent due to magnetic attraction reducing the usable surface area. Initial application of native BMPs for this purpose revealed that further modification was necessary. Organosilane compounds can be covalently bound to iron oxides, introducing a positive ionic charge beneficial for restoring dispersion and facilitating ionic interactions between DNA molecules. BMPs were covalently modified with 3-[2-(2-aminoethyl)-ethylamino]-propyltrimethoxysilane (AEEA) which contains three amine groups (Yoza et al. 2002). DNA extraction using AEEA-modified BMPs showed higher DNA recoveries than other commercially available kits. The DNA binding efficiency increased with increasing numbers of amino groups. Moreover, BMPs were modified with polyamidoamine (PAMAM) dendrimer to increase the number of amino groups, allowing enhanced extraction of DNA from fluid suspension (Yoza et al. 2003b). A PAMAM dendrimer forms a dense outer amine shell through a cascade type of polymer generation (Fig. 12). BMPs at successive dendrimer generations were investigated for DNA binding abilities. The amine numbers double with every layer generated on the BMP surface. On the other hand, the number of amino groups generated on the same size of artificially synthesized magnetic particles (AMPs) is one tenth of that on the BMP at generation six. Modified magnetic particles were mixed with 25 μg of calf thymus DNA. The amount of DNA binding by dendrimer-modified BMPs increased with every generation and 24.83 ± 1.61 μg of DNA bound to 100 μg BMPs at generation six. Furthermore, there was a linear correlation with the amount of DNA removed from solution and increased amino charge as dendrimer generations also increased. This occurrence further confirms the removal of DNA from solution because of cationic charge contribution. DNA recovery was obtained by incubating the DNA complex at 50 °C for 30 min resulting in 87% release (21.70 ± 1.59 μg). DNA recovery with dendrimer-modified BMPs is about six times higher than that with dendrimer-modified AMPs. Furthermore, small quantities of dendrimer-modified BMPs were used to extract DNA from blood.

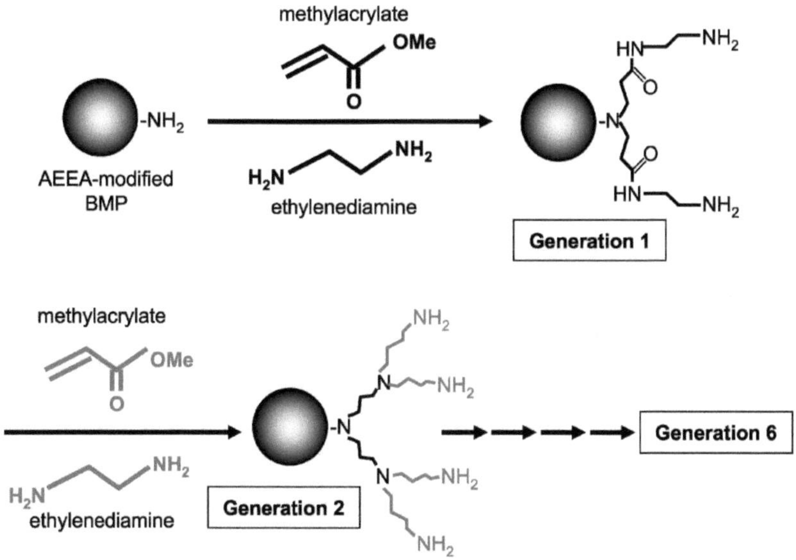

Fig. 12 Synthesis of polyamidoamine dendrimer on BMPs used for DNA extraction. Dendrimer was generated with amine groups derived from AEEA-modified BMPs. Stepwise growth was repeated until the desired number of generations was obtained

The efficiency of DNA recovery was consistently about 30 μg of DNA using 2–10 μg of dendrimer-modified BMPs. This process was incorporated into a fully automated system using a newly developed liquid handling procedure (Yoza et al. 2003a).

5 Future Prospects

Our comprehensive molecular studies on genes and proteins involved in BMP formation in *M. magneticum* AMB-1 have led to the postulation of the step-by-step events of magnetite biomineralization. The results obtained from these fundamental studies have allowed us to develop novel molecular constructs on the BMP surface. The constructed functional BMPs generated by both chemical processes and genetic manipulations are useful for various biotechnological applications in immunoassays, receptor binding assays, DNA extraction, and cell separation. Moreover, the application of functional BMPs in a fully automated system provides precise, rapid, and high-throughput analyses. Further studies will facilitate the development of more advanced biomaterials, such as site-specific immobilization of functional molecules on magnetite crystals, and quantum size effects in magnetic particles with well-controlled size and shape.

Acknowledgements This work was funded in part by a Grant-in-Aid for Specially Promoted Research 2, no. 13002005, from the Ministry of Education, Culture, Sports, Science, and Technology of Japan.

References

Amemiya Y, Tanaka T, Yoza B, Matsunaga T (2005) Novel detection system for biomolecules using nanosized bacterial magnetic particles and magnetic force microscopy. J Biotechnol 120:308–314

Arakaki A, Webb J, Matsunaga T (2003) A novel protein tightly bound to bacterial magnetic particles in *Magnetospirillum magneticum* strain AMB-1. J Biol Chem 278:8745–8750

Bazylinski SA, Garratt-Reed AJ, Frankel RB (1994) Electron microscopic studies of magnetosomes in magnetotactic bacteria. Microsc Res Tech 27:389–401

Calugay RJ, Miyashita H, Okamura Y, Matsunaga T (2003) Siderophore production by the magnetic bacterium *Magnetospirillum magneticum* AMB-1. FEMS Microbiol Lett 218:371–375

Calugay RJ, Okamura Y, Wahyudi AT, Takeyama H, Matsunaga T (2004) Siderophore production of a periplasmic transport binding protein kinase gene defective mutant of *Magnetospirillum magneticum* AMB-1. Biochem Biophys Res Commun 323:852–857

Clackson T, Hoogenboom HR, Griffiths AD, Winter G (1991) Making antibody fragments using phage display libraries. Nature 352:624–628

Frankel RB, Bazylinski DA, Schüler D (1998) Biomineralization of magnetic iron minerals in bacteria. Supramol Sci 5:383–390

Fukuda Y, Okamura Y, Takeyama H, Matsunaga T (2006) Dynamic analysis of a genomic island in *Magnetospirillum* sp. strain AMB-1 reveals how magnetosome synthesis developed. FEBS Lett 580:801–812

Gorby YA, Beveridge TJ, Blakemore RP (1988) Characterization of the bacterial magnetosome membrane. J Bacteriol 170:834–841

Grünberg K, Wawer C, Tebo BM, Schüler D (2001) A large gene cluster encoding several magnetosome proteins is conserved in different species of magnetotactic bacteria. Appl Environ Microbiol 67:4573–4582

Jogler C, Schüler D (2006) Genetic Analysis of Magnetosome Biomineralization. Microbiol Monogr, in this volume

Kawaguchi R, Burgess JG, Sakaguchi T, Takeyama H, Thornhill RH, Matsunaga T (1995) Phylogenetic analysis of a novel sulfate-reducing magnetic bacterium, RS-1, demonstrates its membership of the delta-Proteobacteria. FEMS Microbiol Lett 126:277–282

Knol J, Sjollema K, Poolman B (1998) Detergent-mediated reconstitution of membrane proteins. Biochemistry 37:16410–16415

Komeili A, Vali H, Beveridge TJ, Newman DK (2004) Magnetosome vesicles are present before magnetite formation, and MamA is required for their activation. Proc Natl Acad Sci USA 101:5779–5790

Komeili A (2006) Cell Biology of Magnetosome Formation. Microbiol Monogr, in this volume

Komeili A, Li Z, Newman DK, Jensen GJ (2006) Magnetosomes are cell membrane invaginations organized by the actin-like protein MamK. Science 311:242–245

Kuhara M, Takeyama H, Tanaka T, Matsunaga T (2004) Magnetic cell separation using antibody binding with protein A expressed on bacterial magnetic particles. Anal Chem 76:6207–6213

Lee JS, Shin KS, Pan JG, Kim CJ (2000) Surface-displayed viral antigens on Salmonella carrier vaccine. Nat Biotechnol 18:645–648

Lowenadler B, Jansson B, Paleus S, Holmgren E, Nilsson B, Moks T, Palm G, Josephson S, Philipson L, Uhlen M (1987) A gene fusion system for generating antibodies against short peptides. Gene 58:87–97

Maruyama K, Takeyama H, Nemoto E, Tanaka T, Yoda K, Matsunaga T (2004) Single nucleotide polymorphism detection in aldehyde dehydrogenase 2 (ALDH2) gene using bacterial magnetic particles based on dissociation curve analysis. Biotechnol Bioeng 87:687–694

Matsunaga T, Kamiya S (1987) Use of magnetic particles isolated from magnetotactic bacteria for enzyme immobilization. Appl Microbiol Biotechnol 26:328–332

Matsunaga T, Tadokoro F, Nakamura N (1990) Mass culture of magnetic bacteria and their application to flow type immunoassays. IEEE Trans Magnet 26:1557–1559

Matsunaga T, Sakaguchi T, Tadokoro F (1991) Magnetite formation by a magnetic bacterium capable of growing aerobically. Appl Microbiol Biotechnol 35:651–655

Matsunaga T, Nakamura C, Burgess JG, Sode K (1992) Gene transfer in magnetic bacteria: transposon mutagenesis and cloning of genomic DNA fragments required for magnetosome synthesis. J Bacteriol 174:2748–2753

Matsunaga T, Tsujimura N, Kamiya S (1996a) Enhancement of magnetic particle production by nitrate and succinate fed-batch culture of *Magnetospirillum* sp. AMB-1. Biotechnol Tech 10:495–500

Matsunaga T, Kawasaki M, Yu X, Tsujimura N, Nakamura N (1996b) Chemiluminescence enzyme immunoassay using bacterial magnetic particles. Anal Chem 68:3551–3554

Matsunaga T, Higashi Y, Tsujimura N (1997) Drug delivery by magnetoliposomes containing bacterial magnetic particles. Cellular Eng 2:7–11

Matsunaga T, Sato R, Kamiya S, Tanaka T, Takeyama H (1999) Chemiluminescence enzyme immunoassay using protein A-bacterial magnetite complex. J Magn Magn Mater 194:126–134

Matsunaga T, Togo H, Kikuchi T, Tanaka T (2000) Production of luciferase-magnetic particle complex by recombinant *Magnetospirillum* sp. AMB-1. Biotechnol Bioeng 70:704–709

Matsunaga T, Arakaki A, Takahoko M (2002) Preparation of luciferase-bacterial magnetic particle complex by artificial integration of MagA-luciferase fusion protein into the bacterial magnetic particle membrane. Biotechnol Bioeng 77:614–618

Matsunaga T, Ueki F, Obata K, Tajima H, Tanaka T, Takeyama H, Goda Y, Fujimoto S (2003) Fully automated immunoassay system of endocrine disrupting chemicals using monoclonal antibodies chemically conjugated to bacterial magnetic particles. Anal Chim Acta 475:75–83

Matsunaga T, Okamura Y, Fukuda Y, Wahyudi AT, Murase Y, Takeyama H (2005) Complete genome sequence of the facultative anaerobic magnetotactic bacterium *Magnetospirillum* sp. strain AMB-1. DNA Res 12:157–166

Mirzabekov T, Kontos H, Farzan M, Marasco W, Sodroski J (2000) Paramagnetic proteoliposomes containing a pure, native, and oriented seven-transmembrane segment protein, CCR5. Nat Biotechnol 18:649–654

Murayama E, Takagi Y, Ohira T, Davis JG, Greene MI, Nagasawa H (2002) Fish otolith contains a unique structural protein, otolin-1. Eur J Biochem 269:688–696

Nakamura N, Hashimoto K, Matsunaga T (1991) Immunoassay method for the determination of immunoglobulin G using bacterial magnetic particles. Anal Chem 63:268–272

Nakamura N, Burgess JG, Yagiuda K, Kudo S, Sakaguchi T, Matsunaga T (1993) Detection and removal of *Escherichia coli* using fluorescein isothiocyanate conjugated monoclonal antibody immobilized on bacterial magnetic particles. Anal Chem 65:2036–2039

Nakamura N, Matsunaga T (1993) Highly sensitive detection of allergen using bacterial magnetic particles. Anal Chim Acta 281:585–589

Nakamura C, Burgess JG, Sode K, Matsunaga T (1995a) An iron-regulated gene, *magA*, encoding an iron transport protein of *Magnetospirillum* sp. strain AMB-1. J Biol Chem 270:28392–28396

Nakamura C, Kikuchi T, Burgess JG, Matsunaga T (1995b) Iron-regulated expression and membrane localization of the *magA* protein in *Magnetospirillum* sp. strain AMB-1. J Biochem (Tokyo) 118:23–27

Nakayama H, Arakaki A, Maruyama K, Takeyama H, Matsunaga T (2003) Single-nucleotide polymorphism analysis using fluorescence resonance energy transfer between DNA-labeling fluorophore, fluorescein isothiocyanate, and DNA intercalator, POPO-3, on bacterial magnetic particles. Biotechnol Bioeng 84:96–102

Nyren P, Karamohamed S, Ronaghi M (1997) Detection of single-base changes using a bioluminometric primer extension assay. Anal Biochem 244:367–373

Okamura Y, Takeyama H, Matsunaga T (2000) Two-dimensional analysis of proteins specific to the bacterial magnetic particle membrane from *Magnetospirillum* sp. AMB-1. Appl Biochem Biotechnol 84-86:441–446

Okamura Y, Takeyama H, Matsunaga T (2001) A magnetosome-specific GTPase from the magnetic bacterium *Magnetospirillum magneticum* AMB-1. J Biol Chem 276:48183–48188

Okamura Y, Takeyama H, Sekine T, Sakaguchi T, Wahyudi AT, Sato R, Kamiya S, Matsunaga T (2003) Design and application of a new cryptic-plasmid-based shuttle vector for *Magnetospirillum magneticum*. Appl Environ Microbiol 69:4274–4277

Okuda Y, Denda K, Fukumori Y (1996) Cloning and sequencing of a gene encoding a new member of the tetratricopeptide protein family from magnetosomes of *Magnetospirillum magnetotacticum*. Gene 171:99–102

Parra E, Wingren AG, Hedlund G, Kalland T, Dohlsten M (1997) The role of B7-1 and LFA-3 in costimulation of CD8+ T cells. J Immunol 158:637–642

Rinaldi AC, Mangoni ML, Rufo A, Luzi C, Barra D, Zhao H, Kinnunen PK, Bozzi A, Di Giulio A, Simmaco M (2002) Temporin L: antimicrobial, haemolytic and cytotoxic activities, and effects on membrane permeabilization in lipid vesicles. Biochem J 368:91–100

Sakaguchi T, Burgess JG, Matsunaga T (1993) Magnetite formation by a sulphate-reducing bacterium. Nature 365:47–49

Sakaguchi T, Arakaki A, Matsunaga T (2002) *Desulfovibrio magneticus* sp. nov., a novel sulfate-reducing bacterium that produces intracellular single-domain-sized magnetite particles. Int J Syst Evol Microbiol 52:215–221

Scheffel A, Gruska M, Faivre D, Linaroudis A, Plitzko JM, Schüler D (2006) An acidic protein aligns magnetosomes along a filamentous structure in magnetotactic bacteria. Nature 440:110–114

Schübbe S, Kube M, Scheffel A, Wawer C, Heyen U, Meyerdierks A, Madkour MH, Mayer F, Reinhardt R, Schüler D (2003) Characterization of a spontaneous nonmagnetic mutant of *Magnetospirillum gryphiswaldense* reveals a large deletion comprising a putative magnetosome island. J Bacteriol 185:5779–5790

Schultheiss D, Handrick R, Jendrossek D, Hanzlik M, Schüler D (2005) The presumptive magnetosome protein Mms16 is a poly(3-hydroxybutyrate) granule-bound protein (phasin) in *Magnetospirillum gryphiswaldense*. J Bacteriol 187:2416–2425

Sode K, Kudo S, Sakaguchi T, Nakamura N, Matsunaga T (1993) Application of bacterial magnetic particles for highly selective mRNA recovery system. Biotechnol Tech 7:688–694

Sudo S, Fujikawa T, Nagakura T, Ohkubo T, Sakaguchi K, Tanaka M, Nakashima K, Takahashi T (1997) Structures of mollusc shell framework proteins. Nature 387:563–564

Suzuki T, Okamura Y, Calugay RJ, Takeyama H, Matsunaga T (2006) Global gene expression analysis of iron-inducible genes in *Magnetospirillum magneticum* AMB-1. J Bacteriol 188:2275–2279

Takeyama H, Tsuzuki H, Chow S, Nakayama H, Matsunaga T (2000) Discrimination between Atlantic and Pacific subspecies of northern bluefin tuna (*Thunnus thynnus*) by magnetic-capture hybridization using bacterial magnetic particles. Mar Biotechnol (NY) 2:309–313

Tanaka T, Matsunaga T (2000) Fully automated chemiluminescence immunoassay of insulin using antibody-protein A-bacterial magnetic particle complexes. Anal Chem 72:3518–3522

Tanaka T, Maruyama K, Yoda K, Nemoto E, Udagawa Y, Nakayama H, Takeyama H, Matsunaga T (2003) Development and evaluation of an automated workstation for single nucleotide polymorphism discrimination using bacterial magnetic particles. Biosens Bioelectron 19:325–330

Tanaka T, Takeda H, Kokuryu Y, Matsunaga T (2004a) Spontaneous integration of transmembrane peptides into a bacterial magnetic particle membrane and its application to display of useful proteins. Anal Chem 76:3764–3769

Tanaka T, Takeda H, Ueki F, Obata K, Tajima H, Takeyama H, Goda Y, Fujimoto S, Matsunaga T (2004b) Rapid and sensitive detection of 17beta-estradiol in environmental water using automated immunoassay system with bacterial magnetic particles. J Biotechnol 108:153–159

Wahyudi AT, Takeyama H, Matsunaga T (2001) Isolation of *Magnetospirillum magneticum* AMB-1 mutants defective in bacterial magnetic particle synthesis by transposon mutagenesis. Appl Biochem Biotechnol 91–93:147–154

Wahyudi AT, Takeyama H, Okamura Y, Fukuda Y, Matsunaga T (2003) Characterization of aldehyde ferredoxin oxidoreductase gene defective mutant in *Magnetospirillum magneticum* AMB-1. Biochem Biophys Res Commun 303:223–229

Yang CD, Takeyama H, Tanaka T, Matsunaga T (2001a) Effects of growth medium composition, iron sources and atmospheric oxygen concentrations on production of luciferase-bacterial magnetic particle complex by a recombinant *Magnetospirillum magneticum* AMB-1. Enzyme Microb Technol 29:13–19

Yang CD, Takeyama H, Matsunaga T (2001b) Iron feeding optimization and plasmid stability in production of recombinant bacterial magnetic particles by *Magnetospirillum magneticum* AMB-1 in fed-batch culture. J Biosci Bioeng 91:213–216

Yoshino T, Takahashi M, Takeyama H, Okamura Y, Kato F, Matsunaga T (2004) Assembly of G protein-coupled receptors onto nanosized bacterial magnetic particles using Mms16 as an anchor molecule. Appl Environ Microbiol 70:2880–2885

Yoshino T, Kato F, Takeyama H, Nakai M, Yakabe Y, Matsunaga T (2005) Development of a novel method for screening of estrogenic compounds using nanosized bacterial magnetic particles displaying estrogen receptor. Anal Chim Acta 532:105–111

Yoshino T, Matsunaga T (2005) Development of efficient expression system for protein display on bacterial magnetic particles. Biochem Biophys Res Commun 338:1678–1681

Yoshino T, Matsunaga T (2006) Efficient and stable display of functional proteins on bacterial magnetic particles using mms13 as a novel anchor molecule. Appl Environ Microbiol 72:465–471

Yoza B, Matsumoto M, Matsunaga T (2002) DNA extraction using modified bacterial magnetic particles in the presence of amino silane compound. J Biotechnol 94:217–224

Yoza B, Arakaki A, Maruyama K, Takeyama H, Matsunaga T (2003a) Fully automated DNA extraction from blood using magnetic particles modified with hyperbranched polyamidoamine dendrimer. J Biosci Bioeng 95:21–26

Yoza B, Arakaki A, Matsunaga T (2003b) DNA extraction using bacterial magnetic particles modified with hyperbranched polyamidoamine dendrimer. J Biotechnol 101:219–228

Microbiol Monogr (3)
D. Schüler: Magnetoreception and Magnetosomes in Bacteria
DOI 10.1007/7171_046/Published online: 30 August 2006

Paleomagnetism and Magnetic Bacteria

Michael Winklhofer (✉) · Nikolai Petersen

Department of Earth and Environmental Science,
Ludwig-Maximilians-University of Munich, Theresienstr. 41/IV, 80333 Muenchen, Germany
michaelw@lmu.de

Abstract This contribution focuses on bacterial magnetite as a possible archive of paleomagnetic information. Bacterial magnetite can be formed intracellularly or epicellularly (extracellularly), depending on the type of bacterium. Whilst intracellularly synthesized magnetite (magnetosomes) has magnetic properties suitable for retaining paleomagnetic information, the extracellularly precipitated magnetite is mostly superparamagnetic and therefore not a reliable carrier of paleomagnetic information. We discuss the chemical conditions under which fossil magnetosomes can be preserved and what rock magnetic parameters can be used to detect their presence or absence in sediments. In the last section we briefly review the debate on whether or not the magnetite crystals observed in the Martian meteorite ALH84001 represent fossil traces of early life on Mars.

1 Paleomagnetism

Paleomagnetism deduces the magnetic information that is stored in igneous and sedimentary rocks. This information is carried by the ferromagnetic minerals common in most rocks as accessory constituents. Although the concentration of ferromagnetic minerals in common rocks is normally small, hardly exceeding 5 vol %, its information can nevertheless conveniently be measured due to the high sensitivity of modern magnetometers.

Paleomagnetic information is contained in the natural remanent magnetization (NRM) of rocks. From its direction and intensity, the character of the Earth's magnetic field at the time of formation of the rock can be determined. By these measurements it becomes clear that the dipole axis of the magnetic field has always been parallel to the rotational axis of the Earth throughout geological history. This allows an important application in paleomagnetism: the measurement of a possible displacement of the rock with respect to the magnetic field.

Although the dipole axis of the geomagnetic field is parallel to the Earth's rotational axis, its polarity has changed sign frequently and in an irregular way. The intervals of polarity change last between one and several thousand years, and are characterized by a drop of magnetic field intensity to about 10% its normal value (Merrill et al. 1998). This behaviour is recorded in many sedimentary sequences as "magnetostratigraphy". With the pattern of field reversal known, the measurement of magnetostratigraphy can be used for age determinations, another important application of paleomagnetism.

A more recent field of paleomagnetism, termed "environmental magnetism" (Thompson and Oldfield 1986; Maher and Thompson 1999; Evans and Heller 2003), has evolved by the analysis of certain magnetic properties of sedimentary deposits and their relation to environmental conditions during formation. This has been shown to be a very effective way of analysis, as the magnetic properties of most natural magnetic minerals react extremely sensitively to environmental changes.

2 Carrier of Magnetic Information in Rocks

The dominating magnetic mineral in most common rocks is magnetite, and thus it is also the most important carrier of paleomagnetic information. Magnetite forms in igneous rocks as an early precipitate at high temperatures. In sedimentary rocks it is frequently present as detrital mineral, i.e. as a relic of pre-existing rocks, but it can also form authigenically (in situ) in sediments with slightly reducing conditions. Thus, magnetite was considered a product of inorganic processes until Lowenstam (1962) discovered biogenic magnetite in the teeth of chitons. We know now that biogenic magnetite is widespread in nature and also common in many sedimentary deposits. Here the most important contribution is bacterially formed magnetite.

Lowenstam (1981) recognized that biogenic magnetite forms by processes with varying degrees of control of the organism over the mineralization process. He distinguishes between biologically induced mineralization (BIM) and organic matrix-mediated mineralization; the latter is also called biologically controlled mineralization (BCM) or biologically organized mineralization (BOM). BIM refers to processes with no biological control, while BCM/BOM refers to processes with strict metabolic and genetic control.

Following this distinction we recognize two groups of bacteria that are responsible for the occurrence of biogenic magnetite in sediments and soils. The first group are the so-called magnetotactic bacteria, discovered in 1975 by Richard Blakemore. These organisms synthesize intracellular particles of magnetite (Fe_3O_4) or greigite (Fe_3S_4), usually arranged in chains, which give the bacterium the characteristic property of a swimming compass needle. The magnetic crystals, so-called magnetosomes, are magnetically speaking stable single-domain particles (Sect. 3) characterized by a size such that the particles have minimum induced magnetization and maximum permanent magnetic moment. They are an example of biologically controlled processes (BCM).

The second group of bacteria forms iron minerals epicellularly or extracellularly without directly influencing the crystal size and chemistry (Lovley et al. 1987). Here, iron minerals typically precipitate outside the bacterium on the cell wall through reaction of extraneous ions with metabolic products extruded across or into the cell wall (Mann 2001). The mineral products are crystallochemically heterogeneous and have a wide range of grain sizes. It is not clear if there is a function to this kind of biomineral or if they are mere by-products of metabolic activity. Magnetite formed by this process (BIM) is magnetically different to the magnetosomes found inside the magnetotactic bacteria. To understand this difference in magnetic properties of one and the same mineral, and also to assess the importance for paleomagnetism, we will first discuss the relationship between magnetic properties of magnetite particles and grain size.

3
Grain Size Dependence of Magnetic Properties

The magnetic properties of magnetite crystals, such as magnetic remanence and coercive force, are largely controlled by domain state, which in turn depends upon grain size and aspect ratio. The two domain states relevant for our consideration are single domain (SD) and pseudo-single domain (PSD); superparamagnetic particles will be dealt with separately in the section below. In an ideal SD particle, all the elementary magnetic dipoles (four elementary Bohr magnetons per magnetite unit cell) are aligned parallel to form a uniform magnetization structure (Fig. 1a). In more realistic SD particles, slight departures from the uniform magnetization structure arise near the corners and edges of the particle, which become larger with increasing grain size (Fig. 1b). These quasi-uniform magnetization configurations are called flower states and are energetically most favourable in magnetite cubes (octahedra) as long as the edge length is smaller than 70 nm (91 nm), corresponding to a critical diameter d_0^{min} = 88 nm when expressed as diameter of a sphere of the same volume (Fabian et al. 1996; Witt et al. 2005). Significantly, the flower state can persist as a metastable magnetization configuration in crystals with grain sizes well above d_0^{min}, although it would be energetically more favourable to collapse into a non-uniform state such as a magnetic vortex (Fig. 1c). In a magnetic vortex, the magnetization structure is curl-

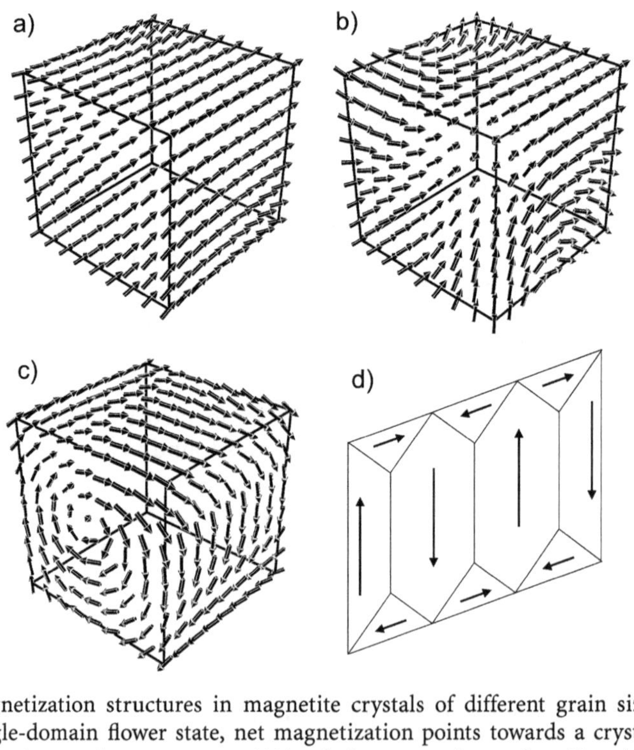

Fig. 1 Magnetization structures in magnetite crystals of different grain size *d*. **a** d = 60 nm, single-domain flower state, net magnetization points towards a crystallographic <111> direction and amounts to 99% of the saturation value M_{sat} (480 Gauss). **b** d = 120 nm, single-domain flower state, $M_{net} = 0.9\ M_{sat}$. **c** d = 140 nm, vortex state, $M_{net} = 0.1\ M_{sat}$. **d** Section through a multi-domain crystal, with lamellar domains of opposite magnetization direction in adjacent domains ($M_{net} = 0$). The small closure domains at the *top* and *bottom* efficiently reduce the pole density on the particle surface. **a–c** were obtained by micromagnetic modelling using a resolution of 16 × 16 × 16 cells

ing, thereby efficiently reducing the stray field outside the particle, however, at the expense of the magnetic moment. Magnetite crystals with grain sizes larger than 1 μm accommodate several uniformly magnetized, lamellar domains (multi-domain, or "MD") with adjacent domains in the interior having opposite polarity (Fig. 1d). As a rule, the number of domains increases with particle volume. For particles with grain sizes above the upper SD threshold size d_0^{max}, the magnetic behaviour changes gradually from SD-like to truly MD-like. These are referred to as PSD particles and comprise crystals that host one or several vortices or a few domains.

While SD particles carry the maximum magnetic moment per particle volume, which here is equal to the saturation magnetization, the specific magnetic moment of a vortex decreases rapidly with grain size, and is nearly zero in MD particles. For the purpose of magnetotaxis it therefore is most efficient for a microorganism to produce SD particles. This can be achieved by limiting the particle size to 140 nm and by increasing the particle elongation. It therefore comes as no surprise that magnetite crystals in magnetic bacteria usually have grain sizes below 140 nm and are elongated (Petersen et al. 1989). It is true that vortex states are energetically more favourable in magnetite crystals greater than 70 nm, but one has to bear in mind that the magnetization in growing magnetosomes already occupies a SD state as the particle size exceeds the threshold size. Up to a critical size of 140 nm, there are energy barriers preventing the (metastable) SD state from collapsing into a vortex. Magnetosomes typically have rounded corners (Balkwill et al. 1980; Buseck et al. 2001), which effectively inhibits nucleation of the vortex states and so stabilizes the SD state in realistic magnetosomes even further compared to "angular" particle shapes like cubes or octahedra (Witt et al. 2005). The uniform magnetization in magnetosomes has convincingly been confirmed by transmission electron holography (McCartney et al. 2001). Likewise, it is necessary to emphasize that the magnetic phase diagrams were calculated for isolated crystals in zero external field. As a rule, magnetosomes are arranged in chains and so strongly magnetically interact. The interactions in a linear particle chain are such that the SD state will be stabilized.

4 Superparamagnetism

If, on the other hand, the particle edge-length becomes less than some 50 nm, magnetite cubes cannot retain a temporally stable SD magnetization at room temperature because the magnetization structure is constantly buffeted by thermal fluctuations, leading to frequent spontaneous magnetization reversals in the particle. Such a behaviour is called Néel superparamagnetism (SP). In a magnetic field, an assemblage of SP particles carries a magnetization, which will decay with time after the magnetic field is switched off. At low temperatures, SP particles become magnetically blocked and then behave as stable magnetic single domains (see also Fig. 2 in Magnetite-Based Magnetoreception in Higher Organisms, for the magnetic detection of SP particles in the beak tissue of homing pigeons) (Winklhofer 2006, in this volume).

Although superparamagnetism does not represent a true domain state but rather describes a thermal relaxation phenomenon, it is sensible to include it in the magnetic phase diagram, defining the lower limit of SD stability (Fig. 2). A few magnetosomes plot into the SP regime. These "outliers" are nascent magnetosomes, newly forming at the chain ends.

It is important to note that Néel superparamagnetism refers to the stability of magnetic remanence with respect to the coordinate system of the particle. If the particles are not embedded (as in a solid rock matrix) but dispersed in a fluid, then the magnetic remanence of an assemblage of particles will decay through Brownian motion in the fluid. Magnetically stable SD particles, which are able to rotate freely in a viscous medium, are therefore subject to the Brownian type of superparamagnetism. Using magnetic relaxometry, one can distinguish Néel and Brownian relaxation by their different relaxation time characteristics (Matz et al. 1998).

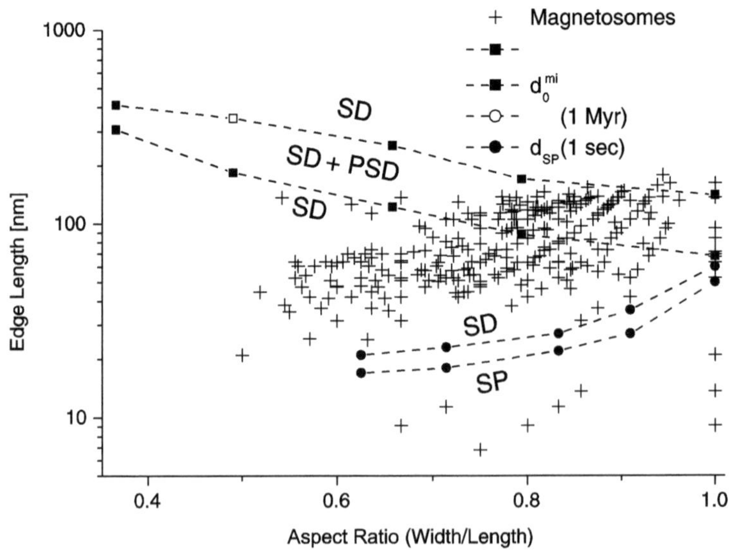

Fig. 2 Magnetic phase diagram for rectangular (parallelepiped shaped) magnetite nanocrystals as a function of the long edge length and aspect ratio, obtained by three-dimensional micromagnetic modelling (SD/PSD boundaries from Fabian et al. 1996, SP boundaries from Winklhofer et al. 1997). Magnetosome data are from Petersen et al. (1989). In the SD/PSD region, high-remanence (SD) states coexist with low-remanence (PSD) states. The superparamagnetic limit depends on the timescale. Single-domain particles with stable magnetization at laboratory timescale ($t = 1$ s) are magnetically unstable over geological timescales ($t = 1$ Myr)

As a rule, magnetite produced by non-magnetic bacteria (epi- and extracellularly; e.g. *Shewanella putrifaciens, Geobacter metallireducens*) falls into the superparamagnetic grain-size range (Moskowitz et al. 1989; Hanzlik et al. 1996) and therefore plays no role in the paleomagnetic signature of sediments. Nevertheless, the BIM magnetite may represent the major source of ultrafine-grained magnetite in recent sediments and, particularly, in soils (Maher 1991).

5 Chemical Stability of Magnetite

The quality of paleomagnetic information not only depends on the physical stability of the natural remanent magnetization of the magnetite particles, but also on their chemical stability. It has to be kept in mind that the chemical environment felt by the magnetosome inside the bacterium will be different after death and destruction of the protective cell membrane. The chemical

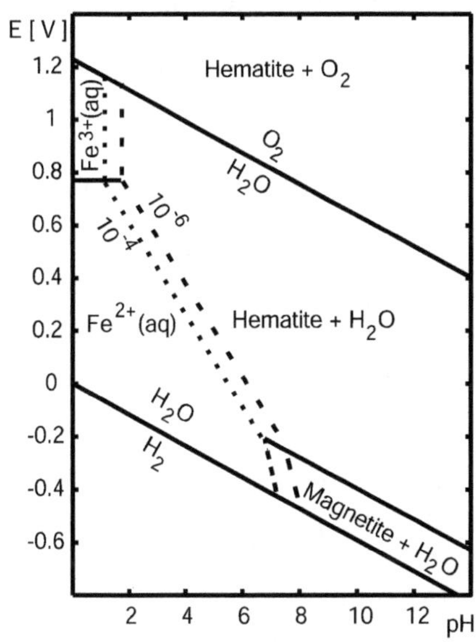

Fig. 3 Eh–pH diagram showing equilibrium conditions for magnetite plus hematite (Fe_2O_3) and an Fe solution containing 10^{-4} and 10^{-6} mol free iron (Fe^{2+} or Fe^{3+}) at 1 bar and 25 °C (*dotted* and *dashed lines*). The left boundaries (*dotted* and *dashed*, respectively) represent equilibrium of the iron oxides with the iron solution. At higher concentration of free iron, the boundary moves to the left, increasing the stability fields of magnetite and hematite. Note the narrow Eh–pH range over which magnetite is stable. The stability field of magnetite shrinks when CO_2 is present, thus allowing siderite ($FeCO_3$) to precipitate (not shown). See Garrels and Christ (1965) for the coexistence between iron oxides with iron sulphides and carbonates

stability and survival of the magnetosome will then depend on the actual environment.

Magnetite forms inorganically in a slightly basic and reducing (weakly anoxic, suboxic) environment, i.e. the pH must be greater than 7 and the redox potential (Eh value) below – 200 mV. As can be seen in Fig. 3, the stability field of magnetite under normal conditions is relatively narrow. When exposed to the open air, magnetite is prone to oxidation. At normal ambient conditions, however, the oxidation rate is such that larger magnetite particles will survive over geological times. The smaller the grain sizes (increasing surface/volume ratio), on the other hand, the more important a possible oxidation becomes.

In the following we will discuss two exemplary cases of chemical instability of fossil bacterial magnetite: case one, the effect of a strongly reducing environment; case two, an oxidizing environment (Sect. 9).

6 Occurrence of Magnetotactic Bacteria and Contribution to the Natural Magnetization of Sediments

Magnetotactic bacteria are present in freshwater and marine environments, where they usually concentrate in the uppermost centimetres of unconsolidated sediment. They are found in both the northern and southern hemispheres, with opposite magnetic polarities. As a rule, their polarity is such that they swim downwards along the field lines of the Earth's magnetic field which brings them back in the shortest time to their habitat when whirled up by currents. In this context it is surprising that magnetotactic bacteria are also found at the geomagnetic equator, where the field lines are parallel to the (water) surface.

Due to their magnetic alignment, magnetotactic bacteria after death potentially contribute to the natural magnetization of the growing and consolidating sediment column. As they are on average aligned to the Earth's magnetic field, they will carry paleomagnetic information in the sediment. The magnetosomes are by evolutionary selection single-domain particles, thus forming ideal carriers of paleomagnetic information.

Fossil bacterial magnetite usually occurs in sediments and soils as part of a mixture of different magnetic components, including detrital, authigenic, anthropogenic and diagenetic phases, which can potentially blur the magnetic signal of the biogenic component. Furthermore, owing to their small particle size, and as shown above in Fig. 3, biogenic magnetite particles are also prone to dissolution under reducing conditions or to undergo low-temperature oxidation. Both processes will affect the magnetic properties of the sediment. In particular, the Verwey transition at 125 K, diagnostic for magnetite, will shift to lower temperatures and eventually disappear with increasing degree of oxidation and maghemitization (Aragón et al. 1985) and/or with grain size decreasing below the superparamagnetic limit. It is therefore of interest to understand how chemical alteration, and the mixing of different mineralogical phase, will affect the bulk rock magnetic properties of the sediment.

7 Relative Contribution of Bacterial Magnetite to the Total Magnetization of the Sediment

The surface calcareous sediments of the Alpine foreland Lake Chiemsee (southern Germany) have long been documented to host plenty of live magnetotactic bacteria (MTB). They therefore form an ideal study ground to assess the relative importance of biogenic versus anorganic (detrital) magnetite.

The uppermost 5 cm of the Lake Chiemsee sediments count on the average 10^6 MTB per millilitre. With a bacterial magnetic moment of approximately 10^{-12} emu (10^{-15} Am^2), the maximum bacterial magnetic contribution to the total sediment magnetic moment is of the order of 10^{-6} emu per millilitre. The measured bulk magnetic moment of 1 ml is of the order of 10^{-5} emu. This means that about 10% of the total sediment magnetization is carried by bacterial magnetite. This is an upper limit, usually it will be less.

Nevertheless, due to the optimized magnetic stability, fossil magnetosomes will form an important carrier of paleomagnetic information in many sedimentary sequences.

8 Criteria for the Distinction of Anorganic and Bacterial Magnetite

From transmission electron microscopy and rock magnetic studies a series of criteria has been introduced to test the presence of fossil magnetosomes in sediments:

- Narrow grain size distribution (Fig. 2)
- Stoichiometry and chemical purity of the magnetite crystals
- Perfection of crystal lattice and particle arrangement in chains
- Direction of the crystallographic < 111 > axes and elongation of the particle
- Morphology of the particles
- Various rock magnetic tests based on magnetic stability measurements, low-temperature magnetic properties (Moskowitz et al. 1993; Pan et al. 2005a) and recently by ferromagnetic resonance spectroscopy (Weiss et al. 2004a)

The rock magnetic tests, though relatively quickly performed, become more and more unreliable with increasing degree of chain disruption, oxidation and decreasing concentration of magnetosomes (due to dissolution, or a decrease in bioproductivity, or an increase in the lithogenic component). As we will show in Sect. 9, an identification of magnetosomes in older deposits will also become problematic.

9
Examples of Biogenic Magnetite in Sediments

Of the large number of investigations proving the presence of fossil magnetosomes in sedimentary deposits (e.g. Stolz et al. 1987; Chang et al. 1989; for a compilation see Evans and Heller 2003), we will select two cases that depict in an exemplary way the most important processes responsible for the preservation or destruction of bacterial magnetite.

9.1
Dissolution of Magnetite in a Reducing Environment

As already described above, the habitat of magnetotactic bacteria is within the uppermost centimetres of the unconsolidated sediment. Several rock magnetic studies have now also shown that the fossil magnetosomes are usually not evenly concentrated within a sediment column, but concentrated within the uppermost 5 to 20 cm (e.g. Frederichs et al. 1999; Pan et al. 2005), occasionally to 50 cm (e.g. Snowball 1993; Peck and King 1996). This phenomenon is observed in freshwater as well as marine sediments.

The dissolution of biogenic magnetite is particularly well documented in the sediment record of the upwelling region of the Benguela basin in the South Atlantic Ocean off Namibia (Bleil 2000). Here, due to the high influx of nutrients and thus high biological activity, the microbial degradation of organic material on the sea floor leads to a correspondingly high oxygen consumption, while nitrate builds up in the sediment column.

In the resulting reducing environment the magnetosomes will no longer be stable (compare Fig. 3) when deprived of their protective cell membranes. The magnetosomes will then be prone to gradual dissolution. An example of beginning dissolution is shown in Fig. 4.

Applying rock magnetic criteria for the presence of magnetosomes, Bleil (2000) demonstrated impressively this effect of dissolution. The grain size sensitive parameters σ_{ar}/σ_{ir} (unhysteretic remanence/isothermal remanence) and σ_{rs}/σ_s (saturation remanence/saturation magnetization) clearly document a downward coarsening of the ferrimagnetic minerals from the predominant fine-grained magnetosomes in the top centimetres to a much coarser fraction of presumably detrital origin dominating the deeper strata (Fig. 5a).

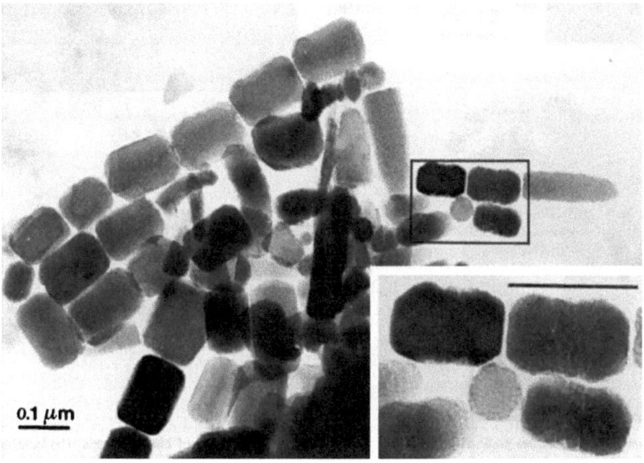

Fig. 4 Magnetic extract from fresh sediment from Lake Ammersee (40 km SW of Munich, Germany) showing incipient dissolution in the form of corrosion. Reprinted from Vali et al. 1987, with kind permission of Elsevier Science Publishers

This becomes particularly evident in the so-called Day diagram (Fig. 5b), where the ratio of σ_{rs}/σ_s is plotted against the ratio of remanence coercivity to normal coercivity B_{cr}/B_c. Interestingly, the smallest grain size closest to stable SD particles is found just below the horizon of the maximum number of living magnetotactic bacteria.

9.2 Low-Temperature Oxidation of Magnetosomes

If the process of sedimentation and early diagenesis is such that the condition of magnetite dissolution is not realized for a sufficiently long time—possibly due to rapid burial in an organic-poor environment—then a slightly oxidizing environment may affect the magnetite particles. This case can be seen in deep-sea sediment cores from the Angola Basin, South Atlantic Ocean, recovered by DSDP Leg 73. The sediment column ranges from Quaternary to Eocene. The composition of the samples varies from almost pure nannofossil ooze to clay-rich material in the Miocene dissolution facies with a clay content well over 50% and $CaCO_3$ less than 10% (Karpoff 1984).

The magnetic mineral fraction of the sediment consists of a mixture of titanomagnetite grains of detrital origin (volcanic ashes) and the much finer grained bacterial magnetosomes, which can be easily distinguished by their magnetic properties (Petersen et al. 1986). Under the electron microscope the magnetofossils can be recognized by their characteristic size and grain shape (Fig. 6). They can be followed down to the bottom of the cores to below 150 m sediment depth. Surprisingly, the fossil magnetosomes are not magnetite any

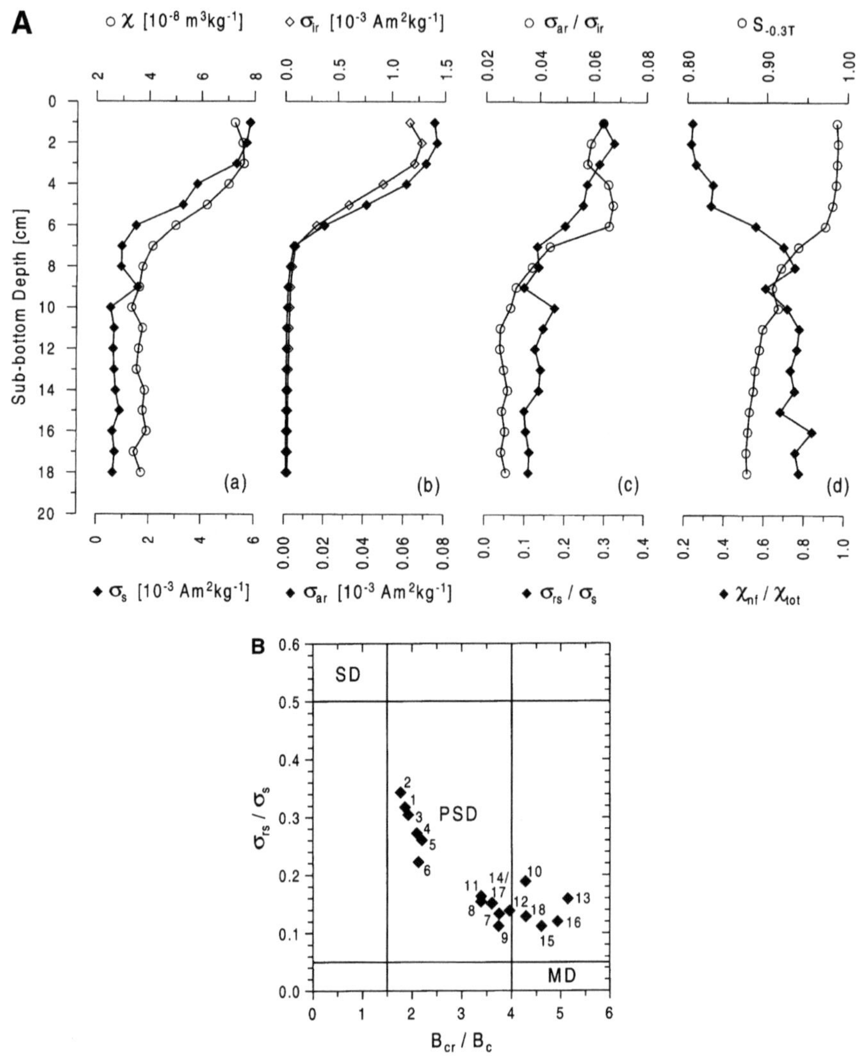

Fig. 5 **a** Rock magnetic downcore profile (Frederichs et al. 1999; Bleil 2000) of a sediment core from the Benguela basin in the South Atlantic Ocean off Namibia, an upwelling region with high biological productivity. (a) Bulk magnetic susceptibility χ (*diamonds*) and saturation magnetization σ_s (*open circles*) delineating variable magnetite contents. (b) Isothermal remanent magnetization σ_{ir} (*open diamonds*) and unhysteretic remanent magnetization σ_{ar} (*full diamonds*) depicting variations in coarser grained (MD/PSD) and finer grained (SD/PSD) magnetite fractions. (c) Magnetite grain size sensitive ratios σ_{ar}/σ_{ir} (*open circles*) and σ_{rs}/σ_s (*full diamonds*). (d) Hematite/goethite index S-0.3T (*open circles*) and relative contribution of non-ferromagnetic sediment matrix constituents to total susceptibility χ_{nf}/χ_{tot} (*full diamonds*). **b** Depth variation of the average magnetite grain size (samples of Fig. 5) as inferred from ratios of saturation remanence to saturation magnetization σ_{rs}/σ_s and coercivity of remanence to coercive field B_{cr}/B_c (Day diagram). The data point numbers refer to the depth of the samples (in cm)

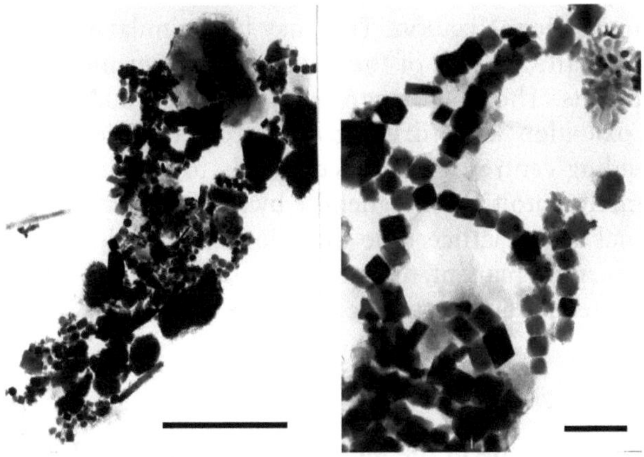

Fig. 6 Magnetic extract from Miocene deep-sea sediments (South Atlantic) showing mostly disrupted magnetosome chains among detrital ferrimagnetic material. *Bars*: 1 μm (*left*), 0.1 μm (*right*). From Petersen et al. (1987) (*left*); from von Dobeneck (1987) (*right*)

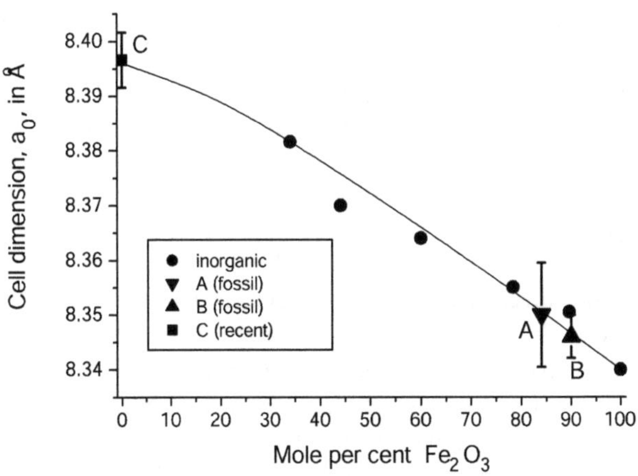

Fig. 7 Fossil (*A*, *B*) and recent (*C*) magnetosomes in the diagram of variation of the cell edge between magnetite (γ-Fe_3O_4) and maghemite (γ-Fe_2O_3) (from Vali et al. 1987)

more, but have completely turned into a different ferrimagnetic mineral, maghemite (γ-Fe_2O_3) (Fig. 7). Interestingly, the process of low-temperature oxidation has not affected the grain shape and size.

From a paleomagnetic point of view it is important to know if the oxidized magnetosome particles kept their original magnetization or were gradually remagnetized during the process of oxidation, acquiring a so-called chemical remanent magnetization. As the magnetostratigraphy measured on the sediment cores gives a reasonably fitting pattern (Tauxe et al. 1984), it seems likely that the magnetosomes retained somehow the direction of the original natural remanent magnetization. This may be a similar process as happens with titanomagnetite, carrier of the natural remanent magnetization of the ocean floor basalts. The titanomagnetite grains also undergo a steady low-temperature oxidation while drifting away from their location of emplacement at spreading centres. This process leads to a gradual transformation of titanomagnetite into titanomaghemite by iron emigration, thereby retaining the original spinel lattice. The direction of the original magnetization obtained as thermoremanent magnetization at the spreading centre is not affected by low-temperature oxidation (Hall 1977).

10
Single-Domain Magnetite in the Mars Meteorite ALH84001 and Possible Life on Mars

In one of the most controversial papers of the past decade, McKay et al. (1996) suggested that the oldest Martian meteorite, ALH84001, with a formation age of $\sim$ 4.5 Ga, harbours relics of early biological activity on Mars. Most prominent among these inferred relics are magnetite crystals of similar morphology and grain size to terrestrial magnetosomes (Thomas-Keptra et al. 2000). Importantly, these magnetite crystals are not associated with the igneous portion of ALH84001, but occur within secondary carbonate globules, which might have recrystallized from shock-melted material (Scott et al. 1997), or precipitated from a low-temperature supersaturated carbonate-rich fluid penetrating fissures and cracks of ALH84001 (Treiman and Romanek 1998; Golden et al. 2000) already crushed by previous asteroid impacts. Despite the uncertainty about the origin of the carbonate phase and particularly about its formation temperature, it was possible after all to determine its formation age as 3.9 billion years ago (Borg et al. 1999), which is contemporaneous with a period in Martian history when the surface is thought to have had flowing water. Thus, in case magnetic bacteria existed on Mars, say in hydrothermal environments similar to Yellowstone Park or deep-sea vents, it is possible that dead cells were suspended in the carbonate-rich fluid. It is not clear, on the other hand, if the physical conditions on Mars 3.9 Gy ago were stable enough for life to develop, given that the early era of heavy meteorite bombardment did not end until 3.8–3.9 Gy ago. For comparison, fossil evidence for life on Earth does not go back beyond 3.6 Ga and chemical indications of yet earlier life, as surmised from carbon isotope studies, remain ambiguous.

Since the magnetite crystals reported in ALH84001 have been considered—not unanimously, however—to be the best piece of evidence to support the hypothesis of pre-terrestrial life, we will sketch here the pros and cons in the debate on the biogenic signature of the ALH84001 magnetites, without advocating a particular hypothesis.

Thomas-Keptra et al. (2000) postulated six criteria that characterize biogenic magnetite: (1) a definite size range and width-to-length ratio, (2) chemically pure magnetite, (3) crystallographic perfection, (4) unusual crystal morphology, (5) elongation of crystals in a $< 111 >$ crystallographic direction, and (6) arrangement of crystals in linear chains.

They argued that the simultaneous presence of all six characteristics should constitute evidence of biological origin and demonstrated the presence of the first five of these characteristics. The last criterion remained elusive, since any information on the in situ spatial arrangement of the magnetite crystals had been erased by the sample preparation: the carbonate had to be dissolved to extract the magnetites for transmission electron microscopy (TEM) studies. Using energy dispersive X-ray analysis (EDXA) and high-power backscattered scanning electron microscopy (SEM), which collects information on the elemental composition from up to 1 micron depth below the sample surface, Friedman et al. (2001) were able to visualize the in situ arrangement of the iron-rich crystals and found them to be arranged in chains. Significantly, just as in terrestrial magnetosome chains, the size and shape of crystals within each chain appeared to be uniform and the chains were often curved. With this technique, however, it is not possible to quantify the overall percentage of magnetite in the sample that is arranged in chains. Recently, Weiss et al. (2004) used three rock magnetic techniques, low-temperature cycling, the Moskowitz test and ferromagnetic resonance spectroscopy, and found that no more than 10% of the magnetite can be present in isolated chains. Nevertheless, the studies by Thomas-Keptra et al. (2000) and Friedman et al. (2001), when taken together, would make a strong case for the biological origin of the magnetite particles, were it not for the fact that the supposed magnetosomes detected in the carbonate globules are associated with Fe-rich carbonates, which offers an alternative and conceptually simpler explanation of the origin of the magnetite crystals: they might as well have formed by thermal decomposition of siderite ($FeCO_3$),

$$3FeCO_3 \rightarrow Fe_3O_4 + CO + 2CO_2 \,,$$

and other ferroan carbonates (Brearley 1998). Experimental evidence for this inorganic formation process was given by Golden et al. (2001), who obtained chemically pure, defect-free, single-domain magnetite crystals by brief heating of Fe-rich carbonates to 470 °C. These synthetic magnetites duplicate all of the shapes reported for the ALH84001 magnetites—cuboid, teardrop, tooth, parallelepiped, hexagonal, irregular and whisker shaped. In contrast to that, the work by Thomas-Keptra et al. (2000) was aimed at explaining the origin of the ALH84001 magnetites with morphologies strictly similar to magnetosomes (e.g. elongated prisms with hexagonal cross section), which constitute only a fourth of the magnetites observed in the carbonate globules.

The finding by Golden et al. (2001) therefore weakens the arguments for a biological origin, but does of course not strictly disprove the biogenicity of the magnetites.

It needs to be mentioned here that magnetosomes in terrestrial magnetic bacteria exhibit a variety of morphologies, including parallelepiped, teardrop, tooth shaped and even whiskers with extended defects (Taylor et al. 2001). The existing morphological evidence is inadequate to support the inference of former life on Mars, as pointed out by Buseck et al. (2001). A decisive argument may come from crystallographic studies. If the magnetite crystals in question are topotactic with their carbonate host, that is, if the carbonate lattice planes continue across the magnetite/carbonate interfaces in three dimensions, then the most plausible explanation would be in situ growth of magnetite by solid-state diffusion (exsolution), as a result of ferroan carbonate decomposition during heating. Barber and Scott (2002) demonstrated that the magnetites that are fully embedded in carbonate are topotactically oriented with respect to the carbonate lattice, and therefore should have formed by exsolution in the solid state. Furthermore, these researchers found the magnetites in the outer, Fe-rich parts of the carbonate globules to be not well oriented and to consist mostly of irregular shaped crystals, which is consistent with in situ growth in a highly strained matrix. These results are in stark contrast to what Thomas-Keptra et al. (2000) observed, namely that some 90% of all magnetite crystals detected, among them the putative magnetosomes, were located in the optically dark rims rather than in the interiors of the globules. Barber and Scott (2001) explain the presence of magnetite chains in ALH84001 by the well-documented mechanism of nucleation and growth on ledges and kink sites on internal carbonate crystal surfaces (e.g. microstructures), which readily yields strings of crystals. These chains may appear curved at the microscopic level because ledges and terraces commonly side step on the atomic scale.

11 Conclusions

Magnetosomes are stable, magnetic, single-domain particles and therefore capable of carrying a high natural remanent magnetization. When preserved over geological time, fossil magnetosomes in sedimentary deposits can be used as reliable archives of paleomagnetic information. On the other hand, because of their small grain size and the narrow stability field of magnetite at ambient temperatures, fossil magnetosomes are prone to rapid dissolution under strongly reducing conditions, that is, in sediment layers rich in organic matter. This can often be seen in downcore rock magnetic profiles, showing a rapid drop in the rock magnetic properties that are diagnostic for bacterial magnetite. The depth at which magnetosomes disappear varies from one sedimentary environment to another and ranges from only a few centimetres to some tens of centometres.

Mildly suboxic conditions are most preferable for the conservation of biogenic magnetite. Oxic conditions lead to progressive maghemitization (oxidation of magnetite), but not necessarily to remagnetization of the original direction of natural magnetization. While maghemitization does blur the low-temperature properties of magnetosomes, it retains their size and morphology.

References

Aragón R, Buttrey DJ, Shepherd JP, Honig JM (1985) Influence of nonstoichiometry on the Verwey transition. Phys Rev B 31:430–436

Balkwill DL, Maratea D, Blakemore RP (1980) Ultrastructure of a magnetotactic spirillum. J Bacteriol 141:1399–1408

Barber DJ, Scott ERD (2002) Origin of supposedly biogenic magnetite in the Martian meteorite Allan Hills 84001. Proc Natl Acad Sci USA 99:6556–6561

Blakemore RP (1975) Magnetotactic bacteria. Science 190:377–379

Bleil U (2000) Sedimentary magnetism. In: Schulz HD, Zabel M (eds) Marine geochemistry. Springer, Berlin Heidelberg New York

Borg LE, Connelly JN, Nyquist LE, Shih CY, Wiseman H, Reese Y (1999) The age of the carbonates in Martian meteorite ALH84001. Science 286:90–94

Brearley AJ (1998) Magnetite in ALH84001: product of decomposition of ferroan carbonate. Lunar Planetary Sci XIXX, Abstract 1451

Buseck PR, Dunin-Borkowski RE, Devouard B, Frankel RB, McCartney MR, Midgley PA, Posfai M, Weyland M (2001) Magnetite morphology and life on Mars. Proc Natl Acad Sci USA 98:13490–13495

Chang S-B, Stolz J, Kirschvink J, Awramik S (1989) Biogenic magnetite in stromatolites. II. Occurrence in ancient sedimentary environments. Precambrian Res 43:305–315

von Dobeneck T (1993) Neue Ansätze zur Messung und Interpretation der magnetischen Hysterese von Tiefseesedimenten. Marie Leidorf, Westfalen

Evans ME, Heller F (2003) Environmental magnetism. Academic, San Diego

Fabian K, Kirchner A, Williams W, Heider F, Leibl T, Hubert A (1996) Three-dimensional micromagnetic calculations for magnetite using FFT. Geophys J Int 124:89–104

Frederichs T, Bleil U, Däumler K, von Dobeneck T, Schmidt A (1999) The magnetic view on the marine paleoenvironment: parameters, techniques and potentials of rock magnetic studies as a key to paleoclimatic and paleoceanographic changes. In: Fischer G, Wefer G (eds) Use of proxies in paleoceanography: examples from the South Atlantic. Springer, Berlin Heidelberg New York, pp 575–599

Friedman EI, Wierzchos J, Ascaso C, Winklhofer M (2001) Chains of magnetite crystals in the meteorite ALH84001: evidence of biological origin. Proc Natl Acad Sci USA 98:2176–2181

Garrels RM, Christ CL (1965) Solutions, minerals, and equilibria. Harper and Row, New York

Golden DC, Ming DW, Schwandt CS, Morris RV, Yang SV, Lofgren GE (2000) An experimental study on kinetically driven precipitation of Ca–Mg–Fe carbonates from solution: implications for the low-temperature formation of carbonates in Martian Allan Hills 84001. Meteorit Planet Sci 35:457–465

Golden DC, Ming DW, Schwandt CS, Lauer HV, Socki RA, Morris RV, Lofgren GE, McKay GA (2001) A simple inorganic process for formation of carbonates, magnetites, and sulfides in Martian meteorite ALH84001. Am Mineral 86:370–375

Hanzlik M, Petersen N, Keller R, Schmidbauer E (1996) Electron microscopy and Fe-57 Mossbauer spectra of 10-nm particles, intermediate in composition between Fe_3O_4 and γ-Fe_2O_3, produced by bacteria. Geophys Res Lett 23:479–482

Hall JM (1977) Does TRM occur in oceanic layer 2 basalts? J Geomagn Geoelectr 29:411–419

Karpoff AM (1984) Miocene red clays of the South Atlantic dissolution facies of calcareous oozes at deep-sea drilling project site 519 to site 523, leg 71. Initial Rep Deep Sea Drill Proj 515–535

Lovley DR, Stolz JF, Nord GI, Phillips EJP (1987) Anaerobic production of magnetite by a dissimilatory iron reducing microorganism. Nature 330:252–254

Lowenstam HA (1962) Magnetite in the denticle capping in recent chitons (Polyplacaphora). Geol Soc Am Bull 73:435–438

Lowenstam HA (1981) Minerals formed by organisms. Science 211:1126–1131

Maher BA (1991) Inorganic formation of ultrafine-grained magnetite. In: Frankel RB, Blakemore RP (eds) Iron biominerals. Plenum, New York, pp 179–191

Maher BA, Thompson R (eds) (1999) Quaternary climates, environments, and magnetism. Cambridge University Press, UK

Mann S (2001) Biomineralization: principles and concepts in bioinorganic materials chemistry. Oxford University Press, UK

Matz H, Drung D, Hartwig S, Gross H, Kötitz R, Müller W, Vass A, Weitschies W, Trahms L (1998) A SQUID measurement system for immunoassays. Appl Supercond 6:577–583

McCartney MR, Lins U, Farina M, Buseck PR, Frankel RB (2001) Magnetic microstructure of bacterial magnetite by electron holography. Eur J Miner 13:685–689

McKay DS, Gibson EK Jr, Thomas-Keptra KL, Vali H, Romanek CS, Clemett SJ, Chillier XDF, Maechling CR, Zare RN (1996) Search for past life on Mars: possible relic biogenic activity in Martian meteorite ALH84001. Science 273:924–930

Merrill RT, McElhinny MW, McFadden PL (1998) The magnetic field of the Earth. Academic, San Diego

Moskowitz BM, Frankel RB, Bazylinski DA, Jannasch HW, Lovley DR (1989) A comparison of magnetite particles produced anaerobically by magnetotactic and dissimilatory iron reducing bacteria. Geophys Res Lett 16:665–672

Moskowitz BM, Frankel RB, Bazylinski DA (1993) Rock magnetic criteria for the detection of biogenic magnetite. Earth Planet Sci Lett 120:283–300

Pan YX, Petersen N, Davila AF, Zhang LM, Winklhofer M, Liu QS, Hanzlik M, Zhu RX (2005) Towards the detection of bacterial magnetite in recent sediments in Lake Chiemsee. Earth Planet Sci Lett 232:109–123

Pan YX, Petersen N, Winklhofer M, Davila AF, Liu QS, Frederichs T, Hanzlik M, Zhu RX (2005a) Magnetic properties of uncultured magnetotactic bacteria. Earth Planet Sci Lett 237:11–325

Peck JA, King JW (1996) Magnetofossils in the sediment of Lake Baikal, Siberia. Earth Planet Sci Lett 140:159–172

Petersen N, von Dobeneck T, Vali H (1986) Fossil bacterial magnetite in deep-sea sediments from the South Atlantic Ocean. Nature 320:611–614

Petersen N, Weiss DG, Vali H (1989) Magnetic bacteria in lake sediments. In: Lowes et al. (eds) Geomagnetism and paleomagnetism, Kluwer, Dordrecht, pp 231–241

Scott ERD, Yamaguchi A, Krot AN (2002) Petrological evidence for shock melting of carbonates in the Martian meterorite ALH84001. Nature 387:377–379

Snowball IF (1994) Bacterial magnetite and the magnetic properties of sediments in a Swedish lake. Earth Planet Sci Lett 126:129–142

Stolz J, Chang S-B, Kirschvink J (1987) Biogenic magnetite in stromatolites. I. Occurrence in modern sedimentary environments. Precambrian Res 43:295–304

Tauxe L, Tucker P, Petersen N, LaBreque (1983) The magnetostratigraphy of leg 73 sediments. Palaeogeogr Palaeoclimatol Palaeoecol 42:65–90

Taylor AP, Barry JC, Webb RI (2001) Structural and morphological anomalies in magnetosomes: possible biogenic origin for magnetite in ALH84001. J Microsc 213:180–197

Thomas-Keprta KL, Bazylinski DA, Kirschvink JL, Clemett SJ, Wentworth DS, Vali H, Gibson EK Jr, Romanek CS (2000) Elongated prismatic magnetite crystals in ALH84001 carbonate globules: potential Martian magnetofossils. Geochim Cosmochim Acta 64:4049–4081

Thompson R, Oldfield F (1986) Environmental magnetism. Allen and Unwin, London

Treiman AH, Romanek CS (1998) Chemical and stable isotopic disequilibrium in carbonate minerals of Martian meteorite ALH84001: inconsistent with high formation temperature. Meterorit Planet Sci 33:737–742

Vali H, Förster O, Amarantidis G, Petersen N (1987) Magnetotactic bacteria and their magnetofossils in sediments. Earth Planet Sci Lett 86:389–400

Weiss BP, Kim SS, Kirschvink JL, Kopp RE, Sankaran M, Kobayashi A, Komeili A (2004) Magnetic tests for magnetosome chains in Martian meteorite ALH84001. Proc Natl Acad Sci USA 101:8281–8284

Weiss BP, Kim SS, Kirschvink JL, Kopp RE, Sankaran M, Kobayashi A, Komeili A (2004a) Ferromagnetic resonance and low-temperature magnetic tests for biogenic magnetite. Earth Planet Sci Lett 224:73–89

Winklhofer M (2006) In: Schüler D (ed) Magnetite-Based Magnetoreception in Higher Organisms, in: Magnetoreception and Magnetosomes in Bacteria, Springer, Berlin Heidelberg New York

Winklhofer M, Fabian K, Heider F (1997) Magnetic blocking temperatures of magnetite calculated with a three-dimensional micromagnetic model. J Geophys Res B102:22695–22709

Witt A, Fabian K, Bleil U (2005) Three-dimensional micromagnetic calculations for naturally shaped magnetite: octahedra and magnetosomes. Earth Planet Sci Lett 233:311–324

Microbiol Monogr (3)
D. Schüler: Magnetoreception and Magnetosomes in Bacteria
DOI 10.1007/7171_047/Published online: 8 September 2006

Formation of Magnetic Minerals by Non-Magnetotactic Prokaryotes

Victoria S. Coker[1] · Richard A. D. Pattrick[1] · Gerrit van der Laan[1,2] · Jonathan R. Lloyd[1] (✉)

[1]School of Earth, Atmospheric & Environmental Sciences
Williamson Research Centre for Molecular Environmental Science,
University of Manchester, Manchester M13 9PL, UK
jon.lloyd@manchester.ac.uk

[2]Magnetic Spectroscopy Group, Daresbury Laboratory, Warrington WA4 4AD, UK

Abstract Magnetic minerals produced by non-magnetotactic microbes include magnetite (Fe_3O_4), greigite (Fe_3S_4) and pyrrhotite (Fe_7S_8). These are produced by a diverse range of Fe(III)-reducing, Fe(II)-oxidising and sulfate-reducing prokaryotes. The mechanisms of formation of magnetic minerals by bacteria are still under investigation, and have been studied most intensively in Fe(III)- and sulfate-reducing bacteria. In the subsurface, the formation of magnetic Fe(II)-bearing minerals can affect the mobility of contaminants through sequestration of toxic metals such as zinc or the reduction of metals such as uranium(VI) or organic contaminants. This review covers the mechanisms of magnetic biomineral synthesis by a diversity of non-magnetotactic prokaryotes, and also discusses their potential biotechnological exploitation in a range of applications including bioremediation and the construction of magnetic recording devices and catalytic systems.

1 Introduction

Magnetic biogenic minerals are produced by microbial activity in a wide range of sub-surface environments. Magnetite, (Fe_3O_4), is produced by both Fe(III)-reducing and Fe(II)-oxidising bacteria, and understanding the formation of biogenic Fe minerals such as magnetite is particularly important because these iron metabolising bacteria generate their energy for growth through oxidation or reduction of iron and as a result are key participants in the global iron cycle (Kappler and Straub 2005; Lovley et al. 1987). These key biogeochemical processes also link to the carbon cycle, and can control trace metal solubilities via a range of passive sorption reactions (Fortin and Langley 2005). As well as magnetite, there are several other important magnetic extracellular biominerals. Greigite (Fe_3S_4) and pyrrhotite (Fe_7S_8) are magnetic sulfides produced by sulfate-reducing bacteria (SRBs) and are formed in hydrothermal as well as sub-surface environments (Watson et al. 2000). There are also several magnetic minerals which can be produced under laboratory conditions with potential biotechnological applications, including magnetic Pd nanoclusters produced by sulfate-reducing bacteria (Macaskie et al. 2005) and magnetic spinels by Fe(III)-reducing bacteria (Coker et al. 2004); these could have applications in catalysis and electronic devices.

The study of the formation of extracellular biogenic minerals, such as those discussed above, may also help provide an understanding of life on early Earth. Those organisms most closely related to the last common ancestor are mostly anaerobic sulfur-reducing hyperthermophiles (Adams 1993; Pace 1991; Stetter 1996; Vargas et al. 1998) and therefore sulfur reduction (producing sulfide available for biomineral precipitation) is considered to have been one of the first forms of microbial respiration. Fe(III) is another candidate as a primary electron acceptor of early global significance, as a range of thermophiles including *Thermotoga maritima* can grow as respiratory organisms when Fe(III) is the only electron acceptor available (Vargas et al. 1998).

From a quantitative perspective magnetite of extracellular origin is probably more abundant than magnetite from magnetosomes, although extracellular magnetite is less stable in changing (geo)chemical conditions making it more susceptible to oxidation (Dekkers 1997). This could be due to magnetite formed extracellularly being more poorly-ordered than magnetite formed by magnetotactic bacteria. Therefore, extracellular magnetite might not be as significant as a palaeomagnetic signal in sediments, as its original abundance might suggest (Dekkers 1997).

2 Mineralogy of Magnetic Minerals

2.1 Magnetite

Magnetite is an important extracellular biogenic magnetic mineral, that has been the focus of numerous studies for more than two decades. The formation of this nanoscale mineral through the bioreduction of Fe(III) coupled to organic matter oxidation, is well established and described in detail in Sect. 3, while recent research has also suggested that magnetite can be formed via biological Fe(II) oxidation (Sect. 4). Magnetite has a cubic spinel structure with one quarter of the tetrahedral (T_d) and one half of the octahedral (O_h) sites filled by iron. The formula for magnetite is $(Fe^{3+})[Fe^{2+}Fe^{3+}]O_4{}^{2-}$, where Fe^{3+} is equally split between O_h (in square brackets) and T_d (in round brackets) sites and Fe^{2+} occupies only O_h sites (Fig. 1). Hence in stoichiometric magnetite the occupancy of the Fe d^6 $O_h : d^5$ $T_d : d^5$ O_h sites is $1:1:1$ and the splitting of the Fe(III) between the two sites means magnetite is classified as an "inverse" spinel. The Fe^{3+} in the O_h and T_d sublattices contain cations

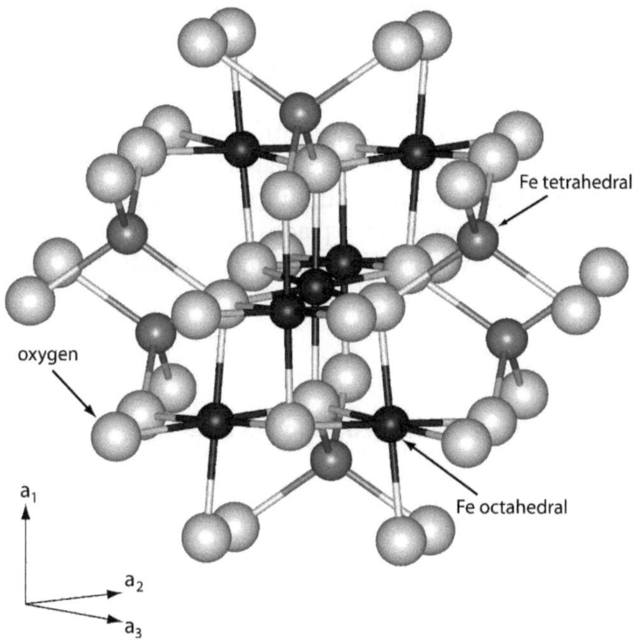

Fig. 1 Magnetite structure showing Fe atoms in the octahedral sites (*small black spheres*) and tetrahedral sites (*small dark grey spheres*). Oxygen atoms are the large *light grey spheres*. Structure created using the programs Cerius2 and Weblab Life, Molecular Simulations Inc (Pattrick et al. 2002)

with antiparallel magnetic moments that cancel each other out but there is a resulting net magnetisation (ferrimagnetism), on the Fe(II) in the O_h site. Magnetite (bio- and geogenic) is the main contributor to magnetism in the geosphere.

Oxidation of Fe^{2+} to Fe^{3+} in the O_h site in magnetite results in a charge imbalance and the release of iron from the structure. The result is neo-vacancies in the structure giving rise to a chemical formula of $(Fe^{3+})[Fe_{1-3\delta}{}^{2+}Fe_{1+2\delta}{}^{3+}\Delta_\delta]O_4{}^{2-} = Fe_{3-\delta}O_4$, where δ is the number of vacancies Δ (Pearce et al. 2006). The end-member of this oxidation process is the mineral maghemite, which has the same structure as magnetite but with no Fe(II) and a formula of γ-Fe_2O_3; some bacteria mentioned below form this non-stoichiometric and more oxidised form of magnetite.

2.2 Iron Sulfides

Pyrrhotite, Fe_7S_8, has a monoclinic structure similar to NiAs, in which cation deficient layers alternate with fully occupied cation layers. Increased iron in the structure filling the vacancies ($Fe_{7+x}S_8$) leads to more complex stacking and hexagonal symmetry. According to Néel (1953), the ferromagnetic behaviour of pyrrhotite is due to the ferromagnetic alignment of cations within iron layers and antiferromagnetic coupling between layers. As there are uncompensated magnetic moments, this results in the ferromagnetic behaviour (Néel 1953). In the series $Fe_{0.875}S$ (Fe_7S_8, monoclinic pyrrhotite) to $Fe_{1.0}S$ (FeS, troilite), phases with $x \geq 0.92$ are antiferromagnetic.

Greigite (Fe_3S_4) is a strongly ferrimagnetic sulfide that can be produced by bacteria. It has a spinel structure (a thiospinel) similar to magnetite (Fig. 1) where O atoms are replaced by S atoms giving a formula of $(Fe^{3+})[Fe^{2+}Fe^{3+}]S_4{}^{2-}$ and it therefore has a net magnetic moment derived in a similar way to magnetite.

3 Magnetite Formation by Fe(III)-Reducing Bacteria

3.1 Diversity of Microbes that Reduce Fe(III)

In the late 1970s micro-organisms were discovered that could conserve energy by reduction of Fe(III) coupled to the oxidation of organic matter (Balashova and Zavarzin 1980). Since that time many different Fe(III)-reducing bacteria have been discovered and they are both phenotypically and taxonomically diverse (Fig. 2) (Lovley et al. 1997; Vargas et al. 1998). *Geobacter metallireducens* (formally GS-15) was isolated from the Potomac River and grown in

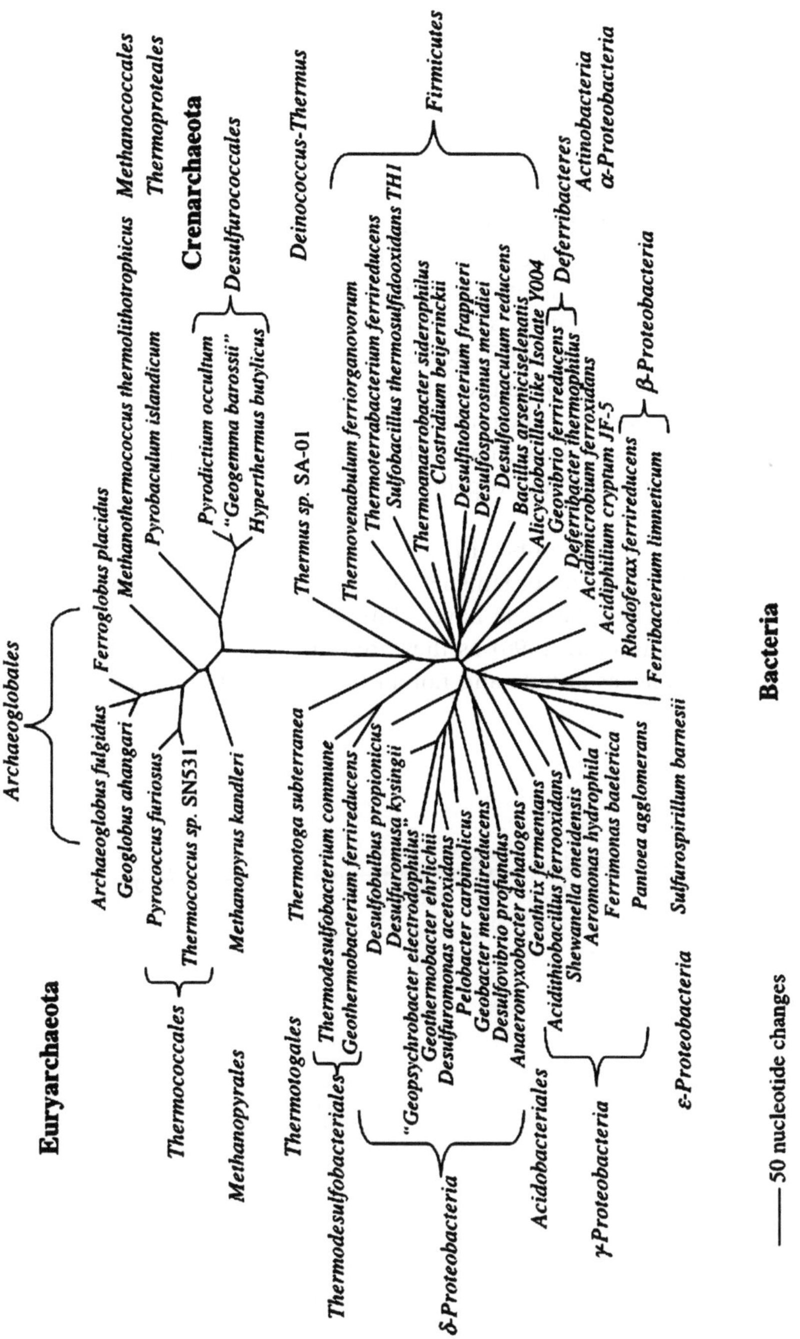

Fig. 2 Phylogenetic tree, based on 16S rRNA gene sequences, of prokaryotes capable of conserving energy to support growth from Fe(III) reduction (reprinted, with permission, from Lovley et al. 2004)

pure culture where it demonstrated the ability to couple the reduction of amorphous ferric oxyhydroxide to the oxidation of organic matter, with fine-grained magnetite formed as a bi-product (Fig. 3) (Lovley et al. 1987). 16S rRNA-based phylogeny placed *G. metallireducens* in the δ-subdivision of the class Proteobacteria (Lovley et al. 1993) where it is closely related to *Desulfuromonas acetoxidans,* a sulfur-reducing bacterium (Pfennig and Biebl 1976; Roden and Lovley 1993), also capable of producing magnetite, along with siderite ($FeCO_3$), when grown on Fe(III)-oxyhydroxides (Nealson and Saffarini 1994). Since then many other closely related Fe(III)-reducers have been isolated including *Geobacter sulfurreducens* (Fig. 3) (Caccavo Jr et al. 1994), *Desulfuromonas palmitatis* (Coates et al. 1995) and *Pelobacter carbinolicus* (Stackebrandt et al. 1989), although the latter can only accomplish Fe(III)-reduction when coupling the reaction to the oxidation of H_2 (Lonergan et al. 1996). All of these species are part of the family *Geobacteraceae* and are capable of complete oxidation of organic compounds coupled to metal reduction (Lonergan et al. 1996).

As well as strict anaerobes in the δ-subdivision of Proteobacteria, there are also Fe(III)-reducing organisms capable of magnetite production in the γ-subdivision. These bacteria have a limited ability to use organic electron donors (Lonergan et al. 1996) as they are only able to oxidise multi-carbon compounds as far as acetate and not all the way to CO_2, unlike the *Geobac-*

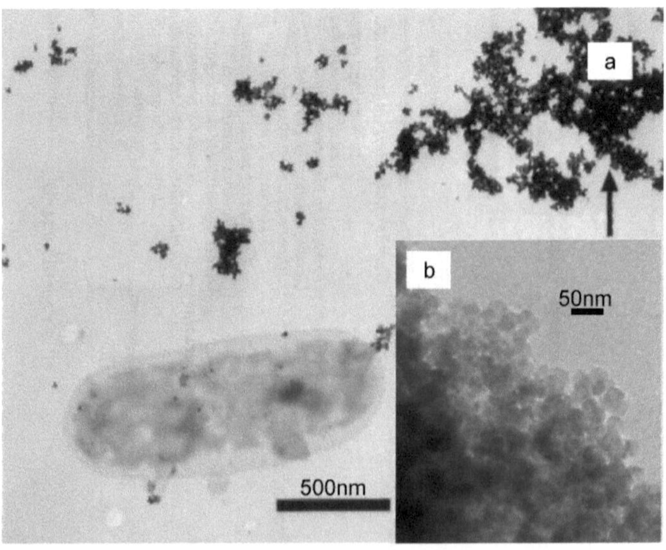

Fig. 3 **a** TEM of air-dried whole-cell preparations showing Tc-containing extracellular magnetite crystals produced by *G. sulfurreducens*, cell just above scale bar (adapted from Lloyd et al. 2000) **b** TEM showing nanocrystals of magnetite produced by *G. metallireducens*

ter species and close relatives (Coates et al. 1999). Included in this group are the facultative anaerobes *Shewanella oneidensis* MR-1 (formally *Alteromonas putrefaciens* and then *Shewanella putrefaciens*) (Lovley et al. 1989) and *S. alga* (Caccavo Jr et al. 1992), both of which are able to produce extracellular magnetite from dissimilatory Fe(III)-reduction.

Although the majority of dissimilatory Fe(III)-reducing bacteria (DIRBs) fall within the class Proteobacteria, the ability to respire using ferric iron and to form extracellular magnetite is also found in other prokaryotic groups. *Geothrix fermentans*, part of the *Holophaga-Acidobacterium* phylum, was initially isolated from aquifer sediments and is capable of producing magnetite through the reduction of Fe(III) coupled to the oxidation of acetate (Coates et al. 1999). Other phylogenetically distinct bacteria that form magnetite include *Geovibrio ferrireducens* (Caccavo Jr et al. 1996), *Deferribacter thermophilus* (Greene et al. 1997), *Anaeromyxobacter dehalogenans* (Petrie et al. 2003) and *Acidiphilium cryptum* JF-5, the latter isolated from acidic sediments (Küsel et al. 1999). *Geothermobacterium ferrireducens* from a hydrothermal environment (Kashefi et al. 2002) and the Gram-positive *Thermoanaerobacter ethanolicus* (previously TOR-39) are also able to precipitate magnetite (Zhang et al. 1998). The latter produce magnetite nanoparticles where the majority are single domain phases which are closer in structure to intracellular magnetite than the more poorly ordered extracellular magnetite formed by strains such as *Geobacter* and *Shewanella* (Frankel and Bazylinski 2003; Zhang et al. 1998). Although the bacteria listed above are thought to form magnetite, analysis of a precipitate from a halotolerant facultative-anaerobe capable of Fe(III)-reduction (Rosello-Mora et al. 1994) using ^{57}Fe Mossbauer, has shown that this bacterium produces nanoparticles that form a spinel with a mineralogy between magnetite and maghemite (Hanzlik et al. 1996), and therefore a slightly oxidised magnetic oxide precipitate. Careful chemical characterisation would be needed to confirm the actual identity of the spinels.

Although the thermophiles mentioned above are part of the bacterial kingdom, a recently discovered Archaeon designated strain-121 is also of interest. This micro-organism has raised the upper temperature limit for life by 8 °C to 121 °C (Kashefi and Lovley 2003). Strain-121 was cultured from sediment collected from an active black smoker hydrothermal vent in the Northeastern Pacific Ocean. Phylogenetic analyses placed the microbe in the Archaeal Kingdom and it is related to members of the genera *Pyrodictium* and *Pyrobaculum*. Strain-121 was cultured on solid amorphous Fe(III) oxide with formate as the electron donor and biogenic extracellular magnetite was produced as the result of incubation at 100 °C (Fig. 4) (Kashefi and Lovley 2003).

Magnetosome-synthesising magnetotactic bacteria are equally diverse and magnetosomes have been found in numerous species of aquatic prokaryotes within the α, δ and *Nitrospira* lineages of Proteobacteria (Schüler 2004). However, they are much more difficult to isolate in pure culture as they must be

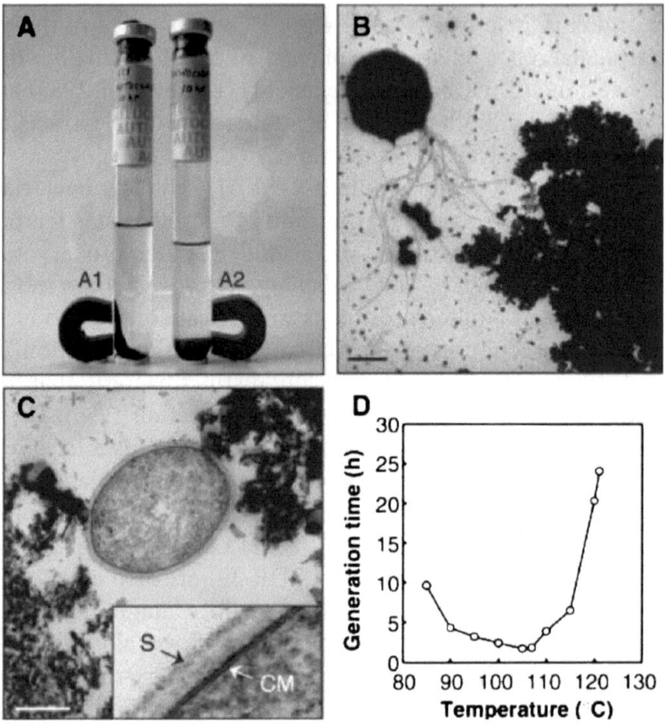

Fig. 4 Magnetite production by the hyperthermophile "Strain 121". **A** Although sterilization indicator tape showed that the inoculated media (A1) should be sterile after 10 hours in an autoclave at 121 °C, Fe(III) continued to be reduced with the formation of magnetite over three subsequent days of incubation at 103 °C. There was no reduction in an uninoculated control (A2). Negatively stained electron micrograph (**B**) and thin section (**C**) of Strain-121 illustrating single layer cell envelope (S) and cytoplasmic membrane (CM). Bar, 1 μm. **D** Time for cells of strain-121 to double at different temperatures (from Kashefi and Lovley 2003 with permission)

grown microaerobically, and in many cases their nutritional requirements remain ill-defined.

3.2 Mechanisms of Bacterial Fe(III) Reduction

Dissimilatory Fe(III)-reducing bacteria have putative ferric iron reductases located on the outer membrane of the cell (Lloyd 2003; Myers and Myers 1992, 1997) that may act as the terminal reductases of an electron transport chain that is linked to the cytoplasmic membrane. The terminal reductases have not yet been unequivocally identified in any organism, but are thought to involve *c*-type cytochromes in electron transport (Lloyd 2005). Electrons are transferred through the chain to the surface of the cell where the exposed reductase transfers them to the Fe(III)-bearing mineral surface. This process conserves energy through the generation of ATP (Wilkins et al. 2006).

For the reduction of solid Fe(III) to occur there needs to be a method for the bacteria to transfer electrons from the reductase to the mineral. There are three ways in which this may be achieved:

- Direct contact between the bacterial cell surface and Fe(III)-bearing mineral.
- Release of chelating compounds to increase the solubility of the Fe(III)-bearing mineral to allow reduction.
- Release and/or utilisation of exogenous extracellular electron shuttling compounds that act as electron transfer mediator compounds between bacterial cell surface and mineral.

Current evidence suggests *Geobacter* species require direct contact between the cell and the Fe(III)-containing mineral (Fig. 5). In addition to direct electron transfer mediated by outer membrane cytochromes, other mechanisms have been identified in *Geobacter* species. For example, a recent study showed that *G. sulfurreducens* produced pili when grown on insoluble ferric iron but

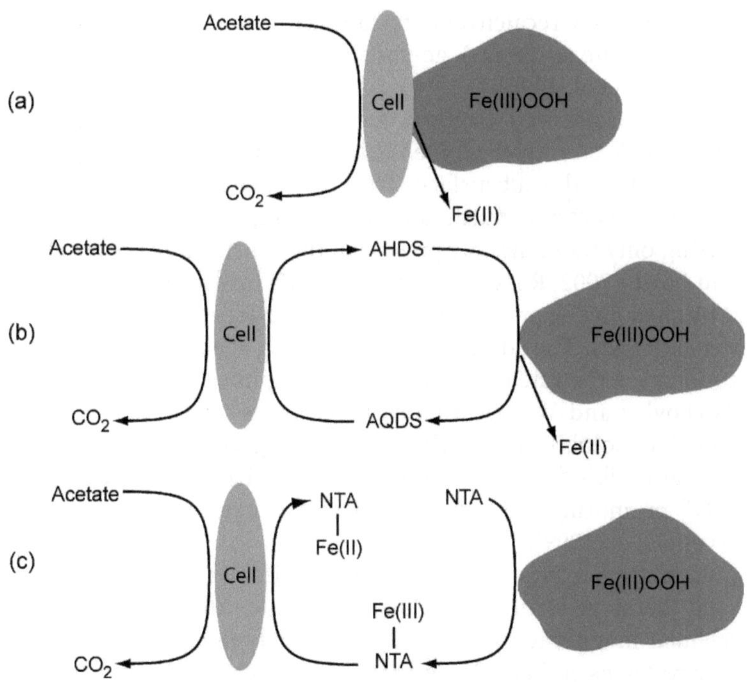

Fig. 5 Mechanisms of reduction of insoluble Fe(III) oxides, via **a** direct contact with the surface of the cell, **b** an extracellular electron shuttle or **c** a chelating agent (adapted from Lloyd 2003)

not when grown on soluble ferric citrate (Reguera et al. 2005). The study showed that the pili were useful not just for attachment to the mineral substrate, but were also highly conductive and could be used for the transfer of electrons by acting as "nanowires" (Reguera et al. 2005). Closely related *G. metallireducens* has also been shown to use flagella and pili to aid movement and attachment to insoluble Fe(III) oxyhydroxides and is chemotactic towards ferric oxides (Childers et al. 2002), facilitating access to metal oxides in anoxic environments (Lovley et al. 2004).

Electron shuttling compounds alleviate the need for direct attachment to the mineral surface. In nature, humics and other extracellular quinone containing molecules are utilised as electron acceptors and are enzymatically reduced by bacteria and the reduced hydraquinone moieties can transfer electrons iotically to Fe(III)-containing minerals (Lovley et al. 1996b). The oxidised humic is then recycled to be reduced again by the bacterial cell, resulting in very efficient metal reduction via a very small amount of electron shuttle (Fig. 5). For example, 100 nM of the humic analogue anthraquinone-2,6-disulfonate (AQDS), has a large effect on the rate of reduction and therefore promotes a faster formation rate of new minerals (Lloyd et al. 1999). Electron shuttles also provide a method for bacteria to access Fe(III)-oxides that are physically inaccessible, for example sub-micron scale fractures in rocks, or to reduce relatively recalcitrant crystalline forms of Fe(III)-bearing minerals, such as hematite (α-Fe_2O_3), goethite (α-FeOOH) and the Fe-bearing clay, smectite (Lovley et al. 1998).

Shewanella species and *Geothrix fermentans* have developed alternative strategies to *Geobacter* to overcome the low solubility, and hence bioavailability of ferric iron in the subsurface. They have been shown able to produce an extracellular electron shuttle and a chelating agent, respectively, rather than utilising only those already present in the subsurface (Lovley et al. 2004; Nevin and Lovley 2002; Rosso et al. 2003). Fe(III) oxides are thought to be solubilised by chelating agents making Fe(III) more accessible to the cell (Fig. 5) (Lovley et al. 1996a). The addition of the chelating agent nitriloacetic acid was found to increase the rate of reduction of amorphous and crystalline Fe(III) minerals (Lovley and Woodward 1996). Chelators may, however, also operate, rather than solubilising Fe(III), by removing Fe(II) from the surface of minerals that inhibit further reduction of the Fe(III) below the surface, thus "indirectly" promoting the reduction of Fe(III) oxides (Roden and Urrutia 1999; Urrutia et al. 1999).

Magnetotactic bacteria use a completely different mechanism to reduce Fe(III) and therefore form intracellular magnetite. Although the mechanism of intracellular magnetite formation is not entirely understood, it is assumed that prokaryotes assimilate Fe by first solubilising the Fe(III) by secreting high-affinity chelating agents called siderophores, which are then taken up by the cell (Andrews et al. 2003). Fe(III) then forms a ferrihydrite-like mineral in the periplasm (Ofer et al. 1984), where an opening between the periplasm and the magnetosome membrane might allow transport of Fe(III) between the compartments (Komeili et al. 2006). Individual magnetite crystals form chains within the bacterial cell enveloped by a trilaminate structure, the magnetosome membrane (Balkwill et al. 1980; Schüler 2004). Magnetite formation requires the presence of mixed-valence iron complexes in solution. Therefore, biomineralisation of this material depends on precise regulation of iron supersaturation and both redox potential and pH (Schüler 2004).

3.3 Magnetite Formation by Fe(III)-Reducing Bacteria

The formation of magnetite by dissimilatory Fe(III)-reducing bacteria (DIRB) is usually achieved by the reductive transformation of a poorly crystalline/amorphous Fe(III) oxyhydroxide, known as ferrihydrite, which has a chemical formula close to $5Fe_2O_3 \cdot 9H_2O$ (Waychunas 1991). Amorphous ferrihydrite, when analysed by X-ray diffraction (XRD), gives only two broad peaks and is therefore often known as 2-line ferrihydrite (Waychunas 1991) or, alternatively, hydrous ferric oxide (HFO) (Schwertmann et al. 1999). HFO is the most bio-available of the Fe(III)-containing minerals present in the subsurface (Glasauer et al. 2003; Lovley and Phillips 1986; Wilkins et al. 2006), and can make up more than 20% of the total Fe in sediments (Thamdrup 2000). HFO is formed abundantly as the initial product of the hydrolysis and precipitation of dissolved iron in natural geologic systems and is a strong scavenger of metal ions from aqueous solutions due to its small particle size (2–3 nm) (Schwertmann et al. 1999) and large surface area (Waychunas 1991).

Ferrihydrite is thermodynamically unstable and will transform to the crystalline goethite and hematite ferric phases (Schwertmann et al. 1999). Crystalline Fe(III)-bearing minerals are, therefore, generally more abundant in sedimentary environments than amorphous equivalents (Schwertmann and Taylor 1989). *S. oneidensis* is capable of reducing Fe(III) in the crystalline phases goethite and hematite as well as ferrihydrite in pure culture experiments (Fredrickson et al. 1998; Roden and Zachara 1996; Zachara et al. 1998), although reduction in "static" systems does not go to completion due to passivation of Fe(III) reduction by Fe(II) surface sorption to either the cell or the Fe(III)-mineral (Roden and Zachara 1996). When similar experiments are carried out under advective flow conditions in a column environment, reduction and dissolution of crystalline goethite can go towards completion (Roden and Urrutia 1999; Roden et al. 2000). Zachara et al. (1998) found that during reduction of naturally occurring crystalline ferric oxides and synthetic equivalents by *S. oneidensis* CN32, geological samples were more reducible by the bacterium. This could have been due to the geological samples being either less crystalline or containing defects that allow bacteria access, in comparison to their synthetic counterparts (Zachara et al. 1998). The reduction of crystalline Fe(III) phases in the environment remains to be proved.

The formation of extracellular magnetite requires suitable regulation of redox, pH and Fe chemistry (Mann et al. 1990; Schwertmann and Fitzpatrick 1992). Schwertmann and Fitzpatrick (1992) suggest that a requirement of microbial magnetite formation is a favourable rate of Fe(II) and Fe(III) supply to the growing crystal, combined with effective buffering capacity of the (natural or synthetic) system that allows bacteria to thrive. For any mineral to form, the solution must be supersaturated with respect to that mineral (Fredrickson et al. 1998; Roh et al. 2003; Schwertmann and Fitzpatrick 1992) and rapidly changing conditions in natural systems may lead to sporadic magnetite formation. Factors that will effect whether magnetite will be precipitated include atmospheric composition, buffer type, incubation temperature and time. For example, siderite instead of magnetite was produced using cultures of *S. oneidensis* under a head-space of $CO_2 - H_2$ instead of N_2 (Roh et al. 2003). Also, if phosphate, instead of carbonate, is used in the buffering system then vivianite ($Fe_3(PO_4)_2$) forms (Islam et al. 2005). Hansel et al. (2003) found that in a flow-through system that the first stage of mineralisation was effected by bacterial metabolism, solution chemistry and kinetically-regulated solid-phase precipitation, whereas the conversion of minerals formed during the first stage to magnetite was primarily effected by flow-regulated Fe(II) concentration, which was the main reason for the difference in mineral end-products (Hansel et al. 2003).

Cooper et al. (2000) found that during experiments with lepidocrocite (γ-FeOOH) and goethite (α-FeOOH), *S. oneidensis* CN32 was able to convert lepidocrocite, but not goethite, to magnetite through dissimilatory Fe(III)-reduction. The exact mechanism of magnetite formation from reduction of Fe(III) oxides is unclear. However, Cooper et al. (2000), building on the previous work of Tamaura et al. (1983), suggest that the layered zigzag structure of lepidocrocite (Schwertmann and Cornell 1991) is "open" enough to allow the formation of inner sphere complexes between large ferrous hydroxide ligands and the Fe(III) ions located within the crystal structure (Tamaura et al. 1983). The comparatively "closed" goethite structure would inhibit such a process.

Since magnetite contains both Fe(III) as well as Fe(II) in a ratio of 2 : 1 it is possible to chemically reduce magnetite using dissolved sulfide but microbial reduction of the mineral was not thought to be thermodynamically possible. However, at temperatures and pHs found in marine and aquatic systems it is possible for *S. oneidensis* to couple growth to magnetite reduction in pure culture (Kostka and Nealson 1995). Dong et al. (2000) conducted experiments in which *S. oneidensis* MR-1 and strain CN32 reduced Fe(III) in magnetite, precipitating siderite or vivianite, depending on whether the reduction was carried out in a carbonate or phosphate buffer. Thus, this results in a dissolution precipitation reaction depending on the Eh and pH conditions (Frankel and Bazylinski 2003); this could have implications for the interpretation of palaeomagnetic data if proven to be environmentally relevant.

4 Other Forms of Metabolism that Mediate the Production of Extracellular Magnetic Minerals

4.1 Magnetite Formation by Fe(II)-Oxidising Bacteria

Bacterial oxidation of Fe(II) is possible using both aerobes and anaerobes however, only the anaerobic Fe(II)-oxidising bacteria, which couple the oxidation of Fe(II) to nitrate reduction mediate the formation of biogenic magnetite (Benz et al. 1998; Chaudhuri et al. 2001). The first observation of this metabolism was made with a lithotrophic enrichment culture (Straub et al. 1996). Most Fe(II)-oxidising nitrate-reducers isolated so far need an organic substrate for growth, which means they only grow mixotrophically with Fe(II) (Benz et al. 1998; Kappler and Straub 2005; Straub et al. 1996). However, *Azospira oryzae* (formally *Dechlorosoma suillum*) is one of the few species where it has been proven that the oxidation of ferrous iron (in this case $FeCl_2$), coupled to dissimilatory nitrate reduction, supports growth. The formation of the resulting biogenic magnetite was via Fe(II)Fe(III)-hydroxides, identified as carbonate containing green rusts (Chaudhuri et al. 2001). Green rust is unstable and known to react with nitrate to form magnetite, so it is possible that the biogenically formed green rust was subsequently abiotically converted to magnetite (Chaudhuri et al. 2001). *A. oryzae* can also oxidise the Fe(II) in siderite and almandine which might also form magnetite as the end-product (Chaudhuri et al. 2001).

Weber and colleagues (2006) found that a wetlands sediment enrichment culture, that had previously reduced Fe(III) in goethite was then able, within the same experiment, to subsequently oxidise Fe(II), when Fe(II) oxidation was promoted by the addition of nitrate. Surprisingly, the bacteria that were the most likely candidates for the Fe(II) oxidation were the same as those that had caused the original Fe(III) reduction and were members of the family *Geobacteraceae* (Weber et al. 2006).

The mechanism of anaerobic Fe(II) oxidation is poorly understood, with important unanswered questions including the cellular location of Fe(II) oxidation, or how the bacteria cope with the poor solubility of the resultant ferric minerals (Kappler and Straub 2005).

4.2 Production of Magnetic Sulfide Minerals by Sulfate-Reducing Bacteria (SRBs)

Iron sulfides are common by-products of the metabolic reactions of sulfate-reducing bacteria (SRBs) in the subsurface. SRBs encountered in sedimentary environments are often members of δ-Proteobacteria including *Desulfovibrio*, *Desulfococcus*, *Desulfobacter*, *Desulfobulbus* and *Desulfobotulus* species, although the ability to respire sulfate is distributed throughout the prokaryotic world (Wagner et al. 1998). For example, other well-studied model organisms include the Gram-positive bacterium *Desulfotomaculum* and *Archeaglobus*, a member of the Archaea.

In the subsurface sulfate is reduced in a three-stage process rather than directly from sulfate to sulfide. First, stable sulfate is activated with ATP to give adenosine phosphosulfate (APS). Then APS is reduced to sulfite by the APS reductase, and finally sulfite is reduced to sulfide by the sulfite reductase. ATP is generated in these last two stages of the process. The sulfate is reduced in the cytoplasm, but a magnetotactic bacterium, *Desulfovibrio magneticus* strain RS-1, can also reduce sulfate to form both intracellular and extracellular sulfide precipitates simultaneously (Sakaguchi et al. 1993).

Usually, the first sulfide mineral to be produced in an aqueous environment from dissolved sulfide is disordered mackinawite ($Fe_{1+x}S$) and, with time, this can react to form greigite and ultimately pyrite or pyrrhotite (Wolthers et al. 2003). Greigite and pyrrhotite are both ferromagnetic iron minerals. In abiotic systems, the actual sulfide mineral assemblage that forms is very sensitive to solution chemistry (Benning et al. 2000; Gilbert and Banfield 2005). Eh and pH are also critical in determining the composition of $Fe_{1-x}S$ within biotic systems, as Eh effects the oxidation state of the Fe and the pH controls the concentrations of sulfide ions in solution (Watson et al.

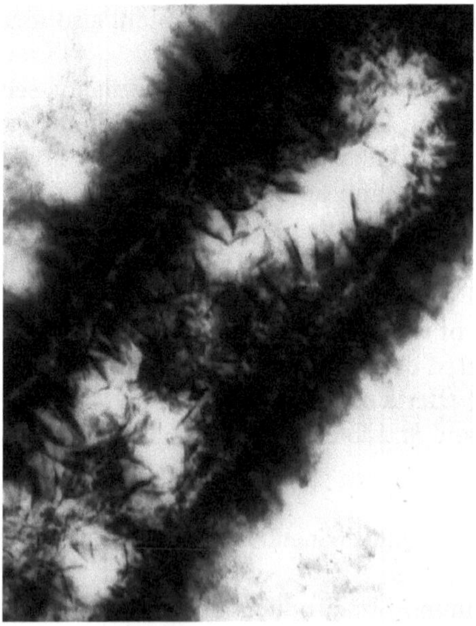

Fig. 6 A transmission electron microscope micrograph of a longitudinal section through a sulfate-reducing bacterium. The bacterium is coated with a finely divided iron sulfide precipitate (Watson et al. 2000)

2000). If iron is present in the growth medium, then iron sulfide precipitates on the surface of the bacterial cell (Fig. 6) (Watson et al. 2000). The mechanism for formation of biogenic magnetic sulfides is poorly understood. It is possible sorption of ferrous iron to functional groups on the cell surface provides sites for heterogeneous nucleation but the sorption and precipitation processes may be separate (Gilbert and Banfield 2005).

5 Implications of Extracellular Biogenic Magnetite Formation

5.1 Banded Iron Formations

It is well known that during the Precambrian, the Earth and its atmosphere were anoxic and life was restricted to prokaryotes (Ahn and Buseck 1990; Canfield 1998; Chaudhuri et al. 2001). Deep ocean waters contained large amounts of Fe(II) from the chemical weathering of the continental crust and/or the sub-seafloor hydrothermal convection processes (Ahn and Buseck 1990; Canfield 1998). This era also marked the formation of banded iron formations (BIFs) which contain the largest concentration of iron resources in the world. These are giant deposits comprising iron-rich layers, which include the minerals hematite, magnetite, chamosite, siderite and pyrite (Garrels et al. 2001), alternating with iron-poor layers of silicates. For iron deposition to happen, oxidation of dissolved Fe(II) needed to occur in the prevailing anoxic environment at the time (Fortin and Langley 2005).

One of the most likely processes that produced the iron-rich layers in BIFs is the microbial activity of both reducing and oxidative capabilities. Fe(II)-oxidising bacteria are capable of producing the huge amounts of ferric iron oxide found in the BIFs; some of this could have been used by dissimilatory Fe(III)-reducing bacteria to form secondary Fe(II)-containing minerals, like magnetite, that are also found in the BIFs (Glasauer et al. 2003). This could also account for the presence of isotopically light carbon in carbonate minerals in BIFs (Walker 1984), since Fe(III)-reducing bacteria co-metabolise and completely oxidise organic substrates such as acetate (Chaudhuri et al. 2001).

5.2 Metal Sequestration by Biogenic Magnetite

Biogenic magnetite can incorporate metals by replacement of the Fe cations in the spinel structure, which has natural and anthropogenic environmental relevance. Arsenic present in aquifers used for drinking and irrigation water by millions of people in Bangladesh and West Bengal has been the focus of intensive research in order to identify the critical controls on arsenic mobility (Akai et al. 2004; Islam et al. 2004; Nickson et al. 1998; Smedley and Kinniburgh 2002; van Geen et al. 2004). Arsenic is present in ground waters as oxyanions of As(V) and As(III) and both are often found associated with Fe oxides and other mineral phases in aquifer sediments (Smedley and Kinniburgh 2002). The relationship between microbial Fe(III)-reduction and As mobilisation is complex (Islam et al. 2004). However, it has been found that microbial formation of nano-magnetite by Fe(III)-reducing bacteria can "lock-up" arsenic, incorporating it into its spinel structure, but only if it is present as As(V) (Coker et al. 2006; Islam et al. 2005). When As(V) sorbed ferrihydrite was reduced by *G. sulfurreducens* in a laboratory study, the As(V) became bound within the magnetite structure replacing some of the Fe(III) in the tetrahedral coordination, immobilising the "problem" contaminant (Coker et al. 2006; Islam et al. 2005). In Bengal, arsenic is found adsorbed to a variety of different minerals including amorphous ferric oxides (Lloyd 2005; Smedley and Kinniburgh 2002), providing an environmental context for these results.

Zinc is another common metal contaminant that can be bioremediated by sequestration within the structure of magnetite (Cooper et al. 2000) rather than by reduction and alteration as described for redox active metals in Sect. 5.4. Zn substitution into the spinel structure can be extensive and $Zn^{2+}Fe_2{}^{3+}O_4$ is an example of a "normal" spinel phase. Zn is associated with a very wide range of anthropological activity from mining and manufacturing to waste disposal and contaminates groundwater, soils and underlying sediments (Riley and Zachara 1992). Cooper et al. (2000) found that magnetite was formed from the dissimilatory Fe(III)-reduction of lepidocrocite (γ-FeOOH) sorbed with Zn(II). It was hypothesised that Zn(II) replaced Fe(II) in the magnetite structure which immobilised the contaminant, consistent with synthetic studies that found Zn replaces Fe(III) on the tetrahedral sites in the spinel (Pearce et al. 2006). Zn(II) speciation was investigated using X-ray absorption spectroscopy (XAS) in a second study in which clay minerals were added to Zn(II)-sorbed ferrihydrite during microbial reduction. Interestingly, the presence of the non-reducible clay phase greatly mitigated the effect that microbial reduction of ferrihydrite had on Zn(II) speciation (Cooper et al. 2005). Therefore, studies of metal mobility related to ferrihydrite reduction in natural environments may not always be extrapolated to natural environments containing significant quantities of clay minerals (Cooper et al. 2005).

5.3 Contaminant Reduction by Biogenic Magnetite

Instead of incorporating metals within biogenic magnetite "locking" them up, many can be reduced and altered to a less toxic or less mobile form using the same biogenic nanomagnetite produced by DIRBs. Many studies have been undertaken on the high reactivity of ferrous iron when bound to ferric oxide minerals. Many reducible pollutants including nitroaromatic compounds, polyhalogenated alkanes, chromium(VI), technetium(VII), uranium(VI), oxamyl and related carbamate pesticides and disinfectants such as monochloramine can be reduced by Fe(II)-bearing minerals such as magnetite, siderite, green rust and troilite (Elsner et al. 2004). Since Fe(II) is abundant in subsurface environments, there is potential for in situ bioremediation via the formation of Fe(II)-bearing biominerals such as magnetite (Elsner et al. 2004 and references therein).

Fe(II)-mediated Tc(VII) reduction is one example of the positive environmental impact of extracellular biogenic nano-magnetite (Lloyd et al. 2000). ^{99}Tc is a fission product of uranium and has been released into the environment during weapons testing and the disposal of radioactive waste. ^{99}Tc is a particular problem due to its long half-life and high mobility in the environment as the stable pertechnetate anion (TcO_4^-) with its subsequent uptake into the food chain as an analogue to sulfate (Cataldo et al. 1989; Lloyd et al. 2000). Reduction of pertechnetate to the insoluble TcO_2 can alleviate these environmental problems. There are several abiotic and biotic methods to accomplish this, including reduction by biogenic nano-magnetite. Lloyd et al. (2000) found that a low rate of Tc(VII) reduction was possible during the reduction of insoluble Fe(III) by *G. sulfurreducens,* but when the electron shuttle AQDS was included in the experiment the rate of reduction of Tc(VII) rose significantly, linked to the rapid formation of magnetite. A similar result was obtained from an enrichment culture isolated from radioactive contaminated soil, leading to the conclusion that stimulation of Fe(III)-reduction could aid remediation of Tc(VII) environmental contamination (Lloyd et al. 2000).

Abiotic magnetite has also been shown to remove uranium from solution as UO_2^{2+} and, after the surface absorption of the uranyl ions, Fe(II) in the magnetite reduces the U(VI) to U(IV) to form a surface coating of UO_2 (Scott et al. 2002). Since nanoparticles are highly reactive and biogenic magnetite has been shown to reduce other metals, it should be a very effective extractor of U(VI) from solution. There is also evidence that uranium can become incorporated into the structure of magnetite as U(VI) (Duff et al. 2002) which would be an even more effective method of accumulating and immobilising aqueous uranyl. Uranium can also be directly reduced from U(VI) to U(IV) by the sulfate-reducing bacteria *Desulfosporosinus* spp. and by the Fe(III)-reducing bacteria *G. metallireducens* and *S. oneidensis* (Lovley et al. 1991; Renshaw et al. 2005).

Ferrites such as magnetite also have the potential to be used in drinking water treatment as the positive charge occurs on the surface of magnetite in acidic conditions allow negatively charged impurities, for example nitrite, to sorb to the surface (Hencl et al. 1995); biogenically produced ferrites are abundant and highly reactive therefore it is possible that these could be used for cleaning drinking water. As well as metallic species, biomagnetite can reduce and remediate toxic aromatics, for example nitroaromatics and chlorinated solvents (Heijman et al. 1993, 1995).

As well as biogenic magnetite, biogenic iron sulfides produced by sulfate-reducing bacteria could be extremely useful for bioremediation, as biogenic iron sulfides are excellent absorbents for a wide range of heavy metals. Watson et al. (2000) tested a strongly and a weakly magnetic iron sulfide with copper and cadmium solutions and found that the highest concentrations of metal absorbed were 2.18 mmol g^{-1} and 0.97 mmol g^{-1}, respectively. Since sulfate-reducing bacteria can produce much larger quantities of nano-scale greigite than the respective magnetotactic bacteria, it would be feasible to use the iron sulfide they produce for bioremediation of heavy metals. Also, the conversion of sulfate to sulfide under anaerobic conditions by sulfate-reducing bacteria forms the basis of a commercial metal biotreatment process at the Budelco Zinc Refinery, Netherlands, where sphalerite (ZnS) is formed as the main by-product (Lloyd and Lovley 2001).

5.4 Other Potential Uses for Magnetic Biominerals

Since extracellular magnetic particles can be made in large quantities and are in the nano-size range they have many potential technological applications in the biosciences, information storage, colour imaging, bioprocessing, magnetic refrigeration and ferrofluids. Many trace metal cations, including cobalt and nickel, in common with arsenic mentioned above, have a high affinity to iron oxides (Fredrickson et al. 2001). These and other di-, tri- and tetravalent cations can substitute for iron in oxides, including magnetite (Pearce et al. 2006). Fredrickson et al. (2001) found when *S. oneidensis* reduced Ni-substituted ferrihydrite (5 mol % Ni) coupled to the oxidation of lactate in either PIPES or $NaHCO_3$ buffered solution magnetite formation occurred (Fredrickson et al. 2001). Ni^{2+} was thought to have substituted into the spinel structure of magnetite replacing some of the Fe(II) ions on the octahedral sites (Fredrickson et al. 2001). This could have biotechnological applications as the substitution of Fe ions in magnetite for other transition metal cations can effect the magnetic properties of the spinel depending on the quantity, placement and valence of the dopant (Pearce et al. 2006). Coker et al. (2004) found that when transition metal substituted magnetite was produced using non-growing cells of *G. sulfurreducens*, large quantities of Co or Ni could be fitted into the magnetite structure by the bacteria. A useful technique in discovering which of the three Fe cation environments in magnetite the dopant has substituted into is X-ray magnetic circular dichroism (XMCD) (Pattrick et al. 2002). This is a synchrotron radiation-based technique that is element-, valence- and site-specific when used on magnetic materials, such as magnetite. It was found that both Co^{2+} and Ni^{2+} substituted preferably for Fe^{2+} on the octahedral site of magnetite.

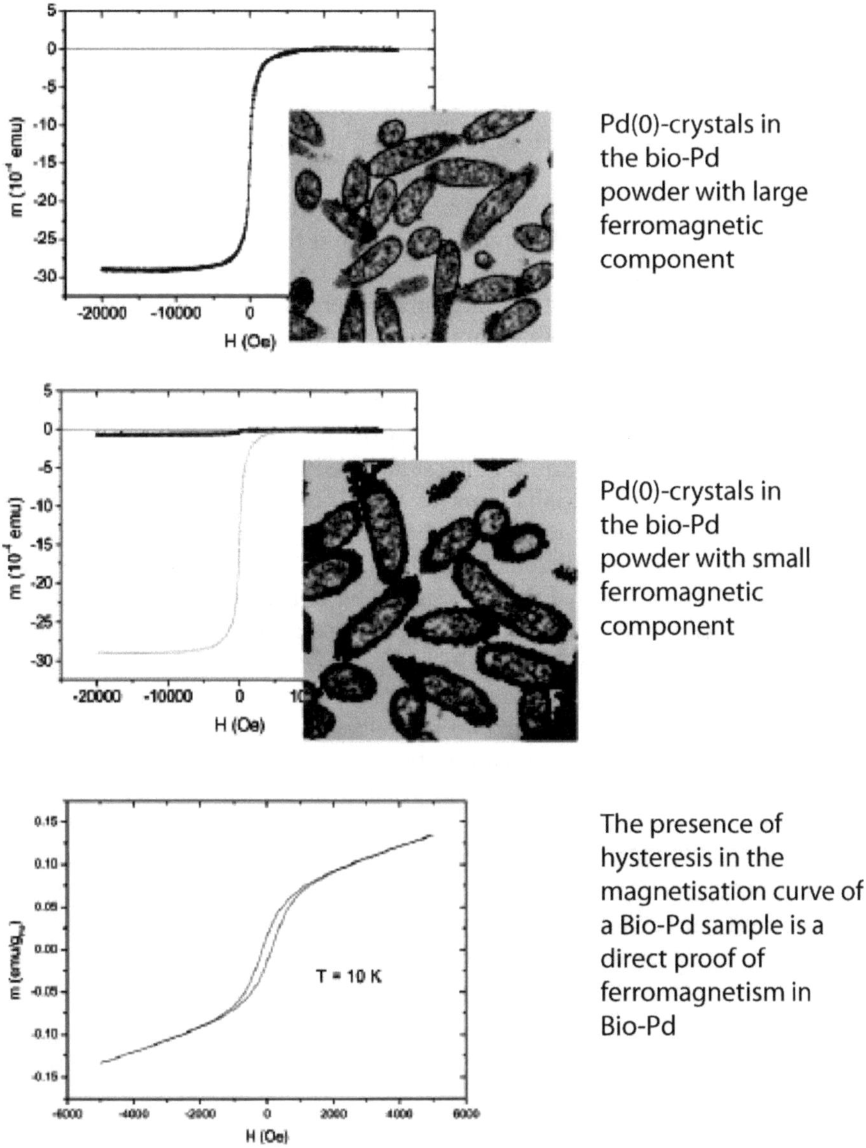

Fig. 7 Illustration of Bio-Pd^0 made on *D. desulfuricans* in two ways to give (*top panel*) small deposits with a high ferromagnetic component or (*middle panel*) larger deposits with a small ferromagnetic component as shown by the magnitude of magnetic moment versus applied magnetic field. The *bottom panel* shows proof of the ferromagnetic property of Bio-Pd^0. The magnetic properties were examined using a vibrating sample magnetometry and superconducting quantum interference device measurements (Macaskie et al. 2005)

Magnetic spinels can also potentially be produced from the reduction of waste materials forming an environmentally friendly method of recycling material that would otherwise be land filled at large commercial and environmental cost. Fe(III)-rich colloidal waste is produced at acid mine drainage (AMD) sites and also from the water polishing steps in the water industry. The Fe(III)-reducing bacterium *G. sulfurreducens* has been shown to couple the reduction of Fe(III) oxyhydroxides in these wastes to the oxidation of sodium acetate producing a magnetic spinel which could then be made into a useful product (Coker et al. 2004).

Finally, an unusual magnetic biogenic mineral with biotechnological applications produced by sulfate- and metal-reducing bacteria is nanoscale elemental palladium. Palladium is a precious metal that is used primarily in catalytic converters by the automotive industry (Baxter-Plant et al. 2002). The demand for palladium is high and recovery of this metal from waste materials is desirable. It has been found that the sulfate-reducing bacteria *Desulfovibrio desulfuricans* can recover palladium (as Pd(II)) from solution and also from automotive catalyst waste leachate by bacterial reduction coupled to H_2 oxidation (Baxter-Plant et al. 2002; Lloyd et al. 1998; Yong et al. 2002). Pd(II) first sorbs to the surface of the cells and then is reduced to produce nanoparticles of elemental palladium, which form clusters on both the cell surface and within the periplasm (Fig. 7). Pd^0 is a catalyst and cells with a layer of nano-Pd^0 on the surface are also effective catalysts due to the small size of the particles and the loading of them on the cells (Baxter-Plant et al. 2002). Bio-Pd^0 can be made by *D. desulfuricans* in two ways to give smaller deposits with a high ferromagnetic component or larger deposits with a smaller ferromagnetic component as shown by the magnitude of the magnetic moment versus applied magnetic field (Fig. 7). In this way it is possible to effect the catalytic efficacy of the Bio-Pd^0. Bio-Pd^0 has been shown to bioremediate solutions containing toxic metals such as Cr(VI) and also polychlorinated biphenyls, which are recalcitrant to microbial or chemical remediation (Baxter-Plant et al. 2002; Macaskie et al. 2005).

6 Conclusions

Magnetic minerals produced extracellularly by prokaryotes are produced under a wide range of environmental and laboratory conditions; the identity of the biomineral formed being dependant on the prevailing geochemical conditions and the bacterial metabolism producing the mineral. Nano-sized magnetite particles are produced by a wide range of bacteria and Archaea and have potential uses in environmental remediation strategies. They are able to remediate toxic and radioactive elements and also have potential uses in electronic applications. Magnetic sulfides produced by SRBs also have applications in remediation of toxic metals, while Bio-Pd^0 is an exotic magnetic biomineral that has significant potential as a catalyst. Magnetic biominerals are an important area of study as they improve knowledge of the biogeochemistry of early Earth and also on the environmental cycling of metals in the subsurface in contemporary Earth systems.

Acknowledgements Research support was provided by a CASE studentship from the EPSRC with the CCLRC and Miniwaste Faraday Programme, grant no. GR/P03230/01.

References

Adams MWW (1993) Enzymes and proteins from organisms that grow near and above 100 °C. Annu Rev Microbiol 47:627–658

Ahn JH, Buseck PR (1990) Hematite nanospheres of possible colloidal origin from the Precambrian banded iron formation. Science 250:111–113

Akai J, Izumi K, Fukuhara H, Masuda H, Nakano S, Yoshimura T, Ohfuji H, Anawar HM, Akai K (2004) Mineralogical and geomicrobiological investigations on groundwater arsenic enrichment in Bangladesh. Appl Geochem 19:215–230

Andrews SC, Robinson AK, Rodriguez-Quinones F (2003) Bacterial iron homeostasis. FEMS Microbiol Rev 27:215–237

Balashova VV, Zavarzin GA (1980) Anaerobic reduction of ferric iron by hydrogen bacteria. Microbiology 48:635–639

Balkwill D, Maratea D, Blakemore R (1980) Ultrastructure of a magnetotactic spirillum. J Bacteriol 141:1399–1408

Baxter-Plant VS, Mabbett AN, Macaskie LE (2002) Bacteria, their precious metal armour, and a new weapon against waste. Microbiol Today 29:80–81

Benning LG, Wilkin RT, Barnes HL (2000) Reaction pathways in the Fe-S system below 100 °C. Chem Geol 167:25–51

Benz M, Brune A, Schink B (1998) Anaerobic and aerobic oxidation of ferrous iron at neutral pH by chemoheterotrophic nitrate-reducing bacteria. Arch Microbiol 169:159–165

Caccavo F Jr, Blakemore RP, Lovley DR (1992) A hydrogen-oxidizing, Fe(III)-reducing microorganism from the Great Bay Estuary, New Hampshire. Appl Environ Microbiol 58:3211–3216

Caccavo F Jr, Lonergan DJ, Lovley DR, Davis M, Stolz JF, McInerney MJ (1994) *Geobacter sulfurreducens* sp. nov., a hydrogen and acetate-oxidising dissimilatory matal reducing. Appl Environ Microbiol 60:3752–3759

Caccavo F Jr, Coates JD, Rosello-Mora RA, Ludwig W, Schleifer KH, Lovley DR, McInerney MJ (1996) *Geovibrio ferrireducens*, a phylogenetically distinct dissimilatory Fe(III)-reducing bacterium. Arch Microbiol 165:370–376

Canfield DE (1998) A new model for proterozoic ocean chemistry. Nature 396:450–452

Cataldo DA, Garland TR, Wildung RE, Fellows RJ (1989) Comparative metabolic behaviour and interrelationships of Tc and S in soybean plants. Health Phys 57:281–288

Chaudhuri SK, Lack JG, Coates JD (2001) Biogenic magnetite formation through anaerobic biooxidation of Fe(II). Appl Environ Microbiol 67:2844–2848

Childers SE, Ciufo S, Lovley DR (2002) *Geobacter metallireducens* accesses insoluble Fe(III) oxide by chemotaxis. Nature 416:767–768

Coates JD, Ellis DJ, Gaw CV, Lovley DR (1999) *Geothrix fermentans* gen. nov., sp. nov., a novel Fe(III)-reducing bacterium from a hydrocarbon-contaminated aquifer. Int J Syst Bacteriol 49:1615–1622

Coates JD, Lonergan DJ, Phillips EJP, Jenter H, Lovley DR (1995) *Desulfuromonas palmitatis* sp. nov., a marine dissimilatory Fe(III) reducer that can oxidise long-chain fatty acids. Arch Microbiol 164:406–413

Coker VS, Gault AG, Pearce CI, van der Laan G, Telling ND, Charnock JM, Polya DA, Lloyd JR (2006) XAS and XMCD evidence for species dependent partitioning of arsenic during microbial reduction of ferrihydrite to magnetite: implications for arsenic mobilization in reducing aquifers. Environ Sci Technol, in press

Coker VS, Pattrick RAD, van der Laan G, Lloyd JR (2004) Use of bacteria to produce spinel nanoparticles. UK Patent 0424636.9

Cooper DC, Picardal F, Rivera J, Talbot C (2000) Zinc immobilization and magnetite formation via ferric oxide reduction by *Shewanella putrefaciens* 200. Environ Sci Technol 34:100–106

Cooper DC, Neal AL, Kukkadapu RK, Brewe D, Coby A, Picardal F (2005) Effects of sediment iron mineral composition on microbially mediated changes in divalent metal speciation: importance of ferrihydrite. Geochim Cosmochim Acta 69:1739–1754

Dekkers MJ (1997) Environmental magnetism: an introduction. Geologie en Mijnbouw 76:163–182

Duff MC, Coughlin JU, Hunter DB (2002) Uranium co-precipitation with iron oxide minerals. Geochim Cosmochim Acta 66:3533–3547

Elsner M, Schwarzenbach RP, Haderlein SB (2004) Reactivity of Fe(II)-bearing minerals toward reductive transformation of organic contaminants. Environ Sci Technol 38:799–807

Fortin D, Langley S (2005) Formation and occurrence of biogenic iron-rich minerals. Earth-Sci Rev 72:1–19

Frankel RB, Bazylinski DA (2003) Biologically induced mineralization by bacteria. In: Dove PM, de Yoreo JJ, Weiner S (eds) Biomineralization. The Mineralogical Society of America, Washington, DC, p 95–114

Fredrickson JK, Zachara JM, Kukkadapu RK, Gorby YA, Smith SC, Brown CF (2001) Biotransformation of Ni-substituted hydrous ferric oxide by an Fe(III)-reducing bacterium. Environ Sci Technol 35:703–712

Fredrickson JK, Zachara JM, Kennedy DW, Dong H, Onstott TC, Hinman NW, Li S (1998) Biogenic iron mineralization accompanying the dissimilatory reduction of hydrous ferric oxide by groundwater bacterium. Geochim Cosmochim Acta 62:3239–3257

Garrels RM, Perry EA, MacKenzie FT (2001) Genesis of Precambrian iron-formations and the development of atmospheric oxygen. Econ Geol 68:1173–1179

Gilbert B, Banfield JF (2005) Molecular-scale processes involving nanoparticulate minerals in biogeochemical systems. In: Banfield JF, Cervini-Silva J, Nealson KN (eds) Molecular Geomicrobiology. Mineralogical Society of America, The Geochemical Society, p 109–155

Glasauer S, Weidler PG, Langley S, Beveridge TJ (2003) Controls on Fereduction and mineral formation by a subsurface bacterium. Geochim Cosmochim Acta 67:1277–1288

Greene AC, Patel BKC, Sheehy AJ (1997) *Deferribacter thermophilus* gen. nov. sp. nov., a novel thermophilic manganese- and iron-reducing bacterium isolated from a petroleum reservoir. Int J Syst Bacteriol 47:505–509

Hansel CM, Benner SG, Neiss J, Dohnalkova A, Kukkadapu RK, Fendorf S (2003) Secondary mineralization pathways induced by dissimilatory iron reduction of ferrihydrite under advective flow. Geochim Cosmochim Acta 67:2977–2992

Hanzlik M, Petersen N, Keller R, Schmidbauer E (1996) Electron microscopy and ^{57}Mossbauer spectra of 10 nm particles, intermediate in composition between Fe_3O_4 and γ-Fe_2O_3, produced by bacteria. Geophys Res Lett 23:479–482

Heijman CG, Greider E, Holliger C, Schwarzenbach RP (1993) Abiotic reduction of 4-chlorobenzene to 4-chloroaniline in a dissimilatory iron-reducing enrichment culture. Appl Environ Microbiol 59:4350–4353

Heijman CG, Greider E, Holliger C, Schwarzenbach RP (1995) Reduction of nitroaromatic compounds coupled to microbial iron reduction in laboratory aquifer columns. Environ Sci Technol 29:775–783

Hencl V, Mucha P, Orlikova A, Leskova D (1995) Utilisation of ferrites for water treatment. Wat Res 29:383–385

Islam FS, Gault AG, Boothman C, Polya DA, Chatterjee D, Lloyd JR (2004) Role of metal-reducing bacteria in arsenic release from Bengal delta sediments. Nature 430:68–71

Islam FS, Pederick RL, Gault AG, Adams LK, Polya DA, Charnock JM, Lloyd JR (2005) Interactions between the Fe(III)-reducing bacterium *Geobacter sulfurreducens* and arsenate, and capture of the metalloid by biogenic Fe(II). Appl Environ Microbiol 71:8642–8648

Kappler A, Straub KL (2005) Geomicrobiological Cycling of Iron. In: Banfield JF, Cervini-Silva J, Nealson KN (eds) Molecular Geomicrobiology. Mineralogical Society of America, The Geochemical Society, p 85–108

Kashefi K, Lovley DR (2003) Extending the upper temperature limit for life. Science 301:934

Kashefi K, Holmes DE, Reysenbach A-L, Lovley DR (2002) Use of Fe(III) as an electron acceptor to recover previously uncultured hyperthermophiles: isolation and characterisation of *Geothermobacterium ferrireducens* gen. nov. sp. nov. Appl Environ Microbiol 68:1735–1742

Komeili A, Li Z, Newman DK, Jensen GJ (2006) Magnetosomes are cell membrane invaginaions organised by actin-like protein mamK. Science 311:242–245

Kostka JE, Nealson KH (1995) Dissolution and reduction of magnetite by bacteria. Environ Sci Technol 29:2535–2540

Küsel K, Dorsch T, Acker G, Stackebrandt E (1999) Microbial reduction of Fe(III) in acidic sediments: isolation of *Acidiphilium cryptum* JF-5 capable of coupling the reduction of Fe(III) to the oxidation of glucose. Appl Envir Microbiol 65:3633–3640

Lloyd JR (2003) Microbial reduction of metals and radionuclides. FEMS Microbiol Rev 27:411–425

Lloyd JR (2005) Mechanisms and environmental impact of microbial metal reduction. In: Gadd GM, Semple KT, Lappin-Scott HM (eds) Micro-organisms and earth systems—advances in geomicrobiology. Cambridge University Press, New York, p 273–302

Lloyd JR, Lovley DR (2001) Microbial detoxification of metals and radionuclides. Curr Opin Microbiol 12:248–253

Lloyd JR, Yong P, Macaskie LE (1998) Enzymatic recovery of elemental palladium by using sulfate-reducing bacteria. Appl Environ Microbiol 64:4607–4609

Lloyd JR, Blunt-Harris EL, Lovley DR (1999) The periplasmic 9.6-kilodalton c-type cytochrome of *Geobacter sulfurreducens* is not an electron shuttle to Fe(III). J Bacteriol 181:7647–7649

Lloyd JR, Sole VA, van Praagh CVG, Lovley DR (2000) Direct and Fe(II)-mediated reduction of technetium by Fe(III)-reducing bacteria. Appl Environ Microbiol 66:3743–3749

Lonergan DJ, Jenter H, Coates JD, Phillips EJP, Schmidt TM, Lovley DR (1996) Phylogenetic analysis of dissimilatory Fe(III)-reducing bacteria. J Bacteriol 178:2402–2408

Lovley DR, Phillips EJP (1986) Availability of ferric iron for microbial reduction in bottom sediments of the freshwater tidal Potomac River. Appl Environ Microbiol 52:751–757

Lovley DR, Woodward JC (1996) Mechanisms for chelator stimulation of microbial Fe(III)-oxide reduction. Chem Geol 132:19–24

Lovley DR, Phillips EJP, Lonergan DJ (1989) Hydrogen and formate oxidisation coupled to dissimilatory reduction of iron or manganese by *Alteromonas putrefaciens*. Appl Environ Microbiol 55:700–706

Lovley DR, Woodward JC, Chapelle FH (1996a) Rapid anaerobic benzene oxidation with a variety of chelated Fe(III) forms. Appl Environ Microbiol 62:288–291

Lovley DR, Holmes DE, Nevin KP (2004) Dissimilatory Fe(III) and Mn(IV) reduction. Adv Microb Phys 49:219–286

Lovley DR, Stoltz JF, Nord GL Jr, Phillips EJP (1987) Anaerobic production of magnetite by a dissimilatory iron-reducing microorganism. Nature 330:252–254

Lovley DR, Phillips EJP, Gorby YA, Landa E (1991) Microbial reduction of uranium. Nature 350:413–416

Lovley DR, Coates JD, Saffarini D, Lonergan DJ (1997) Diversity of dissimilatory Fe(III)-reducing bacteria. In: Winkelman G, Carrano CJ (eds) Iron and related transition metals in microbial metabolism. Harwood Academic Publishers, Chur, Switzerland, p 187–215

Lovley DR, Coates JD, Blunt-Harris EL, Phillips EJP, Woodward JC (1996b) Humic substances as electron acceptors for microbial respiration. Nature 382:445–448

Lovley DR, Fraga JL, Blunt-Harris EL, Hayes LA, Phillips EJP, Coates JD (1998) Humic substances as a mediator for microbially catalyzed metal reduction. Acta Hydrochim Hydrobiol 26:152–157

Lovley DR, Giovannoni SJ, White DC, Champine JE, Phillips EJP, Gorby YA, Goodwin S (1993) *Geobacter metallireducens* gen. nov. sp. nov., a microorganism capable of coupling the complete oxidation of organic coumpounds to the reduction of iron and other metals. Arch Microbiol 159:336–344

Macaskie LE, Baxter-Plant VS, Creamer NJ, Humphries AC, Mikheenko IP, Mikheenko PM, Penfold DW, Yong P (2005) Applications of bacterial hydrogenases in waste decontamination, manufacture of novel bionanocatalysts and in sustainable energy. Biochem Soc Trans 33:76–79

Mann S, Sparks NHC, Wade VJ (1990) Crystallochemical control of iron oxide biomineralisation. In: Frankel RB, Blakemore RP (eds) Iron Biominerals. Plenum Press, New York, p 21–49

Myers CR, Myers JM (1992) Localisation of cytochromes to the outer membrane of anaerobically grown *Shewanella putrefaciens* MR-1. J Bacteriol 174:3429–3438

Myers CR, Myers JM (1997) Outer membrane cytochromes of *Shewanella putrefaciens* MR-1: spectral analysis, and purification of the 83-kDa c-type cytochrome. Biochim Biophys Acta 1326:307–318

Nealson KH, Saffarini D (1994) Iron and manganese in anaerobic respiration: Environmental significance, physiology, and regulation. Annu Rev Microbiol 48:311–343

Néel L (1953) Some new results on antiferromagnetism and ferromagnetism. Rev Mod Phys 25:58–63

Nevin KP, Lovley DR (2002) Mechanisms for accessing insoluble Fe(III) oxide during dissimilatory Fe(III) reduction by *Geothrix fermentans*. Appl Environ Microbiol 68:2294–2299

Nickson R, McArthur J, Burgess W, Ahmed KM, Ravenscroft P, Rahman M (1998) Arsenic poisoning of Bangladesh groundwater. Nature 395:338

Ofer S, Nowik I, Bauminger ER, Papaefthymiou GC, Frankel RB, Blakemore RP (1984) Magnetosome dynamics in magnetoactic bacteria. J Biophys 46:57
Pace NR (1991) Origin of life-facing up to the physical setting. Cell 65:531–533
Pattrick RAD, van der Laan G, Henderson CMB, Kuiper P, Dudzik E, Vaughan DJ (2002) Cation site occupancy in spinel ferrites studied by X-ray magnetic circular dichroism: developing a method for mineralogists. Eur J Mineral 14:1095–1102
Pearce CI, Henderson CMB, Pattick RAD, van der Laan G, Vaughan DJ (2006) Direct determination of cation site occupancies in natural ferrite spinels by $L_{2,3}$ X-ray absorption spectroscopy and X-ray magnetic circular dichroism. Am Mineral 91:880–893
Petrie L, North NN, Dollhopf SL, Balkwill DL, Kostka JE (2003) Enumeration and characterisation of iron(III)-reducing microbial communities from acidic subsurface sediments contaminated with uranium(VI). Appl Environ Microbiol 69:7467–7479
Pfennig N, Biebl H (1976) *Desulfuromonas acetoxidans* gen. nov. and sp. nov., a new anaerobic sulfur reducing, acetate oxidising bacterium. Arch Microbiol 110:3–12
Reguera G, McCarthy KD, Mehta T, Nicoll JS, Tuominen MT, Lovley DR (2005) Extracellular electron transfer via microbial nanowires. Nature 435:1098–1101
Renshaw JC, Butchins LJC, Livens FR, May I, Charnock JM, Lloyd JR (2005) Bioreduction of uranium: environmental implications of a pentavalent intermediate. Environ Sci Technol 39:5657–5660
Riley RG, Zachara JM (1992) Identification of the most common contaminant classes. US Department of Energy, Washington, DC DOE/ER-0547T:15–17
Roden EE, Lovley DR (1993) Dissimilatory Fe(III) reduction by the marine microorganism, *Desulfuromonas acetoxidans*. Appl Environ Microbiol 59:734–742
Roden EE, Zachara JM (1996) Microbial reduction of crystalline iron (III) oxides: Influence of oxide surface area and potential for cell growth. Environ Sci Technol 30:1618–1628
Roden EE, Urrutia MM (1999) Ferrous iron removal promotes microbial reduction of crystalline iron(III) oxides. Environ Sci Technol 33:1847–1853
Roden EE, Urrutia MM, Mann CJ (2000) Bacterial reductive dissolution of crystalline Fe(III) oxide in continuous-flow column reactors. Appl Environ Microbiol 66:1062–1065
Roh Y, Zhang C-L, Vali H, Lauf RJ, Zhou J, Phelps TJ (2003) Biogeochemical and environmental factors in Fe biomineralisation: magnetite and siderite formation. Clays & Clay Minerals 51:83–91
Rosello-Mora RA, Caccavo F Jr, Osterlehner K, Springer N, Schüler D, Ludwig W, Amann R, Vannacanneyt M, Schleifer KH (1994) Isolation and taxonomic characterisation of a halotolerant facultative anaerobic iron-reducing bacterium. Syst Appl Microbiol 17:569–573
Rosso KM, Zachara JM, Fredrickson JK, Gorby YA, Smith SC (2003) Nonlocal bacterial electron transfer to hematite surfaces. Geochim Cosmochim Acta 67:1081–1087
Sakaguchi T, Burgess JG, Matsunaga T (1993) Magnetite formation by a sulfate-reducing bacterium. Nature 365:47–49
Schüler D (2004) Molecular analysis of a subcellular compartment: the magentosome membrane in *Magnetospirillum gryphiswaldense*. Arch Microbiol 181:1–7
Schwertmann U, Taylor RM (1989) Iron oxides: Minerals in Soil Environments. Soil Science Society of America, Madison, Wisconsin
Schwertmann U, Cornell RM (1991) Iron oxides in the laboratory. VCH Publishers Inc, New York

Schwertmann U, Fitzpatrick RW (1992) Iron minerals in surface environments. In: Skinner HCW, Fitzpatrick RW (eds) Biomineralisation, Processes of Iron and Manganese. Catena Verlag, Destedt, Germany, p 7–30
Schwertmann U, Freidl J, Stanjek H (1999) From Fe(III) ions to ferrihydrite and then to hematite. J Colloid Interface Sci 209:215–223
Scott TB, Allen GC, Heard PJ, Randell MG (2002) Reduction of U(VI) to U(IV) on the surface of magnetite. Geochim Cosmochim Acta 69:5639–5646
Smedley PL, Kinniburgh DG (2002) A review of the source, behaviour and distribution of arsenic in natural waters. Appl Geochem 17:517–568
Stackebrandt E, Wehmeyer U, Schink B (1989) The phylogenetic status of *Pelobacter acidigallici, Pelobacter venetianus*, and *Pelobacter carbinolicus*. Syst Appl Microbiol 11:257–260
Stetter KO (1996) Hyperthermophilic procaryotes. FEMS Microbiol Rev 18:149–158
Straub KL, Benz M, Schink B, Widdel F (1996) Anaerobic, nitrate-dependant microbial oxidation of ferrous iron. Appl Environ Microbiol 62:1458–1460
Tamaura Y, Ito K, Katsura T (1983) Transformation of γ-FeO(OH) to magnetite by adsorption of iron(II) ion on γ-FeO(OH). J Chem Soc Dalton Trans 2:189–194
Thamdrup B (2000) Microbial manganese and iron reduction in aquatic sediments. Adv Microb Ecol 16:41–83
Urrutia MM, Roden EE, Zachara JM (1999) Influence of aqueous and solid-phase Fe(II) complexants on microbial reduction of crystalline iron(III) oxides. Environ Sci Technol 33:4022–4028
van Geen A, Rose J, Thoral S, Garnier JM, Zheng Y, Bottero JY (2004) Decoupling of As and Fe release to Bangladesh groundwater under reducing conditions. Part II: Evidence from sediment incubations. Geochim Cosmochim Acta 68:3475–3486
Vargas M, Kashefi K, Blunt-Harris EL, Lovley DR (1998) Microbiological evidence for Fe(III) on early Earth. Nature 395:65–67
Wagner M, Roger AJ, Flax JL, Brusseau GA, Stahl D (1998) Phylogeny of dissimilatory sulfite reductases supports an early origin of sulfate respiration. J Bacteriol 180:2975–2982
Walker JCG (1984) Suboxic diagenesis in banded iron formations. Nature 309:340–342
Watson JHP, Cressey BA, Roberts AP, Ellwood DC, Charnock JM, Soper AK (2000) Structural and magnetic studies on heavy-metal-absorbing iron sulphide nanoparticles produced by sulphate-reducing bacteria. J Magn Magn Mater 214:13–30
Waychunas GA (1991) Crystal chemistry of oxides and oxyhydroxides. In: Lindsley DH (ed) Oxide Minerals: petrologic and magnetic significance. Mineralogical Society of America, Chelsea, Michigan, p 11–68
Weber KA, Urrutia MM, Churchill PF, Kukkadapu RK, Roden EE (2006) Anaerobic redox cycling of iron by freshwater sediment microorganisms. Environ Microbiol 8:100–113
Wilkins MJ, Livens FR, Vaughan DJ, Lloyd JR (2006) The impact of Fe(III)-reducing bacteria on uranium mobility. Biogeochemistry 78:125–150
Wolthers M, van der Gaast SJ, Rickard D (2003) The structure of disordered mackinawite. Am Mineral 88:2007–2015
Yong P, Rowson NA, Farr JPG, Harris IR, Macaskie LE (2002) Bioreduction and biocrystallization of palladium by *Desulfovibrio desulfuricans* NCIMB 8307. Biotechnol Bioeng 80:369–379
Zachara JM, Fredrickson JK, Li S, Kennedy DW, Smith SC, Gassman PL (1998) Bacterial reduction of crystalline Fe^{3+} oxides in single phase suspensions and subsurface materials. Am Mineral 83:1426–1443
Zhang C, Vali H, Romanek CS, Phelps TJ, Liu SV (1998) Formation of single-domain magnetite by a thermophilic bacterium. Am Mineral 83:1409–1418

Microbiol Monogr (3)
D. Schüler: Magnetoreception and Magnetosomes in Bacteria
DOI 10.1007/7171_048/Published online: 30 August 2006

Magnetite-Based Magnetoreception in Higher Organisms

Michael Winklhofer

Department of Earth and Environmental Science,
Ludwig-Maximilians-University of Munich, Theresienstr. 41/IV, 80333 München, Germany
michaelw@lmu.de

Abstract This chapter gives an introduction to one of the last unresolved issues in sensory biology, the physical and neurophysiological basis of geomagnetic-field sensitivity in higher organisms. Two plausible biophysical hypotheses for the magnetoreception mechanism and experimental evidence to support either or both of them will be discussed. The emphasis is on the magnetite hypothesis, according to which biogenic magnetite forms the core of a magnetic field receptor. The structural candidates for a vertebrate magnetoreceptor that have been identified already under the magnetite hypothesis in homing pigeons and rainbow trout will be described. A final discussion concerns the problem of magnetite biomineralization in vertebrates.

1 Introduction

There is ample evidence from behavioural experiments that a large number of animals across all major animal phyla use the magnetic field of the Earth for orientation, long-distance migration and homing (see Wiltschko and Wiltschko 1995 for a comprehensive review). This is a remarkable capability given the fact that the geomagnetic field is comparatively weak compared to technical magnetic fields. Despite the profusion of behavioural data, astonishingly little is known about the nature of the underlying magnetic sense, the main reason for its elusiveness being that magnetic sensory cells—the postulated morphological correlates of the magnetoreceptor—have not yet been identified with certainty. It was only in recent years that candidate magnetoreceptor structures were detected, on which hypotheses can now be tested and specific theoretical models be elaborated to answer the following questions:

1. What is the nature of magnetic sensory cells?
2. By what physical mechanism is the external magnetic field coupled into the organism (magnetoreception)?
3. How sensitive is the mechanism to small changes in the magnetic field (detection threshold)?
4. What physical mechanisms and/or chemical pathways convert the received magnetic energy into a nervous signal (transduction)?

Current research into magnetoreception is driven by two different hypotheses, which were first introduced in the late 1970s (Schulten et al. 1978; Walcott et al. 1979) and have been developed further recently (Shcherbakov and Winklhofer 1999; Ritz et al. 2000). The "radical-pair hypothesis" (Schulten et al. 1978; Ritz et al. 2000) invokes magnetically sensitive biochemical reactions involving spin-correlated radical pairs produced by photoexcitation in the retina. The magnetic field interacting with the radical pair controls the reaction yields and so is transduced into a chemical stimulus. The "magnetite hypothesis" (Walcott et al. 1979; Shcherbakov and Winklhofer 1999), on the other hand, assumes that the external magnetic field interacts with inclusions of magnetite (Fe_3O_4) in tissue, which convert the received magnetic energy into a mechanical stimulus (strain) to be detected by adjacent mechanoreceptors, which eventually generate a nervous signal (receptor potential). Thus magnetoreception is connected to chemoreception in the first case and to mechanoreception in the second case. The two hypotheses are equally plausible at this stage of experimental evidence, and it is due to the focus of the book that this review is mainly focussed on the magnetite hypothesis. Relying on completely different physical principles, the two hypotheses do not mutually exclude each other. On the contrary, there is good experimental evidence that both types of magnetoreceptor principles may be realized, even in one and the same animal, although the primary magnetic information provided by each mechanism appears to be used differently. Examples will be given below.

2 Radical-Pair Mechanism

Even weak magnetic fields can influence biochemical reactions involving spin-correlated radical pairs (SCRP) by altering the dynamics of transitions between spin states (Schulten 1978). This principle may be exploited to devise a chemical compass, as sketched in the following.

2.1 The Biochemical Compass

The biochemical compass includes the following steps (see Ritz et al. 2000 for details). (1) A photoexcited donor molecule D^* transfers an electron to an acceptor molecule A, resulting in a SCRP $D^+ + A^-$. Depending on the mutual orientation of the electron spins, the correlated spin state is referred to as singlet (spins aligned antiparallel) or triplet (parallel). The hyperfine field produced by the nuclear magnetic moment induces transitions between the singlet and triplet states. (2) A weak external magnetic field (compared to the hyperfine field) alters the singlet–triplet conversion rate. The lifetime of the radical pair has to be long enough to allow singlet–triplet conversion to occur and to be modulated by the magnetic field. (3) In the final step, singlet and triplet pairs will decompose into chemically distinct products. This way, the problem of magnetoreception is reduced to the problem of chemoreception.

The magnetic-field effects might well be amplified beyond the level of random fluctuations in a sufficiently large array of receptors, thus making the radical-pair mechanism a feasible basis of a biological compass. Under the assumption that evolutionary pressure has optimized the system, an array of some 4×10^8 and 4×10^{10} receptors is needed to detect field changes of the order of 0.01 and 0.001 Oe, respectively (Weaver et al. 2000). In order to obtain a coherent output, it is necessary that the receptors be oriented in a coherent fashion. Likewise, the cofactors D and A would have to embedded in an immobilized protein to have a stable orientation with respect to the external magnetic field.

Although a chemically based biological compass is theoretically feasible, the specific transducing processes are not understood nor have the receptive structures been identified yet. Cryptochromes in the retina have been proposed as radical pair magnetoreceptors (Ritz et al. 2000), but there is as yet little or no experimental evidence for the hypothesis. While it is known that cryptochromes contain the highly redox-active cofactor flavin adenine dinucleotide (FAD) no radical pairs involving cryptochromes have been found that could provide suggestions for the identity of the second cofactor in the radical pair (Cintolesi et al. 2003).

2.2 Testing the Radical-Pair Hypothesis

Photoexcitation is a way to produce SCRP and therefore is in accord with the fact that magnetic compass orientation in some migratory birds depended on the wavelength of the light they were offered in behavioural experiments.

While red light alone impaired the orientation, green or blue light did not (Wiltschko et al. 1993).

Nevertheless, Ritz et al. (2004) have provided good evidence that the radical-pair mechanism is realized in migratory birds. They devised a series of experiments to selectively affect the radical-pair mechanism by using monochromatic 565-nm green light combined with radio-frequency (RF) fields of low intensity (0.5 mOe) superimposed on the local geomagnetic field (0.5 Oe). Theory (Cintolesi et al. 2003) predicts that an RF field in resonance with the splitting between radical-pair states (singlet and triplet) can directly drive transitions between the spin states, thereby perturbing a radical-pair mechanism. Since this effect is anisotropic, it can be tested experimentally by varying the direction of the dc field with respect to the RF magnetic-field vector. Indeed, when the RF field (7 MHz) was aligned parallel with the dc field, birds displayed normal migratory orientation; however, in the RF field aligned at a 24 or 48° angle to the geomagnetic field, the birds were disoriented, indicating that the RF field interfered with magnetoreception (Ritz et al. 2004). These findings are in agreement with theoretical predictions about resonance effects of oscillating magnetic fields in a radical-pair mechanism, and support the assumption of a radical-pair mechanism underlying the processes mediating magnetic compass information in the birds.

3 Magnetite-Based Magnetoreception

The magnetite hypothesis assumes that some specialized sensory cells contain accumulations of ferrimagnetic material, such as biogenic magnetite, through which the external magnetic field is coupled into the nerve system. In its simplest realization, the ferromagnetic material would act like a compass needle, being rotated into the direction of the magnetic field and thereby exerting a torque on the neighbouring tissue. A torque produces mechanical deformation which, in combination with a mechanoreceptor such as a Pacinian corpuscle, can create a receptor potential and hence trigger a nerve signal. Such a torque mechanism is theoretically well-understood (Yorke 1979) and of course plausible as it is already realized in magnetotactic bacteria (Blakemore 1975; Frankel and Blakemore 1980), which can be considered as microscopic compass needles, swimming along magnetic field lines. What makes this hypothesis so attractive is the fact that magnetic single-domain (SD) crystals of magnetite have indeed been found in animal tissue and closely resemble bacterial magnetite crystals. The most impressive example in this context is magnetite crystals extracted from ethmoid tissue of the sockeye salmon, with grain sizes between 25 and 60 nm and a mean of 48 nm; the crystal morphology was described as cubo-octahedral (Mann et al. 1988). Such a narrow grain-size range is typical of biologically controlled mineralization processes. Due to the extraction procedure, however, there was no information left on the in situ disposition of crystals and their histological context. Chains of crystals were observed in the magnetic extracts but most likely are artefacts from the extraction procedure: ethmoid tissue was ground and dissolved; the released magnetic particles were centrifuged, washed, aggregated magnetically and resuspended ultrasonically (Mann et al. 1988). Likewise, it is not known if the ethmoid tissue is involved in the magnetic sense of sockeye salmon. So far, however, chains of magnetite crystals have not been identified in situ.

3.1 Pitfalls with the Magnetite Hypothesis

The presence of SD magnetite in tissue does not automatically make a case for magnetite-based magnetoreception. SD magnetite has been identified in radula teeth in recent chitons (Mollusca, Polyplacophora), where it forms denticle cappings (Lowenstam 1962). Apart from being magnetic, magnetite is also relatively hard (between calcite and quartz), thus allowing chitons to scrape encrusting algae from intertidal rocks without having their tooth cusps abraded. As an ore mineral, magnetite has a high density (5 g/cc), which can be utilized to design more sensitive gravity receptors than on the basis of calcium carbonate (3 g/cc), of which otoconia in otolithic organs are normally made. Indeed, magnetite particles in sand ingested by rays (guitarfish) were incorporated as otoconia in the vestibular organ, alongside calcitic otoconia (O'Leary et al. 1981). Even though the magnetite particles found in guitarfish are exogenous, they may well interact with the geomagnetic field and produce a mechanical torque on the sensory cells in the otolithic organs (Vilches-Troya et al. 1984).

There may be additional physiological or metabolic functions of endogenously mineralized magnetite that have not been unearthed yet. Magnetite has not only been identified in the human brain, meninx (Kirschvink et al. 1992) and hippocampus (Grassi and Dobson 1999), but also in heart, spleen and liver (Grassi et al. 1997). Magnetite may therefore be an iron dump for the body, or a by-product from iron metabolism, or may even be diagnostic for uncontrolled metabolism in tumour cells. Magnetic remanence measurements on two mouse tumours have revealed large concentrations of ferromagnetic mineral, presumably magnetite (Kirschvink et al. 1982).

Taken together, these findings prompted the necessity of new approaches to verify the magnetite hypothesis. First, magnetite (or some other ferrimagnetic) crystals have to be localized in situ in or next to nerve fibres which convey magnetic-field-modulated impulses to the brain. Detection of magnetic remanence can only be taken as a first hint of where to search more closely. After describing the disposition of the crystals with respect to each other and to the cellular elements, a concrete biophysical model can be developed to the point of making quantitative predictions testable by experiment.

3.2
Superparamagnetic Magnetite in Homing Pigeons

The only examination thus far that met all these requirements was on the upper beak skin of homing pigeons, where a putative magnetite-based magnetoreceptor has been identified and its subcellular organization and ultrastructure characterized (Hanzlik et al. 2000; Winklhofer et al. 2001; Fleissner

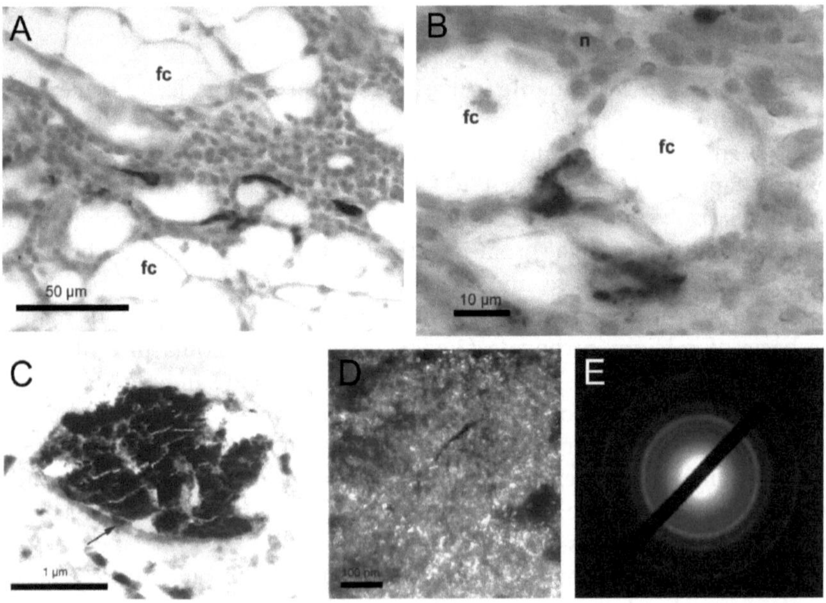

Fig. 1 Clusters of magnetite in the upper-beak skin of homing pigeons. **a** Magnetite clusters as Prussian Blue precipitates (*dark spherules*) under the light microscope. The tissue section was stained with hexacyanoferrate(II) to specifically map ferric iron concentrations; *fc* = fat cells, *n* = cell nucleus (stained with Kernechtrot). **b** Magnetite clusters from linear groups of 10 to 20 clusters were always found in free nerve endings (Fleissner et al. 2003). **c** Transmission electron micrograph of a magnetite cluster at low magnification (*bright field*) and **d** at high magnification (*dark field*). The magnetite particles in the dark field micrograph appear as *white dots*. The cluster size is 1 μm, the crystal size on average is 5 nm. The ultrathin sections (80 nm) for the TEM studies were not stained to avoid changes in composition and ultrastructure. **e** Selected-area electron-diffraction pattern of the cluster in **c**. The diffraction pattern is characteristic of ultrafine-grained magnetite crystals (*line broadening*). The intensity of the diffraction rings follows a uniform angular distribution, implying that the crystallographic axes of the magnetite particles are randomly oriented. *Scale bars*: 50, 10, 1 and 0.1 μm. Pictures from Winklhofer et al. (2001) with kind permission of Schweizerbart Science Publishers

et al. 2003; Davila et al. 2003). The putative magnetoreceptive structures were consistently found at six particular sites in the subcutis, and contained groups of clusters of ultrafine-grained magnetite crystals with grain sizes around 5 nm (Fig. 1). There was no evidence of magnetite particles larger than 10 nm. Preliminary magnetic measurements on pigeon beaks at low temperature had already pointed to tiny amounts of superparamagnetic ma-

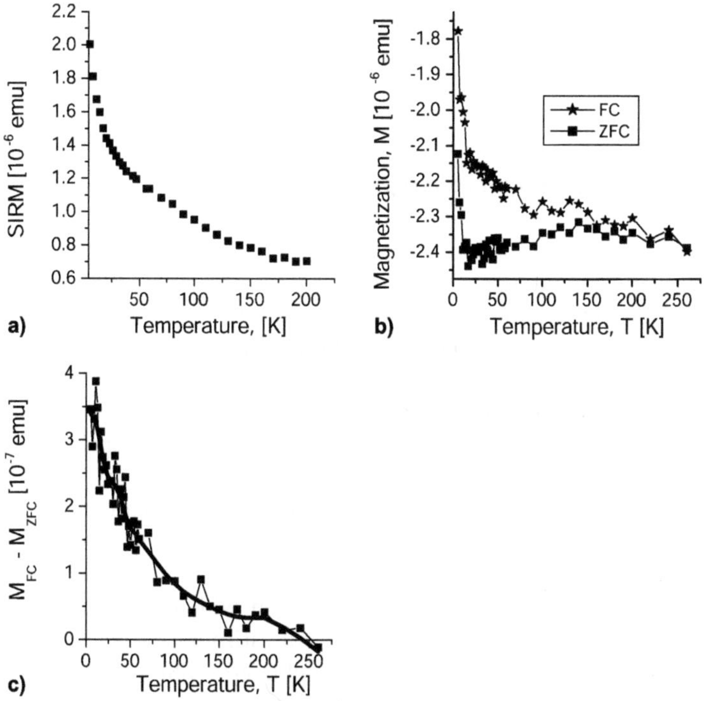

Fig. 2 Low-temperature magnetic measurements indicating the presence of SP crystals in the upper-beak skin of homing pigeon (Hanzlik et al. 2000). **a** Remanence warming curve. A saturation-induced magnetic remanence (SIRM) was imparted on the sample at 5 K using a field of 25 kOe (2.5 T). The SIRM decays strongly with warming, but levels out at 200 K without dropping to zero. This may be an offset due to a small remanant field in the magnetic measurement system. **b** Zero-field-cooled–field-cooled (ZFC–FC) cycling. The sample was cooled to 5 K in a zero field, then the induced magnetization was measured during warming (ZFC, *lower curve*) and subsequent cooling (FC, *upper curve*), both in a field of 50 Oe. The mean concentration of magnetic material in the tissue is so little that the signal is dominated by the diamagnetic background and the variations in the signal are just above the noise level. **c** The resulting remanence curve, obtained by subtracting the ZFC curve from the FC curve (raw data: *squares*; smoothed data: *solid line*), shows that most of the magnetic remanence is blocked well below 100 K. This is consistent with SP particles smaller than 10 nm grain size. Note that the "FC minus ZFC" method, though more susceptible to noise than the SIRM method, is more robust against a small remanant field, by which both curves are equally affected

terial (Fig. 2), which is in accord with the grain sizes determined under the transmission electron microscope (TEM) (Hanzlik et al. 2000). Importantly, the magnetite clusters were found to be contained in free nerve endings (FNE), that is, bare, unmyelinated dendrites as opposed to dendrites enclosed in a connective-tissue capsule. While magnetite was only found in FNE, not all FNE were found to contain magnetite (Fleissner et al. 2003). The magnetite-bearing structures are innervated by the ophthalmic nerve, which conveys the sensory input from the beak skin to the brain (Fleissner et al. 2003). This finding is consistent with electrophysiological recordings showing that the avian ophthalmic nerve carried magnetic information (Beason and Semm 1996).

3.3 Physical Mechanisms of Magnetoreception

The magnetic nanocrystals in the pigeon beak are superparamagnetic (SP), that is, the individual particles cannot carry a magnetization stably fixed in the particle's crystallographic reference frame. They can therefore not be physically twisted by the external magnetic field. Nevertheless, a cluster of SP crystals behaves as a statistical collective in an external magnetic field and takes on an induced magnetization. This way, the external magnetic field can be coupled into the FNE containing the clusters (reception). The next step to be done is to convert the magnetic energy into a physiologically exploitable stimulus that can eventually be transduced into a nerve signal. It is important to recall here that the magnetite clusters occur in FNE. FNE detect temperature, pain and, more importantly, touch, that is, they are sensitive to mechanical stimulation. Thus, a mechanism which converts the received magnetic field energy into deformation will stimulate the FNE. There are several ways of producing deformation in clusters of SP magnetite. An individual cluster will deform into a prolate ellipsoid of revolution with its long axis pointing along the applied magnetic field axis, regardless of whether the SP particles are dispersed in liquid or in a soft elastic matrix like the cytoskeleton (Shcherbakov and Winklhofer 1999; Winklhofer et al. 2001). This behaviour is well known from ferrofluids, a technical representation of a SP system.

A second transducer model is based on magnetic interactions between clusters. As can be seen in Fig. 1, roughly 20 clusters occur in one terminal, loosely arranged in a coherent elongated structure. A detailed analysis shows that the spacing between two adjacent clusters is roughly twice their diameter (Davila et al. 2003). Thus, the clusters will interact magnetically, thereby attracting or repelling each other in dependence on the direction of the magnetic field with respect to the imaginary axis joining the clusters. That way, mechanical forces arise that can stimulate the FNE, too. More precisely, an elongated group of SP clusters will experience a mechanical torque T in an external field applied at an angle ϕ with respect to the long axis of the group,

$$T \approx N \frac{16\pi^2 \chi^2 H^2 R^3}{9\delta} \sin(2\phi) \quad (1)$$

(Davila et al. 2005), where N is the number of clusters in the group, χ is the magnetic susceptibility, H is the strength of the external field, R is the radius of the cluster, and δ is the centre-to-centre distance of two adjacent clusters normalized by R. For the sake of simplicity, it was assumed in the derivation of Eq. 1 that the clusters are all of the same size and arranged in a linear chain. An exact expression for arbitrary dispositions is given in Davila et al. (2005). Anyway, as long as the clusters group in an elongated arrangement, they will behave macroscopically (Fig. 3) like a compass needle and rotate into the axial direction of an applied field. On a microscopic scale, however, the clusters do not rotate but display translational motion ($\boldsymbol{F} = \boldsymbol{m}\nabla\boldsymbol{B}$) caused by the magnetic gradient force $\boldsymbol{F}$ acting on a cluster's dipole moment $\boldsymbol{m}$ in the inhomogeneous field $\boldsymbol{B}(\boldsymbol{r})$ produced by all the other clusters in the group. The clusters in the FNE are of course not completely free to rotate, but confined by the nerve membrane. They will therefore exert a torque on the FNE. The torque in turn will bend the dendrite (Shcherbakov and Winklhofer 2004) and so may trigger transduction (Fig. 3).

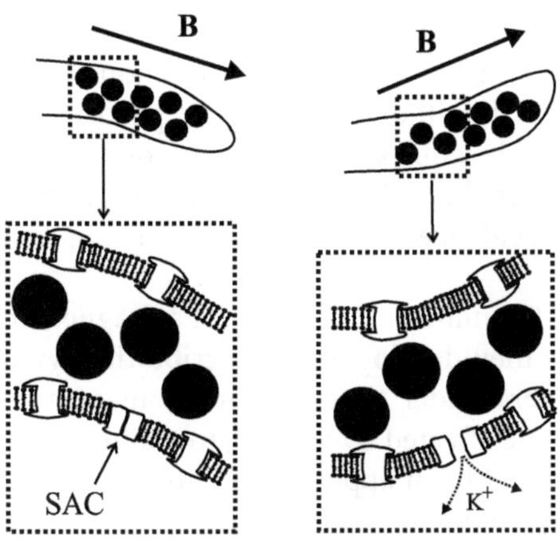

Fig. 3 Physical mechanism by which the magnetic field can be transduced into a nerve signal on the basis of magnetically interacting SP clusters within a free nerve ending (FNE). The magnetic field exerts a torque on the double chain of clusters and, by aligning the chain into the magnetic field, it will bend the FNE. Since FNE are mechanotransducers, the deformation can be converted into a nerve stimulus (Davila et al. 2005). *Insets* show the gating of stretch-activated ion channels (SAC) in the membrane of the FNE (not to scale), controlled by the magnetic-field-induced bending stress. When the membrane around the SAC is under compression (*left*), the SAC is closed; under tension, the SAC opens (*right*), giving rise to an outflow of potassium ions that change the membrane potential. A large number of SAC are required to elicit an action potential

Theoretical calculations show that transducer mechanisms based on clusters of SP magnetite are feasible (Winklhofer et al. 2001; Shcherbakov and Winklhofer 1999; Davila et al. 2003, 2005). Although the models do make quantitative predictions, it is not yet possible (without invoking ad hoc assumptions based on evolutionary optimization arguments) to give a numerical estimate of the field sensitivity of the proposed mechanisms and the resulting detection threshold. This requires measurements of the physical model parameters, such as magnetic susceptibility of SP clusters as well as viscosity and shear modulus of the medium containing the nanocrystals. Likewise, the threshold sensitivity of FNE to mechanical stimuli has to be determined. Still, the FNE containing SP clusters are excellent structural candidates for a magnetoreceptor, on which the theoretical models can be tested.

3.4 Testing the Magnetite Hypothesis with Pulse Experiments

In several behavioural experiments test birds were subjected to a brief but strong magnetic pulse (5000, 3 ms) to specifically test if ferromagnetic material is involved in magnetoreception (see Wiltschko and Wiltschko 1995, for an overview). The pulse intensity was chosen high enough to remagnetize stable single-domain magnetite. The minimum duration of the pulse was dictated by the coil geometry used in the experiments. A duration of less than 5 ms was widely considered short enough to solely remagnetize ferromagnetic material, without affecting SP particles and without triggering any unwanted side effects such as electrodynamically induced nerve pulses. A positive test, i.e. misorientation after pulse treatment, was prematurely taken as evidence of SD magnetite being involved in magnetoreception.

Davila et al. (2005) simulated the effect of a magnetic pulse using the theoretical framework they developed to describe the dynamics of a group of interacting SP clusters under time-dependent magnetic fields. It turns out that SP clusters will be affected by a pulse, too, albeit differently than a chain of SD magnetites would be: if applied at an angle $\phi > 45°$ with respect to the chain axis, a strong magnetic pulse will disrupt the chain, causing temporal impairment of the magnetoreceptor mechanism and misorientation (Davila et al. 2005). Importantly, the chain will re-assemble on a timescale of hours to days after the pulse treatment, which is in good agreement with behavioural experiments, showing that normal orientation behaviour is recovered after 4 to 10 days.

Interestingly, the pulse would only affect experienced birds, but not young birds before their first migration. Compass orientation in young birds, on the other hand, is affected by RF magnetic fields. This led to the paradigm that compass orientation is based on a radical-pair mechanism, while magnetite forms the basis of a magnetic-intensity sensor, which is likely to be used in a so-called magnetic map sense for determining geographical position by means of local magnetic field variations, as caused by spatial variations in the magnetization of the Earth's crust.

3.5 Not Only Pigeons Have Magnets

Interestingly, accumulations of SP magnetite particles have also been reported in the abdomen of honey bees (Gould et al. 1978). SP magnetite has been extracted from the abdomen and thorax of two species of termites (Maher 1998) and of *Pachycondyla marginata*, a migratory ant (Acosta Avalos et al. 1999). The presence of SP magnetite in social insects has also been inferred from electron paramagnetic resonance (Esquivel et al. 1999; Wajnberg et al. 2000) and remanence measurements (Kirschvink and Gould 1981; Wajnberg et al. 2001). On the other hand, TEM investigations on bees have not come so far as to reveal magnetite in its histological context; instead of magnetite, hydrous iron oxides (ferrihydrite) were identified (Kuterbach et al. 1982).

It is important to compare the candidate magnetoreceptor in pigeons with a structure described in rainbow trout *Oncorhynchus mykiss* (Walker et al. 1997). Using electrophysiological recordings, Walker et al. (1997) were able to identify single neurons in the ophthalmic branch of the trigeminal nerve that respond to changes in the intensity but not the direction of an imposed magnetic field. To localize the magnetoreceptor, they used a staining technique to trace the magnetically responsive nerves back to the endings of the individual nerve cells. The candidate magnetoreceptor cells were eventually detected in the olfactory lamellae (nose). Using confocal laser scanning microscopy (CLSM), TEM and energy-dispersive X-ray analysis (EDX), they found iron-rich crystals (grain size 50 nm) in low volume concentrations, which by means of magnetic force microscopy were later shown to have a permanent magnetism with magnetic properties similar to those of SD magnetite (Diebel et al. 2000). Surprisingly, the putative SD magnetite particles are located within a cell rather than in unmyelinated dendrites (FNE) or other mechanosensitive nerve structures. This raises questions on the possible transducer mechanism. It is not clear either if the magnetic particles are coupled to any potentially mechanosensitive elements (such as stretch-sensitive ion channels as proposed by Kirschvink 1992) and if they are arranged in the form of a chain or a cluster, which necessitates detailed ultrastructural investigations. Nevertheless, like the SP magnetite-containing FNE in the beak-skin of pigeons, the presumably SD magnetite-bearing nerve cells in the nose of trout are excellent candidates for a magnetoreceptor, too.

4 Biomineralization of Magnetite in Higher Organisms

A question so far unanswered concerns magnetite biomineralization in higher organisms. In the radula teeth of chitons (molluscs), the iron oxy-hydroxide ferrihydrite ($5Fe_2O_3 \cdot 9H_2O$) was identified as a precursor mineral to magnetite (Towe and Lowenstam 1967). Ferritin may be the key protein in magnetite biomineralization in vertebrates. Ferritins comprise a class of iron storage molecules ubiquitous among living systems. Each molecule of ferritin consists of 24 subunits which are assembled to form an approximately spherical cage-like structure of external diameter 12 nm; the cavity has a diameter of 8 nm (Harrison et al. 1996). Ferritin transforms highly toxic Fe(II) into the less toxic Fe(III), to be sequestered in the cavity in the form of an iron mineral similar to ferrihydrite with varying amounts of phosphate incorporated (StPierre et al. 1989). Interestingly, the SP magnetite particles in homing pigeons have grain sizes below 8 nm and therefore may well have originated in ferritin, with ferrihydrate as precursor to magnetite. At this stage, this is just a working hypothesis and further investigations are clearly warranted to elucidate the cellular and molecular pathways of magnetite biomineralization.

5 Conclusions

The physical and anatomical basis of the magnetic sense in higher animals has long remained elusive. Now that candidate magnetoreceptors have been identified in homing pigeon and rainbow trout, the magnetite hypothesis has become testable by direct micromechanical measurements and electrophysiological recordings. On the other hand, the radical-pair hypothesis, which has become more plausible by behavioural evidence, still lacks a candidate radical-pair magnetoreceptor. It needs to be mentioned that the two hypotheses do not exclude each other and receptors of both kinds may be realized in one and the same animal. In any case, the quest for magnetoreceptors has become a lively field of research, involving specialists from behavioural biology, neurobiology, physiology, biophysics, biochemistry, quantum chemistry and materials science, and it is only a matter of time before the first magnetoreceptor will be identified with certainty.

References

Acosta-Avalos D, Wajnberg E, Oliviera PS, Leal I, Farina M, Esquivel DMS (1999) Isolation of magnetic nanoparticles from *Pachycondyla marginata* ants. J Exp Biol 202:2687–2692

Beason RC, Semm P (1996) Does the avian ophthalmic nerve carry magnetic navigational information? J Exp Biol 199:1241–1244
Blakemore RP (1975) Magnetotactic bacteria. Science 19:377–379
Cintolesi F, Ritz T, Kay CWM, Timmel CR, Hore PJ (2003) Anisotropic recombination of an immobilized photoinduced radical pair in a 50 μT magnetic field: a model avian photomagnetoreceptor. Chem Phys 294:385–399
Davila AF, Fleissner G, Winklhofer M, Petersen N (2003) A new model for a magnetoreceptor in homing pigeons based on interacting clusters of superparamagnetic magnetite. Phys Chem Earth 28:647–652
Davila AF, Winklhofer M, Shcherbakov VP, Petersen N (2005) A magnetic pulse affects a putative magnetoreceptor mechanism. Biophys J 89:55–63
Diebel CE, Proksch R, Green CR, Neilson P, Walker MM (2000) Magnetite defines a vertebrate magnetoreceptor. Nature 406:299–302
Esquivel DMS, Acosta-Avalos D, El-Jaick LJ, Cunha ADM, Malheiros MG, Wajnberg E, Linhares MP (1999) Evidence for magnetic material in the fire ant solenopsis sp. by electron paramagnetic resonance. Naturwissenschaften 86:30–32
Fleissner G, Holtkamp-Rötzler E, Hanzlik M, Winklhofer M, Fleissner G, Petersen N, Wiltschko W (2003) Ultrastructural analysis of a putative magnetoreceptor in the beak of homing pigeons. J Comp Neurol 458:350–360
Frankel RB, Blakemore RP (1980) Navigational compass in magnetic bacteria. J Magn Magn Mater 15–18:1562–1564
Gould JL, Kirschvink JL, Deffeyes KS (1978) Bees have magnetic remanence. Science 201:1026–1028
Hanzlik M, Heunemann C, Holtkamp-Rötzler E, Winklhofer M, Petersen N, Fleissner G (2000) Superparamagnetic magnetite in the upper-beak tissue of homing pigeons. Biometals 13:325–331
Harrison PM, Arioso P (1996) The ferritins: molecular properties, iron storage function and cellular regulation. Biochim Biophys Acta 1275:161–203
Kirschvink JL (1992) Comment on constraints on biological effects of weak extremely low-frequency electromagnetic fields. Phys Rev A 46:2178–2184
Kirschvink JL, Gould JL (1981) Biogenic magnetite as a basis for magnetic field detection in animals. BioSystems 13:181–201
Kirschvink JL, Tabrah FL, Batkin S (1982) Ferromagnetism in two mouse tumors. J Exp Biol 101:321–326
Kirschvink JL, Kobayashi-Kirschvink A, Woodford BJ (1992) Magnetite biomineralization in the human brain. Proc Natl Acad Sci USA 89:7683–7687
Kuterbach DA, Walcott B, Reeder RJ, Frankel RB (1982) Iron-containing cells in the honey bee (*Apis mellifera*). Science 218:695–697
Lowenstam HA (1962) Magnetite in denticle capping in recent chitons (polyplacophora). Geol Soc Am Bull 73:435–438
Maher BA (1998) Magnetite biomineralization in termites. Proc R Soc Lond B Biol Sci 265:733–737
Mann S, Sparks NH, Walker MM, Kirschvink JL (1988) Ultrastructure, morphology and organization of biogenic magnetite from sockeye salmon, Oncorhynchus nerka: implications for magnetoreception. J Exp Biol 140:35–49
O'Leary DP, Vilches-Troya J, Dunn RF, Campos-Munoz A (1981) Magnets in the guitarfish vestibular receptors. Experientia 37:86–88
Ritz T, Adem S, Schulten K (2000) A model for photoreceptor-based magnetoreception in birds. Biophys J 78:707–718

Ritz T, Thalau P, Phillips JB, Wiltschko R, Wiltschko W (2004) Resonance effects indicate a radical-pair mechanism for avian magnetic compass. Nature 429:177–180

Schulten K, Swenberg CE, Weller A (1978) A biomagnetic sensory mechanism based on magnetic field modulated coherent electron spin motion. Z Phys Chem NF 111:1–5

Schultheiss-Grassi PP, Dobson J (1999) Magnetic analysis of human brain tissue. Biometals 12:67–72

Schultheiss-Grassi PP, Heller F, Dobson J (1997) Analysis of magnetic material in the human heart, spleen and liver. Biometals 10:351–355

Shcherbakov VP, Winklhofer M (1999) The osmotic magnetometer: a new model of a magnetite-based magnetoreceptor in animals. Eur Biophys J 28:380–392

Shcherbakov VP, Winklhofer M (2004) Elastic bending of magnetic filaments. Phys Rev E 70:61803

St Pierre TG, Webb J, Mann S (1989) In: Mann S, Webb J, Willimas RJP (eds) Ferritin and hemosiderin: structural and magnetic studies of the iron core. Biomineralization—chemical and biochemical perspectives. Wiley, Weinheim, pp 295–344

Towe KM, Lowenstam HA (1967) Ultrastructure and development of iron mineralization in the radular teeth of cryptochiton stelleri (mollusca). J Ultrastruct Res 17:1–13

Vilches-Troya J, Dunn RF, O'Leary DP (1984) Relationship of the vestibular hair cells to magnetic particles in the otolith of the guitarfish sacculus. J Comp Neurol 226:489–494

Wajnberg E, Acosta-Avalos D, El-Jaick L, Abraçado LG, Coelho JLA, Bakuzis AF, Morais PC, Esquivel DMS (2000) Electron paramagnetic resonance study of the migratory ant *Pachycondyla marginata* abdomens. Biophys J 78:1018–1023

Wajnberg E, Cernicchiaro G, Acosta-Avalos D, El-Jaick L, Esquivel DMS (2001) Induced magnetization of social insects. J Magn Magn Mater 226–230:2040–2041

Walcott C, Gould JL, Kirschvink JL (1979) Pigeons have magnets. Science 205:1027–1029

Walker MM, Diebel CE, Haugh CV, Pankhurst PM, Montgomery JC (1997) Structure and function of the vertebrate magnetic sense. Nature 390:371–376

Weaver JC, Vaughan TE, Astumian D (2000) Biological sensing of small field differences by magnetically sensitive chemical reactions. Nature 405:707–709

Wiltschko R, Wiltschko W (1995) Magnetic orientation in animals. Springer, Berlin Heidelberg New York

Wiltschko W, Munro U, Ford H, Wiltschko R (1993) Red light disrupts magnetic orientation of migratory birds. Nature 364:363–369

Winklhofer M, Holtkamp-Rötzler E, Hanzlik M, Fleissner G, Petersen N (2001) Clusters of superparamagnetic magnetite particles in the upper-beak skin of homing pigeons: evidence of a magnetoreceptor. Eur J Mineral 13:659–669

Yorke ED (1979) A possible magnetic transducer in birds. J Theor Biol 77:101–105

Subject Index

MIX
Papier aus verantwortungsvollen Quellen
Paper from responsible sources
FSC® C105338

If you have any concerns about our products,
you can contact us on
ProductSafety@springernature.com

In case Publisher is established outside the EU,
the EU authorized representative is:
Springer Nature Customer Service Center GmbH
Europaplatz 3, 69115 Heidelberg, Germany

Printed by Libri Plureos GmbH
in Hamburg, Germany